Corporate Governance

Prof (Dr) PK Ghosh MA LLB MBA PhD
Visiting Professor
Regional College of Management
Bhubaneswar 751 016
Odisha

IKON BOOKS
Publishers and Distributors
New Delhi • Hyderabad

CBS Publishers & Distributors Pvt Ltd

New Delhi • Bengaluru • Chennai • Kochi • Mumbai • Pune
Hyderabad • Kolkata • Nagpur • Patna • Vijayawada

CBS Publishers & Distributors Pvt Ltd
4819/XI, 24 Ansari Road, Daryaganj
New Delhi-110 002 (India)

IKON BOOKS, Publishers and Distributors
B-37, First Floor, Office No. 105
Street No. 14, Madhu Vihar
I.P. Extention, Delhi - 110 092
Telephone: 91-11-43618579
Email: ikonbooks@gmail.com
        cs@ikonbooks.com
        Website: www.ikonbooks.com

CORPORATE
GOVERNANCE

First Edition: 2014

ISBN: 978-81-239-2404-5

Published by Satish Kumar Jain for

CBS Publishers & Distributors Pvt Ltd
4819/XI Prahlad Street, 24 Ansari Road, Daryaganj, New Delhi 110 002, India.
Ph: 23289259, 23266861, 23266867   Website: www.cbspd.com
Fax: 011-23243014   e-mail: delhi@cbspd.com; cbspubs@airtelmail.in.
Corporate Office: 204 FIE, Industrial Area, Patparganj, Delhi 110 092
Ph: 4934 4934   Fax: 4934 4935   e-mail: publishing@cbspd.com; publicity@cbspd.com

Branches

- **Bengaluru:** Seema House 2975, 17th Cross, K.R. Road,
  Banasankari 2nd Stage, Bengaluru 560 070, Karnataka
  Ph: +91-80-26771678/79   Fax: +91-80-26771680   e-mail: bangalore@cbspd.com
- **Chennai:** 20, West Park Road, Shenoy Nagar, Chennai 600 030, Tamil Nadu
  Ph: +91-44-26260666, 26208620   Fax: +91-44-42032115   e-mail: chennai@cbspd.com
- **Kochi:** 36/14 Kalluvilakam, Lissie Hospital Road, Kochi 682 018, Kerala
  Ph: +91-484-4059061/65   Fax: +91-484-4059065   e-mail: kochi@cbspd.com
- **Mumbai:** 83-C, Dr E Moses Road, Worli, Mumbai-400018, Maharashtra
  Ph: +91-22-24902340/41   Fax: +91-22-24902342   e-mail: mumbai@cbspd.com
- **Pune:** Bhuruk Prestige, Sr. No. 52/12/2+1+3/2 Narhe, Haveli
  (Near Katraj-Dehu Road Bypass), Pune 411 041, Maharashtra
  Ph: +91-20-64704058/59, 32392277   Fax: +91-20-24300160   e-mail: pune@cbspd.com

Representatives

- **Hyderabad**   0-9885175004
- **Nagpur**   0-9021734563
- **Kolkata**   0-9831437309, 0-9051152362
- **Patna**   0-9334159340
- **Vijayawada**   0-9000660880

Printed at India Binding House, Noida, UP

PREFACE

Corporate Governance involves the building of a set of relationship between the company, its board, the management, the shareholder and other stakeholders by putting in place a structure and a system through which the established goals of the company may be achieved. The board and apex management of a corporation is responsible for practicing good corporate governance. Governance is the mechanism by which the values, principles, policies and procedures of a corporation are indicated and manifested. The essence of corporate governance is in promoting and maintain integrity, transparency and accountability in the higher Whelan's of management by the last few years. Issues of good governance have been receiving considerable attention in political discourse in India. The term good governance is used so widely now that it inherently requires continuous nurturing and adapting to the dynamic business governance. The recurrent corporate failures and corporate misconduct are a system that corporate governance mechanism has failed to come up to the expectations of various corporate constituencies.

The objective in writing this book is to explain the subject of corporate governance in a systematic and comprehensive manner. The book highlights the fact that strong corporate governance yields good economic prosperity and social development. In this context, the concept of drivers and enabler of goal attainments has been explained in the following chapters.

Chapter 1 – Corporate Governance courses the gamut of activities having a direct or indirect effect on the health of the entity. Starts with definitions concept & nature. Thereafter Further objectives and principles are stated in this chapter in concise terse.

Setting the stage, the chapter focuses on difference between governing of governance dealing with diversity, dynamics, and complexity. Highlights on the Need for Corporate Governance and side by side the Kumurmangalam Birla Committee on Corporate Governance (Sebi Reporting long side) mentions the Framework of OECD principles; Discloser and transparency in governance; and theoretical aspects of Corporate Governance and summary of different theories. Lastly theories discussed in this chapter should be viewed in the light of the business.

It also focuses on the development of Corporate Governance codes and their uses in different countries.

This chapter bears out the mechanism and overview. 4 Ps of Corporate Governance are enunciated with figure and models of Corporate Governance are discussed.

Here the definition of Good Governance and the paradigm shift in current states is discussed.

This chapter brings to bear some features of Corporate Governance and J.J. Irani Committee report in company law.

Lastly this chapter exposes the way leading to wealth creation in Governance and decision in Governance and decision making in investment with wisdom.

Chapter 2 – entirely devotes to board constitution and role and duties of director along with the board structure process and evaluation. In this chapter role of exercise director & independent director their duties, independence, remuneration, etc have been mentioned.

Disclosure and transparency of the board is highlighted to bring adopt the clear picture of the channels for disseminating information. Equitable treatment of shareholder is discussed to bring about the position of minority in the corporate body.

Lastly, performance measures are briefly dealt with to bring about the notion of good or bad corporations.

Chapter 3 – deals with the concept of the Board, Chairman and the CEO and their function. In the second chapter exposes the Audit mechanism and functions of Audit Committee besides an overview of Impact out Corporate Governance Committees.

The demand for Disclosure and Information culture is sine-qua-non for the success of Corporate Governance.

So this chapter gives their important application in PSEs and concomitant issues of the PSEs.

So what are the challenges for Corporate Governance in the Developing world. The answer is given succinctly in this chapter.

Chapter 4 – deals with corporate social responsibility which is concerned with the relationship between a corporation and a local society in which it resides or operates.

The principles of CSR incorporated in the chapter discuss vividly the basic principles together with the rational arguments for its justification.

Further the chapter focuses on CSR in the era of economic liberation and its impact of CG.

Other areas inserted in the chapter are Banking sector and how Corporate Governance system is laid out with reference to Indian Banks.

While discussing the Corporate misconduct and mis-governance, reference is made in Corporate Values; the core knowledge resources of business or industry.

Lastly measurement of Governance is put to expose the nature of corporate excellence.

The last part of this chapter refers to the theories of the firm and related issues leading to the solutions toward the change in good governance.

Chapter 5 – Relevant content relating to different legal elements, economic elements, socio cultural elements etc. Further the chapter pinpoints the excellence in leadership perceived by the corporate.

Some views of corporate governance classical views, etc contemporary views etc. mentioned along with ethical of corporate body.

In this chapter ethical approaches to corporate governance in India are mentioned. Should corporations must act ethically the answer given in this chapter highlighting the corporate philosophy and ethical practices in relation and morality.

Further innovation frame work in the content of corporate governance is mentioned to make the corporate governance system more effecting and offer the basis of modern good business. Information technology plays an important role in business today. So that is new in management information system has been discussed.

Lastly, the value creation in corporate governance is delineated. The same is mentioned in the context of world economic scenario and changing expectation of stakeholders & shareholders.

CONTENTS

CORPORATE GOVERNANCE

INTRODUCTION

Essentially CORPORATE GOVERNANCE, covers the gamut of activities having a direct or indirect effect on the health of the entity. Nobel Laureate MILTON FRIEDMAN stated; "Corporate Governance is to conduct business in accordance with Shareholders' desire while confirming to local laws and customs". More recently the President, World Bank—J. Wolfensohn made a more contemporary definition of the same. "Corporate Governance is all about promoting Corporate Fairness, Transparency and Accountability". How much of this is pertinent and really followed today remains to be seen.

Few Definitions

1. *"Corporate governance is a field in economics that investigates how to secure/ motivate efficient management of corporations by the use of incentive mechanisms, such as contracts, organizational designs and legislation. This is often limited to the question of improving financial performance, for example, how the corporate owners can secure/motivate that the corporate managers will deliver a competitive rate of return". "Corporate governance deals with the ways in which suppliers of finance to corporations assure themselves of getting a return on their investment".*

2. *"Corporate governance is the system by which business corporations are directed and controlled. The corporate governance structure specifies the distribution of rights and responsibilities among different participants in the corporation, such as, the board, managers, shareholders and other stakeholders, and spells out the rules and procedures for making decisions on corporate affairs. By doing this, it also provides the structure through which the company objectives are set, and the means*

of attaining those objectives and monitoring performance", OECD April 1999. OECD's definition is consistent with the one presented by Cadbury (1992).

3.   *"Corporate governance—which can be defined narrowly as the relationship of a company with its shareholders or, more broadly, as its relationship with society".*

Corporate Governance Employees Accounting Quality focuses on or any of the following variables: company accounting policies, disclosure standards, proactive adoption of accounting policy improvements, internal audit and control mechanisms for addressing auditor's queries. The top companies were ranked accordingly. Similarly, for "Value Creation Focus" business strategy (driven by value creation focus), effective use of cash surplus, capital structure, usage of IPO funds, shareholder friendliness are among the key variables. For "Fair policies among actions", the fund managers take the cue from fair treatment of minority shareholders, transparency of trades by top management and ethical behavior with customers, suppliers, tax authorities and government. Similar variables were used for ranking companies based on other parameters.

Historical Perspective

Corporate governance guidelines and best practices have evolved over a period of time. The Cadbury Report on the financial aspects of corporate governance, published in the United Kingdom in 1992, was a landmark. It led to the publication of the Vienot Report in France in 1995. This report boldly advocated the removal of cross-shareholdings that had formed the bedrock of French capitalism for decades. Further, The General Motors Board of Directors Guidelines in the United States and the Dey Report in Canada proved to be influential in the evolution of other guidelines and codes across the world. Over the past decade, various countries have issued recommendations for corporate governance. Compliance with these is generally not mandated by law, although codes that are linked to stock exchanges sometimes have a mandatory content.

The Sarbanes-Oxley Act, which was signed by the U.S. President George W. Bush into law in July 2002, has brought about sweeping changes in financial reporting. This is perceived to be the most significant change to federal securities law since the 1930s. Besides directors and auditors, the Act has also laid down new accountability standards for security analysts and legal counsels.

In November 2003, the SEC approved changes to the NYSE and NASDAQ listing requirements. The changes focused mainly on Board independence, independent committees of the Board, audit committee composition, code of business conduct and ethics and related party transactions.

The Higgs Report on non-executive directors and the Smith Report on audit committees, both published in January 2003, from part of the systematic review of corporate governance being undertaken in the U.K. and Europe. This is in light of recent corporate failures. The recommendations of these two reports are aimed at strengthening the existing framework for corporate governance in the U.K. Enhancing the effectiveness of the non-executive directors and switching the key audit relationship from executive directors to an independent audit committee are part of this. These recommendations are intended as revisions to the Combined Code on Corporate Governance.

In April 2004, the governments of the 30 Organization for Economic Cooperation and Development (OECD) countries approved a revised version of the OECD's Principles of Corporate Governance adding new recommendations for good practice in corporate behavior with a view to rebuilding and maintaining public trust in companies and stock markets. The revised principles call on governments to ensure effective regulatory frameworks and on companies to be more accountable. The principles include increased awareness among institutional investors, enhanced role for shareholders in executive compensation, greater transparency and effective disclosures to counter conflicts of interest.

In India, the Confederation of Indian Industry (CII) took the lead in framing a desirable code of corporate governance in April 1998. This was followed by the recommendations of the Kumar Mangalam Birla Committee on Corporate Governance. This committee was appointed by the Securities and Exchange Board of India (SEBI). The recommendations were accepted by SEBI in December 1999, and are now enshrined in Clause 49 of the Listing Agreement of every Indian stock exchange. SEBI also instituted a committee under the chairmanship of Mr. N. R. Narayana Murthy which recommended enhancements in corporate governance. SEBI has incorporated the recommendations made by the Narayana Murthy Committee on Corporate Governance in clause 49 of the listing agreement.

In addition, the Department of Company Affairs, Government of India, constituted a nine-member committee under the chairmanship of Mr. Naresh Chandra, former Indian ambassador to the U.S., to examine various corporate governance issues. The committee's recommendations are now mandatory.

Governance occurs just as a corporate entity; acquires life and more particularly when ownership of the enterprise is separated from its management. The term "governance" occurred for the first time in the writings of Chaucer. The phrase was not in vogue in the management literature until 1980. Adam Smith recognized the importance of corporate governance long back though he did not use the phrase. According to him :

> "The directors of the companies being managers of other people's money than their own, it cannot well be expected that they should watch

over it with the same anxious vigilance with which the partners in a private coparcenary frequently watch over their own".

19th Century – The century of the entrepreneur – The 19th century witnessed the laying down of the foundations of Modern Corporation. It was truly the century of the entrepreneur. The corporate sector witnessed the establishment of a large number of enterprises.

During 20th Century, there has occurred a phenomenal growth of management theory, consultants and management techniques. Great strides were made in the development of management ideas about marketing, finance, human resource development, operations, etc. However, the functioning of Boards did not receive much attention despite it being the principal governing organ in every corporate entity.

Concept of Corporate Governance

A company is an artificial and unnatural entity concerned with achieving, inter alia, the following goals:

  (i) Continuity, i.e. succession planning whereby a corporate enterprise may continue as a working unit.

 (ii) Stimulating the enterprise so that it is capable of indentifying the opportunities available at a particular juncture of time.

(iii) Facilitating constructive challenges within the business.

(iv) Identification of right challenges so as to make an appropriate allocation of resources.

A company attempts to achieve these objectives through the instrumentality of a group of people known as the board of directors, but the interests of the board may not always match those of the shareholders (owners of the company) on account of various reasons. It is in this context that the need of corporate governance arises. It may, however, be noted that governance is a dynamic concept which cannot be placed in the straitjacket of any single definition. Corporate Governance is drawn from diverse fields like laws, economics, ethics, politics, management, finance, etc. Consequently, a proper understanding of corporate governance requires familiarity with concepts, assumptions and vocabulary of each of these disciplines, besides the capacity to synthesize and transcend them.

Nigel MacDonald, a member of the Cadbury Committee, has warned against corporate governance as a system whereby a company controls its risks. This view leads to a negative approach instead of considering corporate governance as a System that stimulates entrepreneurial drive.

Following arc the different perspectives from which the concept of corporate governance has been defined by different scholars or agencies:

1. Simply stated, corporate governance is about 'performance as well as conformance'. It concerns management's power, responsibility, influence and accountability.

2. According to Ada Demb and Friedrich Neubauer "Corporate Governance is the process by which corporation is made responsive to the rights and wishes of stakeholders".

3. The Australian Committee on Corporate Governance is of the View that the Board's key role is to ensure that corporate management is continuously and effectively striving for above average performance, taking into account the factor risk.

4. Monks and Minow have defined corporate governance as "Relationships among various participants in determining the direction and performance of a corporation." According to them, the primary participants in a corporation are the tripod of shareholders management-led by the CEO, and the Board of Directors

5. Standard and Poor consider corporate governance as "the way a company is organized and managed to ensure that all financial stakeholders (shareholders and creditors) receive their fair share of a company's earnings and assets."

6. OECD has defined the corporate governance to mean "a system by which business corporations are directed and controlled". The corporate governance structure specifies the distribution of rights and responsibilities among different participants in the corporation such as the board, managers, shareholders and other stakeholders: and spells out the rules and procedures for making decisions in corporate affairs. By doing this, it provides the structure through which the company's objectives are set along with the means of attaining these objectives as well as for monitoring performance.

7. The term "corporate governance" connotes a blend of rules, regulations, laws and voluntary practices that enable companies to attract finance and human capital, perform efficiently and thereby maximize long-term value for the shareholders, besides respecting the aspirations of multiple stakeholders including those of the society.

8. Cadbury Committee (U.K.) has defined corporate governance as :

   "(it is) the system by which companies are directed and controlled" It may also be defined as a system of structuring, operating and controlling a company with the following specific aims:

(i) fulfilling long-term strategic goals of the owners,

(ii) taking care of the interests of the employees,

(iii) a consideration for the environment and local community,

(iv) maintaining excellent relations with both customers and suppliers, and

(v) proper compliance with all the applicable legal and regulatory requirements.

9. Corporate governance denotes the process, structure and relationship through which the Board of Directors oversee what the management does. It is also about being answerable to different stakeholders. It is concerned with the ways of bringing the interests of investors and managers in line and ensuring that firms are run for the benefit of investors.

10. According to Confederation of Indian Industry's Code, corporate governance refers to "an economic, legal and institutional environment that allows companies to diversify, grow, restructure and exit and do everything necessary to maximize long-term shareholder value". Commenting on a minimal definition, the CII Code states that "Corporate governance deals with laws, procedures, practices and implicit rules that determine a company's ability to take managerial decisions vis-a-vis its claimants. In particular, its shareholders, creditors, the State and employees".

11. Financial institutions have defined corporate governance as 'that philosophy by which owners and managers arc expected to be perennially responsive to other entities such as minority shareholders, promoters, institutional shareholders, etc.'

On an overall basis, corporate governance is an umbrella term encompassing various issues arising from interaction among senior management, shareholders, board of directors and other corporate stakeholders.

Scope of Corporate Governance

The scope of corporate governance tends to the following:

(i) **Structuring of Boards:** It covers aspects relating to the composition of boards, representation of insiders and outsiders on the Board, role of non-executive and independent directors.

(ii) **Board Procedures:** It covers aspects like convening of board meetings, frequency and attendance at board meetings, fulfilling information requirements of the board for decision making: constitution of variables board committees like audit committee, compensation committee, shareholders grievance committee etc.

(iii) Enhancing of shareholders' participation, disclosure of financial information and fulfilling shareholders' rights.

(iv) Industrial Democracy through co-determination on boards, setting up of ESOPs and representation of institutional investors.

Corporate governance also encompasses ethics values and morals of a corporation and it directors. Since governance is a kind of social contract between the company and its wider constituencies, it morally obliges the company and its directors to take account of other stakeholders' interests. It also involves monitoring and overseeing strategic direction in a social-economic and cultural context.

Thus, corporate governance is a system which involves the distribution of rights and responsibilities among different participants in the corporation such as the board, management, shareholders and other stakeholders. It spells out the rules and procedures for making decisions about corporate affairs. It also includes structures, processes, cultures and systems through which a company sets its objectives, determines the means of attaining those objectives and monitoring its performance.

Nature of Corporate Governance

Human society needs governing. Whenever power is exercised to direct, control and regulate social activity affecting people's legitimate interests, governance comes to the fore. Governance is necessary whether the social or economic group is a nation, State, a professional body or a business corporation. Each has its governing body and particular governance activities. Governance identifies rights and responsibilities, legitimizes actions and determines accountability. It is concerned with derivation, uses and limitations of power.

Corporate governance is concerned with the process by which corporate entities and particularly limited liability companies are governed. Specifically, it is concerned with the exercise of power over the direction of the enterprise, the supervision of executive actions, the acceptance of duty to be accountable, and the regulation of the corporation within the jurisdiction of the State in which it operates. Primarily corporate governance is concerned with the board of directors, its structure, style and processes, their relationships and roles, the linkages and activities, as well as the roles of company's members, auditors and others.

The appointment of Cadbury Committee and the adoption of Cadbury Report in 1992 was done due to concern about the inadequacy of financial control and accountability.

The focus of corporate governance in India is to improve disclosures and compliances, upgrade corporate governance practices, and facilitate the integration of Indian business with their global counterparts, and to revitalize the bourses and so on.

The following points may, however, help in understanding the nature of Corporate Governance :

- Corporate governance is not an end in itself. Rather it is an important element in building up international competitiveness. Each nation needs strong boards, strong corporate managements and strong investors, all working under a deliberately created environment of creative tension so as to ensure good corporate governance.

- The purpose of corporate governance is to achieve a responsible, value-oriented management and control of a corporation.

- Corporate governance is only a part of the larger economic context in which business firms operate.

- Corporate governance is affected by the relationships among participants in the governance system.

- Corporate governance is not a "one size fits all" proposition. There may be a wide diversity of approaches to corporate governance and it is quite appropriate. A corporation's practice evolves as it adapts to changing situations.

- Corporate governance is concerned with formalising, clarifying decision-making process within the organization.

Franks and Mayer (1992) have distinguished between two systems of corporate governance viz. (i) the outsider system and (ii) the insider system as follows :

- Outsider System - Dominant in the UK and the USA. It is characteristic of economies with large number of quoted companies, a liquid capital market and little concentration of shareholdings. It relies on the market and outside investors for corporate control.

- Insider System - Attributed to continental Europe and Japan, it is characterized by a small number of listed companies, an illiquid capital market and a high concentration of shareholding in the hands of corporations, institutions, families, etc. This model uses a system of interlocking networks and committees.

Corporate governance has two aspects which are as follows :

1. **Internal Aspects:** These refer to the set of organisational rules within a company. These aspects comprise of the following :

    (a) Sound internal processes and procedures

    (b) Sound corporate philosophy based on ethical principles

    (c) Good quality leadership by the board and senior management

    (d) The management's mindset imbued by vision, sense of responsibility, respect for law and value

2. **External Aspects :** These aspects are reflected through a corporate entity's focus on profit optimisation rather than maximisation as well as the existence of smooth relation between the company and the various stakeholders. These aspects relate to the assessment of performance of the company through market mechanism. Both the above aspects must ensure the ushering of the corporation into an era where no single person or group could command concentrated decision-making power.

Governance vis-a-vis Management

Corporate governance is concerned with the way the directors control the activities of the company and ensure that the management to whom they delegate many functions are accountable. A good system or corporate governance should enable responsibility to be clearly identified. Directors have the responsibility for governance of their companies. Etymologically, the word "governance" seems to have been derived from the Latin word "governance" which means "to rule", "to steer" and the Greek for steerman. Governance is concerned with the intrinsic nature, purpose, integrity and identity of the institution with primary focus on the entity's relevance, continuity and the fiduciary aspects. It comprises of a process structure and relationship through which the board oversees what the executives do.

Basically management is concerned with supervisory actions involving judicious use of means to accomplish certain ends. It focuses on specific goal attainments over a definite time-frame in given organization. Thus, corporate management means what the executives do to define and achieve the objectives of the company.

The following diagram depicts the relationship between corporate governance and management :

Fig. 1.1: The Activities of Governance and Management Compared

Following are the vital differences between governance and management :

1. *Difference about focus:* Governance has an external focus whereas the management focus is internal.

2. *Difference in assumption:* Governance assumes an open system whereas the management assumes the system to be closed.

3. *Difference in orientation:* Governance is strategy-oriented. Management is task/oriented.

4. *Difference in approach:* Governance relates to where the company is to go. Management is concerned with getting the company to the targeted goal.

5. *Difference in who performs each of the above*: Governance is the job of the board of directors whereas management is the job of the executives.

The four principal activities of corporate governance are :

- Direction : It concerns the formulation of strategic direction for the future of he organization in the long term.
- Executive Action : It signifies involvement in crucial executive decisions
- Supervision : It encompasses the monitoring and overseeing of management performance.
- Accountability : It is a recognition of responsibilities to those making legitimate demand for accountability.

Objectives of Corporate Governance

Good governance is integral to the existence of a company. It inspires and strengthens investor confidence by ensuring company's commitment to higher growth and profits. It aims at the following :

(i) Installation of a properly structured board 'which is capable of taking independent and objective decisions

(ii) Ensuring a properly balanced board having representation of an adequate number of non-executive independent directors capable of taking care of the interests of all the stakeholders

(iii) Adoption of transparent procedures and practices in decision-making by an informed comity of board members

(iv) Effective and regular monitoring of management functioning by the Board

(v) Disclosures to shareholders with a view to help them become informed of the relevant developmentsimpacting the company

(vi) Exercise of effective control on corporate affairs by the board at all times

The overall objectives of governance should be to maximize long-term value and shareholders' wealth.

Fundamental principles of corporate governance

Governance styles may be as different as the nature of companies. It was for this reason that Adrian Cadbury had cautioned against there being any simple style of governance suiting all types of companies. Every company may have a unique governance style of its own. Despite uniqueness of styles, there are some fundamental principles noted below that permeate all kinds of governance styles :

(i) **Transparency:** It involves the explaining of company's policies and actions to those to whom it owes responsibilities. It should lead to the making of appropriate disclosures without jeopardizing company's strategic interests. Internally, transparency means openness in a company's relationship with its employees as well as the conduct of its business in a manner that will bear scrutiny. Unfortunately, total transparency is not the dominant culture in corporations. Transparency requires courage. The destruction in shareholder value in Enron and Marcona did not occur due to wrong business decisions but because they did not share their setbacks with the shareholders.

(ii) **Accountability:** It signifies that the Board of Directors are accountable to the shareholders and management is accountable to the Board of Directors. Both the board and management must be accountable to the shareholders for the performance of tasks assigned to them. Its thrust is to ensure effective management of resources and achievement of results with efficiency coupled with empowerment. Accountability provides impetus to performance.

(iii) **Trusteeship:** Large corporations have both a social and economic purpose. They represent a coalition' of interests of shareholders, lenders of capital as well as business associates and employees. It casts a responsibility of trusteeship on the Board of Directors who must act to protect and enhance shareholder value. The Board must ensure that the company fulfils its obligations and responsibilities to its other stakeholders. Inherent in the concept of trusteeship is the responsibility to ensure equity, namely, that the rights of all shareholders, large or small, are protected.

(iv) **Empowerment:** It signifies that management must have the freedom to drive the enterprise forward. It is a process of actualizing potential of its employees. Empowerment unleashes creativity and innovation throughout the organization by truly vesting decision-making powers at the most appropriate levels in the organizational hierarchy.

(v) **Ethics:** A corporation must set exemplary standards of ethical behaviour both within the organization and in its external relationship. Deviation from ethical principles corrupts organizational culture and underlines stakeholder value.

(vi) **Oversight:** It means the existence of a system of checks and balances. It should prevent misuse of power and facilitate timely management response to change and risks.

(vii) **Fairness to all stakeholders:** It involves a fair and equitable treatment of all participants in the corporate governance structure. There should be no discrimination between two equally situated participants. For example, the voting rights between the same category of shareholders must be uniform.

Factors Affecting the Quality of Corporate Governance

(i) **Definition of Roles and Powers of Board:** The absence of clearly defined roles and powers of the board weakens accountability mechanism and threatens the achievement of organisational goals.

Therefore, the foremost requirement of good governance is the clear identification of powers, roles, responsibilities and accountability of the Board, CEO and Chairman of the Board. The role of the Board should be clearly documented in a Board Charter.

(ii) **Legislation:** Clear and unambiguous legislation and regulations are fundamental to effective corporate governance. Legislation that requires continuing legal interpretation, or that is difficult lo interpret on a day to day basis is prone to deliberate manipulation or inadvertent misunderstanding.

(iii) **Management Environment:** Management environment includes setting clear objectives and appropriate ethical framework; establishing clear cut processes, providing for transparency; clear enunciation of responsibility and accountability, implementing sound business planning, encouraging business risk assessment, having right people and right skills for the jobs, establishing clear boundaries for acceptable behavior, establishing performance measures, evaluating performance and sufficiently recognizing individual and group contribution.

(iv) **Board Skills:** To be able to undertake its functions efficiently and effectively, the Board must possess the necessary blend of qualities, skills, knowledge and experience. Each of the directors should bring unique and valuable quality. A Board should collectively have a mix of the following skills, knowledge and experience :

- Operational or technical expertise including policy skills and leadership experience.
- Financial skills,
- Legal skills, and
- Knowledge of Government and regulatory requirements.

(v) Board Appointments: To ensure that the best people are appointed to the Board, the Board position should be tilted through a wide external search. There should be a well-defined and open procedure for appointment of new directors. The appointment process should meet all statutory and administrative requirements. Consideration should be given to the skill requirements of the Board. All new directors should be provided with a letter of appointment setting out their duties and responsibilities.

(vi) Board Induction and Training: Directors must have a broad understanding of the area of operation or business, corporate strategy and current issue facing the Board.

(vii) Board Independence: An Independent Board is essential to sound corporate governance. This goal can be achieved by having a majority of independent directors. The director's independence is necessary to ensure that there are no actual or perceived conflicts of interest.

(viii) Board Meetings: Directors must devote time and attention necessary to fulfill their obligations. As a minimum, regular director attendance at Board meetings is important. Board meetings provide the forum for decision-making by the members of the Board. These meetings enable directors to discharge their responsibilities.

(ix) Board Resources: Board members should have sufficient resources to enable them to discharge their duties effectively. It includes the facility of access by directors to independent legal and professional advice at the company's expense. The costs of supporting the Board should be transparent and reported.

(x) Code of Conduct: It is essential that the organisation's guiding ethics and code of conduct are clearly understood and followed by each member of the organization and communicated to all stakeholders.

(xi) Strategy Setting: The objectives of the company must be clearly documented in the form of a long term corporate strategy and an annual business plan together with achievable and measurable performance targets and milestones.

(xii) Business and Community Consultation: Though basic activity of a business entity is inherently commercial, yet it must also take care of its community obligations.

(xiii) Financial and Operational Reporting: The Board requires comprehensive, regular, reliable, timely and relevant information in a form and of a quality that is appropriate to discharge its function of monitoring corporate performance. For this purpose, clearly defined performance measures–financial and non-financial, should be established which enable the efficiency and effectiveness of the organisation to be assured.

(xiv) Monitoring the Board performance: The Board must monitor its own performance and also that of the individual directors at periodic intervals. This could be done by using key performance indicators besides peer review. The Board should establish an appropriate mechanism for reporting the results of Board assessments.

(xv) Audit Committee: The Committee is responsible for liaisioning with the management, internal and statutory auditors for reviewing the adequacy of internal control and compliance with significant policies and procedures, and finally reporting to the Board on the audit work.

(xvi) Risk Management: Risk is an important element of corporate governance. There should be a clearly established process of identifying, analyzing and treating risk, the absence of which could prevent the company from effectively achieving its objectives.

Corporate Accountability and Responsibility

* Board of Directors must be independent
* Audit committee gains total control over external audits and auditors
* Audit committee must include a financial expert
* Enhanced whistle blower regulations
* CEOs and CFOs must certify financial statements and the evaluation and effectiveness of internal procedures and controls
* Enhanced insider trading regulations for executives and board members
* New limitations on executive loans
* SEC will adopt new rules to address securities analyst's conflict of interest
* SEC will receive additional funding and conduct additional studies
* Enhanced document retention rules

- Increased criminal fraud accountability with increased criminal and civil penalties
- SEC recommends that CEO sign tax returns

Internal Procedures and Controls

- On an annual basis, CEOs and CFOs must evaluate, document and test internal procedures and controls related to financial reporting
- External auditors must attest to the effectiveness of these controls as part of their annual audit process

Audit and Accounting

- Auditors must register with the SEC
- Auditors must be independent, with new limitations imposed on non-audit work
- All non-audit services must be pre-approved by the audit committee
- Audit partners must limit their time with accounts and rotate new partners

Enhanced Disclosure and Reporting Requirements

- Filing requirements are accelerated
- Enhanced disclosure on internal procedures and controls
- Any non-GAAP financials must be disclosed
- All material of balance-sheet items must be disclosed

Codes and Guidelines

Corporate governance principles and codes have been developed in different countries and issued from stock exchanges, corporations, institutional investors, or associations (institutes) of directors and managers with the support of governments and international organizations. As a rule, compliance with these governance recommendations is not mandated by law, although the codes linked to stock exchange listing requirements may have a coercive effect. For example, companies quoted on the London and Toronto Stock Exchanges formally need not follow the recommendations of their respective national codes. However, they must disclose whether they follow the recommendations in those documents and, where not, they should provide explanations concerning divergent practices. Such disclosure requirements exert a significant pressure on listed companies for compliance.

In contrast, the guidelines issued by associations of directors, corporate managers and individual companies tend to be wholly voluntary. For example, The GM Board Guidelines reflect the company's efforts to improve its own governance capacity. Such documents, however, may have a wider multiplying effect prompting other companies to adopt similar documents and standards of best practice.

1.1. DIFFERENCE OF GOVERNING AND GOVERNANCE

A working definition of 'social-political' or 'interactive' governance, or simply governing and governance – the elements of which will be clarified as follows :

Governing can be considered as the totality of interactions, in which public as well as private actors participate, aimed at solving societal problems or creating societal opportunities; attending to the institutions as contexts for these governing interactions; and establishing a normative foundation for all those activities.

Governance can be seen as the totality of theoretical conceptions on governing.

Social-political governance means using an analytical and normative perspective on any societal governance that is 'collective'. Collective activities as a public task (the 'state') and the private sector (the market), or of the third sector (civil society) are not in isolation, but as a shared set of responsibilities. Public governing came out with an eye to 'private carrying capacities', and private governing carried out with an eye to 'public carrying capacities'. Some of those activities are of a self-governing nature, others are considered to be co-governing, and there is also a place for authoritative or hierarchical governing.

The perspective of this study also can be called interactive governance. Conceptually interaction plays a dominant role and forms, models and modes of governing interactions are key conceptual elements. Interaction as a social phenomenon, and governing interactions as a specific type of them, are a rich source for analyzing and synthesizing insights into many facets of governance. In particular, in governing interactions the diversity, dynamics and complexity of governance issues in modern societies can be expressed. Interaction thus becomes a 'linking pin' between societal attributes and governance qualities by the quality of its governing interactions.

Governance as a concept is not new but currently it is being treated more systematically, and this might be expected to continue. The state is the central governing actor, and state society relations are the focus.

We are still in a period of creative disorder concerning governance, optimistic that governance theory has tremendous potential in opening up alternative ways of looking at political institutions, domestic-global linkages, transnational cooperation, and different forms of public-private exchange. Whether this potential will be fulfilled depends on certain conditions being met in particular a number of boundaries will have to be crossed; conceptual boundaries, boundaries between theory and practice and last but not least between 'world views'.

In order to develop systematic ideas on governance such social science disciplines as political science, public administration/ management, sociology, international relations and some institutional economics, Generally governance is examined through an exploration 'in breadth' of its many aspects and manifestations, rather than as a systematic theoretical exercise 'in depth'. The relation between concepts and theory is perhaps double-side, may be even paradoxical, without theory there are no concepts; without concepts there is no theory. The proper concepts are needed to formulate a good theory, but we need a good theory to arrive at the proper concepts.

The governance approach focuses on the interactions taking place between governing actors within social-political situations. These interactions give human actions as attempts made toward understanding the diversity, complexity and dynamics of these situations. In doing so there is scope for influencing societal features that occur between the 'modern' and the 'post-modern'. A growing awareness and acceptance of different modes of governance, including 'self-governing' capacities of social systems, 'co-governing' arrangements and hierarchical governing; each contributing in single and mixed modes to questions of governability in a broader sense. Openness to difference, a willingness to communicate, and a willingness to learn are important criteria in coping with societal diversity, dynamics and complexity. Substantive criteria on which basis actors are willing and able to interact with each other and accept each other's boundaries are also needed.

Global standardization may be useful and necessary, the same norms apply to caveat local and autonomy. Some insights are more useful than others, and some truths seem better than others. This true statement could make diversity of opinion the lifeblood of social-political science and of social-political governance. If insight and truth is in the eye of the beholder, then what we need are social-political governing processes (and structures) that take both interactions and actors seriously.

Interactions shape actors and actors shape interactions; they are equal as basic units of analysis and theory development. The governance perspective starts from the diversity, dynamics and complexity of the societies to be governed – and the governance themselves. These societies need order, but nothing can change without dynamics. They required similarity to enable communication, but diversity to gain new insights. The standards need to reduce uncertainty risk, and complexity to solve problems and create opportunities.

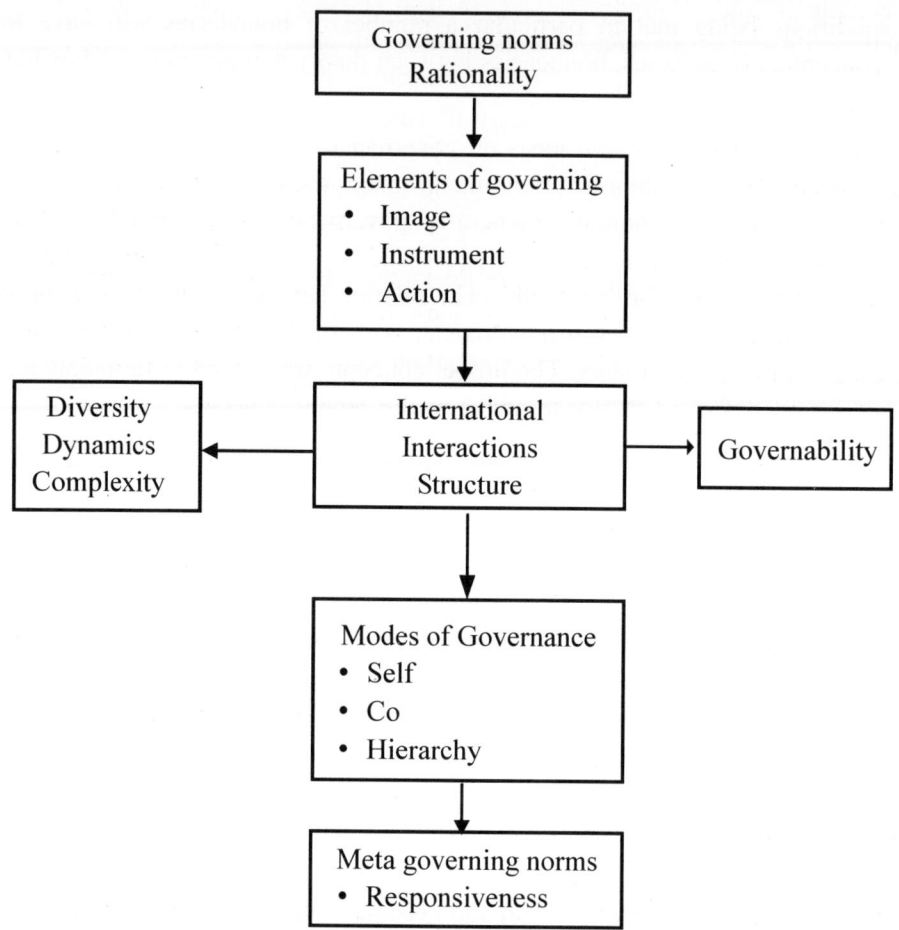

Schemes of analysis :
(Governing as governance : by Jan Kooimen)

Societal diversity, dynamics, complexity and interactions are inevitable

In the governance perspective emphasizing interactions and particularly, governing-as-interaction(s), it is essential not to the lose sight of the actors, diverse as they may be. Indeed, actors cannot be separated from the role they play in the interactions that occur. We are used to considering individuals and other entities as being independent from the interactions they participate in. They interact and, seemingly, can stop at will. But actors are continuously shaped by (and in) the interaction in which they relate to each other.

In the interaction concept tensions between the action and the interaction can be considered to be the expression of societal dynamics.

The complexity of social-political systems can be seen in the fact that a multitude of interactions take place in many different forms and at different intensities. Indeed, most interactions in governing will occur concurrently and inter-dependently of others. Such interactions can only be influenced if aspects of complexity are sufficiently understood. To govern social political problems and opportunities requires clarity about the nature of interactions involved in a problem to be tackled or an opportunity to be created, the way these interactions interrelate, and their characteristic patterns.

In governing interactions insight can be gained in the diversity, dynamics and complexity of modern societies, and the way these features appear in governance issues.

The three types of infractions is also important in relation to modes of governance: interferences to self-governance, interplays to co-governance, and interventions to hierarchical governance. They are the object of governing, but at the same time can be seen as forms of governance themselves; an interplay in a work situation is a form of work organization that thus governs the activities of the workers.

Governing actors have a choice of different types of interactions to participate in at the intentional level of governing interactions; by doing so they explicitly or implicitly contribute to the structure of these types of interactions as modes of governance.

Modes of Governance

Self Governance

Self-governance is an important mode of societal governance. Self governance refers to the capacity of social entities to govern themselves autonomously.

Co-Governance

The implications of what co-governance as a special mode of governance may mean in the broader perspective the governance element between interacting parties and something 'in common' to pursue together, in some way autonomy and identity. Coordination, in my opinion, does not fit this definition. Coordination is arguably a mechanism in which parties adjust their behavior without their identities or autonomy being directly involved.

Collaboration and Co-operation

Collaboration and Co-operation may seem to be two words for the same thing, but it is important to differentiate the two. Collaboration is the less formal of the two and co-operation the more formal. Concerning collaboration, ideas on 'co' are presented from both an action research tradition that paid attention to empirical detail and practical experience, and the second a political-economic tradition, strong in deductive analytical reasoning. Both are important contributions to our insight in 'co' behavior, 'co' processes, and 'co' structures, and both offer a basis for further thinking and conceptualizing 'co' phenomena in modern governance.

Communicative Governance

New patterns of governance are addressed to stimulate learning processes, leading to cooperative behavior and mutual adjustment, in order that responsibility for managing change is shared by all or most involved actors.

Communicative governance suits governing situations where those involved in governing interplays are willing to reach inter-subjective understanding for co-governing purposes.

1.2. THE NEED FOR CORPORATE GOVERNANCE

A corporation is a congregation of various stakeholders, namely, customers, employees, investors, vendor partners, government and society. A corporation should be fair and transparent to its stakeholders in all its transactions. This has become imperative in today's globalized business world where corporations need to access global pools of capital, need to attract and retain the best human capital from various parts of the world, need to partner with vendors on mega collaborations and need to live in harmony with the community. Unless a corporation embraces and demonstrates ethical conduct, it will not be able to succeed.

Corporate governance is about ethical conduct in business. Ethics is concerned with the code of values and principles that enable a person to choose between right and wrong, and therefore, select from alternative courses of action. Further, ethical dilemmas arise from conflicting interests of the parties involved. In this regard, managers make decisions based on a set of principles influenced by the values, context and culture of the organization. Ethical leadership is good for business as the organization is keen to conduct its business in line with the expectations of all stakeholders.

Corporate governance is beyond the realm of law. It stems from the culture and mindset of management, and cannot be regulated by legislation alone. Corporate governance deals with conducting the affairs of a company such that there is fairness to all stakeholders and that its actions benefit the greatest number of stakeholders.

Good Corporate Governance requires the cooperation of all the stakeholders; and such cooperation is enhanced by the corporation adhering to the best corporate governance practices.

Corporate governance is a key element in improving the economic efficiency of a firm. Good corporate governance also helps to ensure that corporation takes into account the interests of a wide range of constituencies, as well as of the communities within which they operate. Further, it ensures that their Boards are accountable to the shareholders. This, in turn, helps enure that corporations operate for the benefit of society as a whole. While large profits can be made taking advantage of the asymmetry between stakeholders in the short run, balancing the interests of all stakeholders along will ensure survival and growth in the long run. This includes, for instance, taking into account societal concerns about labour and the environment.

The failure to implement good governance can have a heavy cost beyond regulatory problems. Evidence suggests that companies that do not employ meaningful governance procedures can pay a significant risk premium when competing for scarce capital in the public markets.

The credibility offered by good corporate governance procedures also helps maintain the confidence of investors both foreign and domestic to attract more "patient", long-term capital, and will reduce the cost of capital. This will ultimately induce more stable sources of financing.

Corporate scandals are the result of financial crises. For instance, the Asian financial crisis brought the subject of corporate governance to the surface in Asia. Further, recent sandals disturbed the otherwise placid and complacent corporate landscape in the US. These scandals, in a sense, proved to be serendipitous. They spawned a new set of initiatives in corporate governance in the US and triggered fresh debate in the European union as well as in Asia. The many instances of corporate misdemeanors have also shifted the emphasis on compliance with substance, rather than form, and brought to sharper focus the need for intellectual honesty and integrity. Proper financial and non-financial disclosures made by any firm are only as good and honest as the rations are governed properly should be allowed to be a part of committees. This includes the Prime Minister and Finance Minister's advisory councils, committees set up by the Confederation of Indian Industry (CII) the Securities and Exchange Board of India ("SEBI"), the Department of Company Affairs, ministries and the boards of large banks and financial institutions.

Corporate governance initiatives in India began in 1998 with the Desirable Code of Corporate Governance a voluntary code published by the CII, and the first formal regulatory framework for listed companies especially for corporate governance, established by the SEBI. The later was made in February 2000, following the recommendations of the Kumarmangalam Birla Committee Report.

The majority of the definitions articulated in the codes relate corporate governance to "control" of the company, of corporate management, or of company conduct or managerial conduct. Perhaps the simplest and most common definition of this sort or that provided by the Cadbury Report (UK), which is frequently quoted or paraphrased; "Corporate governance is the system by which businesses are directed and controlled".

The definition in the preamble of the OECD principle is also all encompassing Corporate governance..... involves a set of relationships between a company's management, its board, its shareholders and other stakeholders. Corporate governance also provides the structure through which the objectives of the company are set and the means of attaining those objectives and monitoring performance are determined.

The most common school of thought would have us believe that if management is about running business, governance is about ensuring that it is run properly. All companies need governing as well as managing. The aim of "Good Corporate Governance" is to enhance the long-term value of the company for its shareholders and all other partners. The enormous significance of corporate governance is clearly evident in this definition, which encompasses all stakeholders. Corporate governance integrates all the participants involved in a process, which is economic, and at the same time social. This definition is deliberately broader than the frequently heard narrower interpretation that only takes account of the corporate governance postulates aimed at shareholder interests.

Studies of corporate governance practices across several countries conducted by the Asian Development Bank (2000), International Monetary Fund (1999), Organization for Economic Cooperation and Development ("OECD") 1999 and the World Bank (1999) reveal that there is no single model of good corporate governance. This is recognized by the OECD code. The OECD code also recognizes that different legal systems, institutional frameworks and traditions across countries have led to the development of a range of different approaches to corporate degree of priority placed on the interest of shareholders, who place their trust in corporations to use their investment funds wisely and effectively. In addition, best managed corporations also recognize that business ethics and the corporate awareness of the environmental and societal interest of the communities within which they operate, can have an impact on the reputation and long-term performance of corporations.

The Kumarmangalam Birla Committee on Corporate Governance

SEBI had constituted a Committee on May 7, 1999 under the chairmanship of Shri Kumarmangalam Biral, then Member of the SEBI Board "to promote and raise the standards of corporate governance". Based on the recommendations of this Committee, a new clause 49 was incorporated in the Stock Exchange Listing Agreements ("Listing Agreements").

The recommendations of the Kumarmangalam Birla Committee on Corporate Governance (the "Recommendations") are set out to this report.

Financial reporting and disclosures

Financial disclosure is a critical component of effective corporate governance. SEBI set up an Accounting Standards Committee, as a Standing Committee, under the chairmanship of Shri Y. H. Malegam with the following objectives:

- To review the continuous disclosure requirements under the listing agreement for listed companies;

- To provide input to the Institute of Chartered Accountants of India ("ICAI") for introducing new accounting standards in India; and

- To review existing Indian accounting standards, where required and to harmonise these accounting standards and financial disclosures at par with international practices.

SEBI has interacted with the ICAI on a continuous basis in the issuance of recent Indian accounting standards on areas including segment reporting, related party disclosures, consolidated financial statements, earnings per share, accounting for taxes on income, accounting for investments in associates in consolidated financial statements, discontinuing operations, interim financial reporting, intangible assets, financial reporting of interests in joint ventures and impairment of assets.

With the introduction of these recent Indian accounting standards, financial reporting practices in India are almost on par with International Accounting Standards.

Compliance with the Code and SEBI's Experience

All companies are required to submit a quarterly compliance report to the stock exchanges within 15 days from the end of a financial reporting quarter. The report has to be submitted either by the Compliance Officer or by the Chief Executive Officer of the company after obtaining due approvals. SEBI has prescribed a format in which the information shall be obtained by the Stock Exchanges from the companies. The companies have to submit compliance status on eight sub-clauses namely:

- Board of Directors;
- Audit Committee;
- Shareholders/Investors Grievance Committee;
- Remuneration of directors;
- Board procedures;

- Management;
- Shareholders; and
- Report on Corporate Governance.

Stock exchanges are required to set up a separate monitoring cell with identified personnel, to monitor compliance with the provisions of the Recommendations. Stock exchanges are also required to submit a quarterly compliance report from the companies as per the Schedule of Implementation. The stock exchanges are required to submit a consolidated compliance report within 30 days of the end of the quarter to SEBI.

The status of compliance with respect to provisions of corporate governance must be enforced.

- The compliance level in respect of requirements relating to Board of Directors, Audit Committee, Shareholders Grievance Committee and Shareholders should be very high.

SEBI observed that the compliance with the requirements in clause 49 of the Listing Agreement is, by and large, satisfactory; however, an analysis of the financial statements of companies and the report on corporate governance discloses that their quality is not uniform. This is observed on parameters such as the nature of qualifications in audit reports, the quality of the corporate governance report itself (which is often perfunctory in nature), and the business transacted and the duration of audit committee meetings. Variations in the quality of annual reports, including disclosures, raises the question whether compliance is in form or in substance; and emphasise the need to ensure that the laws, rules and regulations do not reduce corporate governance to a mere ritual. This question has come under close scrutiny in recent times.

Rationale for a review of the Code

SEBI believes that efforts to improve corporate governance standards in India must continue. This is because these standards are themselves evolving, in keeping with market dynamics. Recent events worldwide, primarily in the United States, have renewed the emphasis on corporate governance. These events have high lighted the need for ethical governance and management, and for the need to look beyond mere systems and procedures. This will ensure compliance with corporate governance codes, in substance and not merely in form.

Again, one of the goals of good corporate governance is investor protection. The individual investor is at the end of a chain of financial information, stretching from corporate accountants and management, through Boards of Directors and audit committees, to independent auditors and stock market analysts, to the Investing public. Many of the links in this chain need to be strengthened or replaced to preserve its integrity.

SEBI, therefore, believed that a need to review the existing code on corporate governance arose from two perspectives (a) to evaluate the adequacy of the existing practices, and (b) to further improve the existing practices.

In the contet of ratinal set out in this report, SEBI believed if necessary to form a committee as corporate Governance comprising representatives from the stock exchangechambers of commerce, investor, professional bodies recommendate.

— Fairness Does the recommendation enhance fairness to all stakeholders, by minimizing asymmetry of benefits?

— Accountability Does the recommendation make corporate management more accountable?

— Transparency Does the recommendation enhance transparency?

— Ease of implementation Is the recommendation easy to implement?

— Verification Is the recommendation objective verifiable?

— Enforcement Can the recommendation be effectively enforced?

The ratings received from members were first aggregated across recommendations and tabulated. Recommendations whose ratings were 7 and above were then aggregated, on each of the seven parameters set out above. The rating score for each such recommendation was aggregated.

Key Issues Discussed and Recommendations

The key issues debated by the Committee and the related recommendations are discussed below.

Audit Committees

Review of information by audit committees

Suggestions were received from members that audit committees of publicly listed companies should be required to review the following information mandatarily :

• Financial Statements;

• Management discussion and analysis of financial condition and results of operations;

Reports relating to compliance with laws and to risk management;

Management letters/letters of internal control weaknesses issued by statutory/ internal auditors; and

Records of related party transactions.

In view of the above deliberations, the Committee makes the following mandatory recommendation :

Audit committees of publicly listed companies should be required to review the following information mandatorily :

— Financial statements and draft audit report, including quarterly /half-yearly financial information;

— Management discussion and analysis of financial condition and results of operations;

— Reports relating to compliance with laws and to risk management;

— Management letters/letters of internal control weaknesses issued by statutory / internal auditors; and

— Records of related party transactions.

— Financial literacy of members of the audit committee.

Non-mandatory recommendation

Companies should be encouraged to move towards a regime of unqualified financial statements. This recommendation should be reviewed at an appropriate juncture to determine whether the financial reporting climate is conducive towards a system of filing only unqualified financial statements.

Board disclosures

The Committee believes that it is important for corporate Boards to be fully aware of the risks facing the business and that it is important for shareholders to know about the process by which companies manage their business risks.

In light of this, it was suggested that procedures should be in place to inform Board members about the risk assessment and minimization procedures. These procedures should be periodically reviewed to ensure that executive management controls risk through means of a properly defined framework. These risks will include global risks; general, economic and political risks; industry risks; and company specific risks.

Nominee directors

Exclusion of nominee directors from the definition of independent directors.

It was suggested that nominee directors should be excluded from the definition of independent directors.

The Committee felt that the institution of nominee directors creates a conflict of interest that should be avoided. Such directors often claim that they are answerable only to the institutions they represent and take no responsibility for the company's management or fiduciary responsibility to other shareholders. It is necessary that all directors, whether representing institutions or otherwise, should have the same responsibilities and liabilities.

Committee further makes recommendation

There shall be no nominee directors.

Where an institution wishes to appoint a director on the Board, such appointment should be made by the shareholders.

An institutional director, so appointed, shall have the same responsibilities and shall be subject to the same liabilities as any other director.

Nominee of the Government on public sector companies shall be similarly elected and shall be subject to the same responsibilities and liabilities as other directors.

Independent Directors

Definition of independent directors

The Committee noted that the definition of independent directors should be clarified in the recommendations. It observed that the definition of independent directors as set out in the code of the International Corporate Governance Network may be referred to. The Committee also noted that the Naresh Chandra Committee report has attempted to define the term "independent director". The Committee was of the view that the same definition may be used to define independent directors.

An issue often raised in the context of independence is whether independent directors are entitled to any material benefits from the company other than sitting fees, remuneration, and travel and stay arrangements. Such benefits include stock options and performance bonuses that executive directors may be entitled to. The central issue is whether such benefits serve as incentives or hindrances to the objectivity of decision-making and hence, compromise its quality. It also needs to be considered that restrictions such as these could disenchant a person from accepting the position of independent director that carries onerous responsibilities without appropriate reward.

The Committee decided that the term "Independent director" shall have the same meaning as contained in report of the Naresh Chandra Committee.

Mandatory recommendation

SEBI should make rules for the following :

- Disclosure in the report issued by a security analyst whether the company that is being written about is a client of the analyst's employer or an associate of the analyst's employer, and the nature of services rendered to such company, if any; and,

- Disclosure in the report issued by a security analyst whether the analyst or the analyst's employer or an associate of the analyst's employer hold or held (in the 12 months immediately preceding the date of the report) or intend to hold any debt or equity instrument in the issuer company that is the subject matter of the report of the analyst.

Recommendations of the Naresh Chandra Committee

This Report states that the Committee would also recommenced that the following mandatory recommendations in the report of the Naresh Chandra Committee, relating to corporate governance, be implemented by SEBI.

This section sets out such recommendations of the Naresh Chandra Committee that were considered by this Committee.

Disclosure of Contingent Liabilities (Naresh Chandra Committee Report)

The Committee makes the following mandatory recommendation :

Management should provide a clear description in plain English of each material contingent liability and its risks, which should be accompanied by the auditor's clearly worded comments on the management's view. This section should be highlighted in the significant accounting policies and notes on accounts, as well as, in the auditor's report, where necessary.

This is important because investors and shareholders should obtain a clear view of a company's cotangent liabilities as these may be significant risk factors that could adversely affect the company's future financial condition and results of operations.

CEO/CFO Certification (Naresh Chandra Committee Report)

The Committee makes the following mandatory recommendation :

For all listed companies, there should be a certification by the CEO (either the Executive Chairman or the Managing Director) and the CFO (whole-time Finance Director or other person discharging this function) which should state that, to the best of their knowledge and belief :

- They have reviewed the balance sheet and profit and loss account and all its schedules and notes on accounts, as well as the cash flow statements and the Directors' Report;

- These statements do not contain any material untrue statement or omit any material fact nor do they contain statements that might be misleading;

- These statements together present a true and fair view of the company, and are in compliance with the existing accounting standards and/ or applicable laws/ regulations;

- They are responsible for establishing and maintaining internal controls and have evaluated the effectiveness of internal control systems of the company; and they have also disclosed to the auditors and the Audit Committee, deficiencies in the design or operation of internal controls, if any, and what they have done or propose to do to rectify these;

- They have also disclosed to the auditors. as well as the Audit Committee, instances of significant fraud, if any, that involves management or employees having a significant role in the company's internal control systems; and

- They have indicated to the auditors, the Audit Committee and in the notes on accounts, whether or not there were significant changes in internal control and/ or of accounting policies during the year.

Naresh Chandra Committee Report

In this Report,

The Committee makes the following mandatory recommendation :

All audit committee members shall be non-executive directors

Independent Director Exemptions (Naresh Chandra Committee Report)

The Committee makes the following recommendation :

Legal provisions must specifically exempt non-executive and independent directors from criminal and civil liabilities under certain circumstances. SEBI should recommend that such exemptions need to be specifically spelt out for the relevant laws by the relevant departments of the Government and independent regulators, as the case may be.

However, independent directors should periodically review legal compliance reports prepared by the company as well as steps taken by the company to cure any taint. In the event of any proceedings against an independent director in connection with the affairs of the company, defense should not be permitted on the ground that the independent director was unaware of this responsibility.

Other Suggestions and the Committee's Response

The Committee also received certain other suggestions relating to corporate governance. These suggestions and the Committee's response/recommendation are set out in the following paragraphs.

Harmonization

It was suggested that SEBI should work towards harmonizing the provisions of clause 49 of the Listing Agreement and those of the Companies Act, 1956.

The Committee noted that major differences between the requirements under clause 49 and the provisions of the Companies Act, 1956 should be identified. SEBI should then recommend to the Government that the provisions of the Companies Act, 1956 be changed to bring it in line with the requirements of the Listing Agreement.

Removal of Independent Directors

It was suggested that companies should inform SEBI/ stock exchanges within five business days of the removal/resignation of an independent director, along with a statement certified by the managing director/ director/ company secretary about the circumstances of such removal resignation (specifically whether there was any disagreement with the independent director that caused such removal/ resignation). Any independent director sought to be removed or who has resigned because of a disagreement with management should have the opportunity to be heard in general meeting.

The Committee noted that under the existing provisions, companies are required to inform the stock exchanges of any changes in directors. The existing safeguards are adequate and hence no further action is required.

Term of Office of Non-Executive Directors

It was suggested that there must be a cap on the term of office of a non-executive director.

The Committee noted that persons should be eligible for the office of non-executive director so long as the term of office did not exceed nine years (in three terms of three years each, running continuously).

The Committee also noted that it would be a good practice for directors to retire after a particular age. Companies may fix the retirement age at either 65 or 70 years.

The Committee recommends that the age limit for directors to retire should be decided by companies themselves. Corporate Boards should have an adequate mechanism of self-renewal, as part of corporate governance best practices.

1.3. FRAME WORK OF OECD PRINCIPLE

I. Ensuring the Basis for an Effective Corporate Governance Framework

The corporate governance framework should promote transparent and efficient markets, be consistent with the rule of law and clearly articulate the division of responsibilities among different supervisory, regulatory and enforcement authorities.

A. The corporate governance framework should be developed with a view to its impact on overall economic performance, market integrity and the incentives it creates for market participants and the promotion of transparent and efficient markets.

B. The legal and regulatory requirements that affect corporate governance practices in a jurisdiction should be consistent with the rule of law, transparent and enforceable.

C. The division of responsibilities among different authorities in a jurisdiction should be clearly articulated and ensure that the public interest is served.

D. Supervisory, regulatory and enforcement authorities should have the authority, integrity and resources to fulfill their duties in a professional and objective manner. Moreover, their rulings should be timely, transparent and fully explained.

II. The Rights of Shareholders and Key Ownership Functions

The corporate governance framework should protect and facilitate the exercise of shareholders' rights.

A. Basic shareholder rights should include the right to: 1) secure methods of ownership registration; 2) convey or transfer shares; 3) obtain relevant and material information on the corporation on a timely and regular basis; 4) participate and vote in general shareholder meetings; 5) elect and remove members of the board; and 6) share in the profits of the corporation.

B. Shareholders should have the right to participate in, and to be sufficiently informed on, decisions concerning fundamental corporate changes such as: 1) amendments to the statutes, or articles of incorporation or similar governing documents of the company; 2) the authorisation of additional shares; and 3) extraordinary transactions, including the transfer of all or substantially all assets, that in effect result in the sale of the company.

C. Shareholders should have the opportunity to participate effectively and vote in general shareholder meetings and should be informed of the rules, including voting procedures, that govern general shareholder meetings :

    1. Shareholders should be furnished with sufficient and timely information

concerning the date, location and agenda of general meetings, as well as full and timely information regarding the issues to be decided at the meeting.

2. Shareholders should have the opportunity to ask questions to the board, including questions relating to the annual external audit, to place items on the agenda of general meetings, and to propose resolutions, subject to reasonable limitations.

3. Effective shareholder participation in key corporate governance decisions, such as the nomination and election of board members, should be facilitated. Shareholders should be able to make their views known on the remuneration policy for board members and key executives. The equity component of compensation schemes for board members and employees should be subject to shareholder approval.

4. Shareholders should be able to vote in person or in absentia, and equal effect should be given to votes whether cast in person or in absentia.

D. Capital structures and arrangements that enable certain shareholders to obtain a degree of control disproportionate to their equity ownership should be disclosed.

E. Markets for corporate control should be allowed to function in an efficient and transparent manner.

1. The rules and procedures governing the acquisition of corporate control in the capital markets, and extraordinary transactions such as mergers, and sales of substantial portions of corporate assets, should be clearly articulated and disclosed so that investors understand their rights and recourse. Transactions should occur at transparent prices and under fair conditions that protect the rights of all shareholders according to their class.

2. Anti-takeover devices should not be used to shield management and the board from accountability.

F. The exercise of ownership rights by all shareholders, including institutional investors, should be facilitated.

1. Institutional investors acting in a fiduciary capacity should disclose their overall corporate governance and voting policies with respect to their investments, including the procedures that they have in place for deciding on the use of their voting rights.

2. Institutional investors acting in a fiduciary capacity should disclose how they manage material conflicts of interest that may affect the exercise of key ownership rights regarding their investments.

G. Shareholders, including institutional shareholders, should be allowed to consult with each other on issues concerning their basic shareholder rights as defined in the Principles, subject to exceptions to prevent abuse.

III. The Equitable Treatment of Shareholders

The corporate governance framework should ensure the equitable treatment of all shareholders, including minority and foreign shareholders. All shareholders should have the opportunity to obtain effective redress for violation of their rights.

A. All shareholders of the same series of a class should be treated equally.

　　1. Within any series of a class, all shares should carry the same rights. All investors should be able to obtain information about the rights attached to all series and classes of shares before they purchase. Any changes in voting rights should be subject to approval by those classes of shares which are negatively affected.

　　2. Minority shareholders should be protected from abusive actions by, or in the interest of, controlling shareholders acting either directly or indirectly, and should have effective means of redress.

　　3. Votes should be cast by custodians or nominees in a manner agreed upon with the beneficial owner of the shares.

　　4. Impediments to cross border voting should be eliminated.

　　5. Processes and procedures for general shareholder meetings should allow for equitable treatment of all shareholders. Company procedures should not make it unduly difficult or expensive to cast votes.

B. Insider trading and abusive self-dealing should be prohibited.

C. Members of the board and key executives should be required to disclose to the board whether they, directly, indirectly or on behalf of third parties, have a material interest in any transaction or matter directly affecting the corporation.

IV. The Role of Stakeholders in Corporate Governance

The corporate governance framework should recognise the rights of stakeholders established by law or through mutual agreements and encourage active cooperation between corporations and stakeholders in creating wealth, jobs, and the sustainability of financially sound enterprises.

A. The rights of stakeholders that are established by law or through mutual agreements are to be respected.

B. Where stakeholder interests are protected by law, stakeholders should have the opportunity to obtain effective redress for violation of their rights.

C. Performance-enhancing mechanisms for employee participation should be permitted to develop.

D. Where stakeholders participate in the corporate governance process, they should have access to relevant, sufficient and reliable information on a timely and regular basis.

E. Stakeholders, including individual employees and their representative bodies, should be able to freely communicate their concerns about illegal or unethical practices to the board and their rights should not be compromised for doing this.

F. The corporate governance framework should be complemented by an effective, efficient insolvency framework and by effective enforcement of creditor rights.

V. Disclosure and Transparency

The corporate governance framework should ensure that timely and accurate disclosure is made on all material matters regarding the corporation, including the financial situation, performance, ownership, and governance of the company.

A. Disclosure should include, but not be limited to, material information on :

1. The financial and operating results of the company.

2. Company objectives.

3. Major share ownership and voting rights.

4. Remuneration policy for members of the board and key executives, and information about board members, including their qualifications, the selection process, other company directorships and whether they are regarded as independent by the board.

5. Related party transactions.

6. Foreseeable risk factors.

7. Issues regarding employees and other stakeholders.

8. Governance structures and policies, in particular, the content of any corporate governance code or policy and the process by which it is implemented.

B. Information should be prepared and disclosed in accordance with high quality standards of accounting and financial and non-financial disclosure.

C. An annual audit should be conducted by an independent, competent and qualified, auditor in order to provide an external and objective assurance to the board and shareholders that the financial statements fairly represent the financial position and performance of the company in all material respects.

D. External auditors should be accountable to the shareholders and owe a duty to the company to exercise due professional care in the conduct of the audit.

E. Channels for disseminating information should provide for equal, timely and cost-efficient access to relevant information by users.

F. The corporate governance framework should be complemented by an effective approach that addresses and promotes the provision of analysis or advice by analysts, brokers, rating agencies and others, that is relevant to decisions by investors, free from material conflicts of interest that might compromise the integrity of their analysis or advice.

VI. The Responsibilities of the Board

The corporate governance framework should ensure the strategic guidance of the company, the effective monitoring of management by the board, and the board's accountability to the company and the shareholders.

A. Board members should act on a fully informed basis, in good faith, with due diligence and care, and in the best interest of the company and the shareholders.

B. Where board decisions may affect different shareholder groups differently, the board should treat all shareholders fairly.

C. The board should apply high ethical standards. It should take into account the interests of stakeholders.

D. The board should fulfil certain key functions, including :

1. Reviewing and guiding corporate strategy, major plans of action, risk policy, annual budgets and business plans; setting performance objectives; monitoring implementation and corporate performance; and overseeing major capital expenditures, acquisitions and divestitures.

2. Monitoring the effectiveness of the company's governance practices and making changes as needed.

3. Selecting, compensating, monitoring and, when necessary, replacing key executives and overseeing succession planning.

4. Aligning key executive and board remuneration with the longer term interests of the company and its shareholders.

5. Ensuring a formal and transparent board nomination and election process.

6. Monitoring and managing potential conflicts of interest of management, board members and shareholders, including misuse of corporate assets and abuse in related party transactions.

7. Ensuring the integrity of the corporation's accounting and financial reporting systems, including the independent audit, and that appropriate systems of control are in place, in particular, systems for risk management, financial and operational control, and compliance with the law and relevant standards.

8. Overseeing the process of disclosure and communications.

E. The board should be able to exercise objective independent judgement on corporate affairs.

1. Boards should consider assigning a sufficient number of non-executive board members capable of exercising independent judgement to tasks where there is a potential for conflict of interest. Examples of such key responsibilities are ensuring the integrity of financial and non-financial reporting, the review of related party transactions, nomination of board members and key executives, and board remuneration.

2. When committees of the board are established, their mandate, composition and working procedures should be well defined and disclosed by the board.

3. Board members should be able to commit themselves effectively to their responsibilities.

F. In order to fulfil their responsibilities, board members should have access to accurate, relevant and timely information.

1.4. CORPORATE GOVERNANCE

Theoretical Aspects

Corporate governance has only relatively recently come to prominence in the business world; the term 'corporate governance' and its everyday usage in the financial press is a new phenomenon of the last fifteen years or so. However, the theories underlying the development of corporate governance, and the areas it encompasses, are drawn from a variety of disciplines including finance, economics, accounting, law, management, and organizational behaviour.

It must be remembered that the development of corporate governance is a global occurrence and, as such, is a complex area including legal, cultural, ownership, and other structural differences, therefore some theories may be more appropriate and relevant to some countries than others, or more relevant at different times depending on what stage an individual country, or group of countries, is at. The stage of development may refer to the evolution of the economy, corporate structure, or ownership groups, all of which affect how corporate governance will develop and be accommodated within its own country setting. An aspect of particular importance is whether the company itself operates within a shareholder framework, focusing primarily on the maintenance or enhancement of shareholder value as its main objective, or whether it takes a broader stakeholder approach, emphasizing the interests of diverse group such as employees, providers of credit, suppliers, customers, and the local community.

Summary of theories

Summary of theories affecting corporate governance development

Agency theory: Agency theory identifies the agency relationship where one party, the principal, delegates work to another party, the agent. In the context of a corporation, the owners are the principal and the directors are the agent.

Transaction cost economics: Transaction cost economics views the firm itself as a governance structure. The choice of an appropriate governance structure can help align the interests of directors and shareholders.

Stakeholder theory: Stakeholder theory takes account of a wider group of constituents rather than focusing on shareholders. Where there is an emphasis on stakeholders, the governance structure of the company may provide for some direct representation of the stakeholder groups.

Stewardship theory: Directors are regarded as the stewards of the company's assets and will be to act in the best interest of the shareholders.

Class hegemony: Directors view themselves as an elite at the top of the company, and will recruit/promote to new director appointments taking into account how appointments might fit into that elite.

Managerial hegemony: Management of a company, with its knowledge of day-to-day open effectively dominate the directors and hence weaken the influence of the directors.

Theories associated with the development of corporate governance.

The main theories that have affected the development of corporate governance – agency theory, transaction cost economics, stakeholder theory, and stewardship theory are discussed in more detail below:

Agency Theory

A significant body of work has built up in this area within the context of the principal-agent framework. The work of Jensen and Meckling (1976) in particular, and of Fama and Jensen (1983), are important. Agency theory identifies the agencyship where one party, the principal, delegates work to another party, the agent. The agency relationship can have a number of disadvantages relating to the opportunism or self interest of the agent, for example, the agent may not act at rechest of the principal, or the agent may act only partially in the best interests of the principal. There can be a number of dimensions to this, for example, the agent misusing his power for pecuniary or other advantage, and the agent not taking appropriate risks in pursuance of the principal's interests because he (the agent) views those risks as not being appropriate (he and the principal may have different attitudes to risk). There is also the problem of information asymmetry whereby the principal and the agent have access to different levels of information; in practice, this means that the principal is at a disadvantage because the agent will have more information.

In the context of corporations and issues of corporate control, agency theory views corporate governance mechanisms, especially the board of directors, as being an essential monitoring device to try to ensure that any problems that may be brought about by the principal agent relationship, are minimized. Blair (1996) states :

Managers are supposed to be the 'agents' of a corporation's 'owners', but managers must be monitored and institutional arrangements must provide some checks and balances to make sure they do not abuse their power. The costs resulting from managers misusing their position, as well as the costs of monitoring and disciplining them to try to prevent abuse, have been called 'agency costs'.

Much of agency theory as related to corporations is set in the context of the separation of ownership and control, as described in the work of Berle and Means (1932). In this context, the agents are the managers and the principals are the shareholders, and this is the most commonly cited agency relationship in the corporate governance context. However, it is useful to be aware that the agency relationship can also cover various other relationships including those of company and creditor, and of employer and employee.

Separation of Ownership and Control

Secondly the Agency theory focuses on the potential problems of the separation of ownership and control were identified in the eighteenth century by Smith (1838): '... the directors of such companies [joint stock companies] however being the managers rather of other people's money than of their own, it cannot well be expected that they should watch over it with the same anxious vigilance [as if it were their own].' Almost a century later, the work of Berle and Means (1932) is often cited as providing one of the fundamental explanations of investor and corporate relationships. Berle and Means' work highlighted that, as countries industrialized and developed their markets, the ownership and control of corporations became separated. This was particularly the case in the USA and the UK where the legal systems have fostered good protection of minority shareholders and hence there has been encouragement for more diversified shareholder bases.

Once shareholders do begin to act like owners again, then they will be able to exercise a more direct influence on companies and their boards, so that boards will be more accountable for their actions and, in that sense, the power of ownership will be returned to the owners (the shareholders). Useem (1996) highlights, however, though that institutional investors will ultimately become accountable to 'the millions of ultimate owners... who may come to question the policies of the new powers that be. Then the questions may expand from whether the professional money managers are achieving maximum private return to whether they are fostering maximum public good. Their demands for downsizing and single-minded focus on shareholder benefits – whatever the costs – may come to constitute a new target of ownership challenge'.

Transaction Cost Economics

Transaction Cost Economics (TCE), as expounded by the work of Williamson (1975, 1984), is often viewed as closely related to agency theory. TCE views the firm as a governance structure whereas agency theory views the firm as a nexus of contracts. Essentially, the latter means that there is a connected group or series of contracts amongst the various players, arising because it is seemingly impossible to have a contract that perfectly aligns the interests of principal and agent in a corporate control situation.

In the discussion of agency theory above, the importance of the separation of ownership and control of a firm was emphasized. As firms have grown in size, whether caused by the desire to achieve economies of scale, by technological advances, or by the fact that natural monopolies have evolved, they have increasingly required more capital, which has to be raised from the capital markets and a wider shareholder base has been established. The problems of the separation of ownership and control and the resultant corporate governance issues have thus arisen. Coase (1937) examines the rationale for firms' existence in the context of a framework of the efficiencies of internal, as opposed to external, contracting. He states the operation of a market costs by forming an organisation and allowing some authority (an 'entrepreneur') to direct the resources, certain marketing costs are saved. The entrepreneur has to carry out his function at less cost, taking into account the fact that he may get factors of production at a lower price than the market transactions which he supersedes.

In other words, there are certain economic benefits to the firm itself to undertake transactions internally rather than externally. In its turn, a firm becomes larger the more transactions it undertakes and will expand up to the point where it becomes cheaper or more efficient for the transaction to be undertaken externally. Coase therefore posits that firms may become less efficient the larger they become; equally, he states that 'all changes which improve managerial technique will tend to increase the size of the firm'.

Williamson (1984) builds on the earlier work of Coase, and provides a justification for the growth of large firms and conglomerates, which essentially provide their own internal capital market. He states that the costs of any misaligned actions may be reduced by 'judicious choice of governance structure rather than merely realigning incentives and pricing them out'.

Stiles and Taylor (2001) point out that 'both theories [TCE and agency] are concerned with managerial discretion, and both assume that managers are given to opportunism (self-interest seeking) and moral hazard, and that managers operate under bounded rationality... [and] both agency theory and TCE regard the board of directors as an instrument of control'. In this context, 'bounded rationality' means that managers will tend to satisfice rather than maximize profit (this, of course, not being in the best interests of shareholders).

Stakeholder Theory

In juxtaposition to agency theory is stakeholder theory. Stakeholder theory takes account of a wider group of constituents rather than focusing on shareholders. A consequence of focusing on shareholders is that the maintenance or enhancement of shareholder value is paramount, whereas when a wider stakeholder group, such as employees, providers of credit, customers, suppliers, government, and the local community, is taken into account the overriding focus on shareholder value becomes less self-evident. Nonetheless, many companies do strive to maximize shareholder value whilst at the some time trying to take into account the interests of the wider stakeholder group. One rationale for effectively privileging shareholders over other stakeholders is that they are the recipients of the residual free cash flow (being the profits remaining once other stakeholders, such as loan creditors, have been paid). This means that the shareholders have a vested interest in trying to ensure that resources are used to maximum effect, which in turn should be to the benefit of society as a whole.

Shareholders and stakeholders may favour different corporate governance structures and also monitoring mechanisms. We can, for example, see differences in the corporate, governance structures and monitoring mechanisms of the so-called Anglo-American model, with its emphasis on shareholder value and a board composed totally of executive and non-executive directors elected by shareholders, compared to the German model, whereby certain stakeholder groups such as employees, have a right enshrined in law: for their representatives to sit on the supervisory board alongside the directors.

•

Stewardship Theory

Stewardship theory draws on the assumptions underlying agency theory and TCE. The work of Donaldson and Davis (1991) cautioned against accepting agency theory; given and introduced an alternative approach to corporate governance Stewards theory.

The thrust of Donaldson and Davis' paper was that agency theory emphasises the control of managerial 'opportunism' by having a board chair independent of the CEO and using incentives to bind CEO interests to those of shareholders. Stewardship theory stresses the beneficial consequences on shareholder returns of facilitative authority structures which unify command by having roles of CEO and chair held by the same person... The safeguarding of returns to shareholders may be along the track, not of placing management under greater control by owners, but of empowering managers to take autonomous executive action.

The Theories in Context

The approach taken in this book is to assume a public corporation business form (that is, a publicly quoted company), unless specifically stated otherwise. Therefore the theories discussed above should be viewed in the light of this type of business form. In the UK, this type of business form generally has a dispersed shareholder base, although there is concentration of shareholdings amongst the institutional investors such as the pension funds and insurance companies. Agency theory, together with the work of Berle and Means, seems particularly relevant in this context.

The theories that have affected the development of corporate governance should also be viewed in conjunction with the legal system and capital market development, as well as the ownership structure. For example, countries like the UK and the USA have a common law system that tends to give good protection of shareholder rights, whilst civil law countries, such as France, tend to have less effective legal protection for shareholder rights, and more emphasis may be given to the rights of certain stakeholder groups.

However, it is clear that companies cannot operate in isolation without having regard to the effect of their actions on the various stakeholder groups. To this end, companies need to be able to attract and retain equity investment, and be accountable to their shareholders, whilst at the same time giving real consideration to the interests of wider stakeholder constituencies.

1.5. DEVELOPMENT OF CORPORATE GOVERNANCE CODES

The Growth in Corporate Governance Codes

The corporate governance codes and guidelines have been issued by a variety of bodies ranging from committees, appointed by government departments, and usually including prominent respected figures from business and industry, representatives from the investment community, representatives from professional bodies, and academics, through to stock exchange bodies, various investor representative groups and professional bodies, such as those representing directors or company secretaries.

As regards compliance with the various codes, compliance is generally on a voluntary disclosure basis, whilst some codes, such as the UK's Combined Code (2003), are on a 'comply or explain basis', that is, either a company has to comply fully with the code and state that it has done so, or it explains why it has not.

In this chapter the development of corporate governance in the UK is covered in some detail, particularly in relation to the Cadbury Report (1992), which has influenced the development of many corporate governance codes globally. Similarly, the OECD Principles are reviewed in detail as these have also formed the cornerstone of many corporate governance codes. The impact of various other international organizations on corporate governance developments, including the World Bank, Global Corporate Governance Forum, International Corporate Governance Network, and Commonwealth Association for Corporate Governance, are discussed. Recent developments in the EU, which have implications both for existing, and potential, member countries' corporate governance, are covered. There is also a brief overview of the Basle Committee recommendations for corporate governance in banking organizations.

Finally, recent corporate collapses in the USA have had a significant impact on confidence in financial markets across the world and corporate governance developments in the USA are discussed in some detail.

Corporate Governance in the UK

The UK has a well developed market with a diverse shareholder base including institutional investors, financial institutions, and individuals. The UK illustrates well the problems that may be associated with the separation of the ownership and control of corporations and hence has many of the associated agency problems, including misuse of corporate assets by directors and a lack of effective control over, and accountability of, directors' actions, contributed to a number of financial scandals in the UK.

As in other countries the development or corporate governance in the UK was initially driven by corporate collapses and financial scandals. The UK's Combined Code

(1998) embodied the findings of a trilogy of codes: the Cadbury Report (1992), the Greenbury Report (1995), and the Hampel Report (1998). Brief mention is made of each of these three at this point to set the context, whilst a detailed review of the Cadbury Report (1992) is given subsequently in chapter because it has influenced the development of many codes across the world. Reference is made to relevant sections of various codes in appropriate subsequent chapters.

The development of corporate governance in the UK represents the Combined Code published in 2006 by the Financial Reporting Council. We can see the various influences since 1998 (the original Combined Code, published in 1998, encompassed the Cadbury, Greenbury, and Hampel report recommendations). These influences can be split into four broad areas. First, there are reports that have looked at specific areas of corporate governance: the Turnbull report on internal controls, the Myners review of institutional investment, the Higgs review of the role and effectiveness of non-executive directors, and the Smith review of audit committees. Secondly, there has been the influence of institutional investors and their representative groups. Thirdly, influences affecting the regulatory framework within which corporate governance in the UK operates have included the UK company law review and the Financial Services Authority review. Fourthly, there have been what might be termed 'external influences' such as the EU review of company law and the US Sarbanes Oxley Act. Each of these is now discussed in turn.

Cadbury Report (1992)

Following various financial scandals and collapses (Coloroll and Polly Peck, to name but two) and a perceived general lack of confidence in the financial reporting of many UK companies, the Financial Reporting Council, the London Stock Exchange, and the accountancy profession established the Committee on the Financial Aspects of Corporate Governance in May 1991. After. the Committee was set up, the scandals at BCCI and Maxwell happened, and as a result, the Committee interpreted its remit more widely and looked beyond the financial aspects to corporate governance as a whole. The Committee was chaired by Sir Adrian Cadbury and, when the Committee reported in December 1992, the report became widely known as 'the Cadbury Report'.

The recommendations covered: the operation of the main board; the establishment, composition, and operation of key board committees; the importance of, and contribution that can be made by, non-executive directors; the reporting and control mechanisms of a business. The Cadbury Report recommended a Code of Best Practice with which the boards of all listed companies registered in the UK should comply, and utilized a 'comply or explain' mechanism. This mechanism means that a company should comply with the code but, if it cannot comply with any particular aspect of it, then it should explain why it is unable to do so. This disclosure gives investors detailed information about any instances of non-compliance and enables them to decide whether the company's non-compliance is justified.

Greenbury Report (1995)

The Greenbury Committee was set up in response to concern at both the size of directors' remuneration packages and their inconsistent and incomplete disclosure in companies' annual reports. It made, in 1995, comprehensive recommendations regarding disclosure of directors' remuneration packages. There has been much discussion about how much disclosure there should be of directors' remuneration and how useful detailed disclosures might be. Whilst the work of the Greenbury Committee focused on the directors of public limited companies, it hoped that both smaller listed companies and unlisted companies would find its recommendations useful.

Central to the Greenbury Report recommendations were strengthening accountability and enhancing the performance of directors. These two aims were to be achieved by (i) the presence of a remuneration committee comprised of independent non-executive directors who would report fully to the shareholders each year about the company's executive remuneration policy, including full disclosure of the elements in the remuneration of individual directors, and (ii) the adoption of performance measures linking rewards to the performance of both the company and individual directors, so that the interests of directors and shareholders were more closely aligned.

Since that time (1995), disclosure of directors' remuneration has become quite prolific in UK company accounts. The main elements of directors' remuneration are considered further in this book.

Hampel Report (1998)

The Hampel Committee was set up in 1995 to review the implementation of the Cadbury and Greenbury Committee recommendations. The Hampel Committee reported in 1998. The Hampel Report said: 'We endorse the overwhelming majority of the findings of the two earlier committees.' There has been much discussion about the extent to which a company should consider the interests of various stakeholders, such as employees, customers, suppliers, providers of credit, the local community, etc., as well as the interests of its shareholders. The Hampel Report stated that 'the directors as a board are responsible for relations with stakeholders; but they are accountable to the shareholders'. However, the report does also state that 'directors can meet their legal duties to shareholders, and can pursue the objective of long-term shareholder value successfully, only by developing and sustaining these stakeholder relationships'.

The Hampel Report, like its precursors, also emphasized the important role that institutional investors have to play in the companies in which they invest (investee companies). It is highly desirable that companies and institutional investors engage in dialogue and that institutional investors make considered use of their shares in other words, institutional investors should consider carefully the resolutions on which they have a right to vote and reach a decision based on careful thought, rather than engage in 'box ticking'.

Combined Code (1998)

The Combined Code drew together the recommendations of the Cadbury, Greenbury, and Hampel reports. It has two sections, one aimed at companies and another aimed at institutional investor. The Combined Code operates on the 'comply or explain' basis mentioned above. In relation to the internal controls of the business, the Combined Code states that 'the board should maintain a sound system of internal control to safeguard shareholders' investment and the company's assets' and that 'the directors should, at least annually, conduct a review of the effectiveness of the group's system of internal control and should report to shareholders that they have done so. The review should cover all controls, including financial, operational, and compliance controls and risk management'. The Turnbull Report issued in 1999 gave directors guidance on carrying out this review.

Turnbull (1999)

The Turnbull Committee, chaired by Nigel Turnbull, was established by the Institute of Chartered Accountants in England and Wales (ICAEW) to provide guidance on the implementation of the internal control requirements of the combined Code. The Turnbull Report confirms that it is the responsibility of the board of directors to ensure that the company has a sound system of internal control, and that the controls are working as they should. The board should assess the effectiveness of internal controls and report on them in the annual report. Of course, a company is subject to new risks both from the outside environment and as a result of decisions that the board makes about corporate strategy and objectives. In the managing of risk, boards will need to take into account the existing internal control system in the company and also whether any changes are required to ensure that new risks are adequately and effectively managed.

Myners (2001)

The Myners Report on institutional investment, issued in 2001 by HM Treasury, concentrated more on the trusteeship aspects of institutional investors and the legal

requirements for trustees, with the aim of raising the standards and, promoting greater shareholder activism. For example, the Myners Report expects that institutional investors should be more proactive, especially in the stance they take with underperforming companies. Some institutional investors have already shown more of a willingness to engage actively with companies to try to ensure that shareholder value is not lost by underperforming companies.

Higgs (2003)

The Higgs Review, chaired by Derek Higgs, reported in January 2003 on the role and effectiveness of non-executive directors. Higgs offered support for the Combined Code whilst making some additional recommendations. These recommendations included: stating the number of meetings of the board and its main committees in the annual report, together with the records of individual directed that a chief executive director should or not also become chairman of the same company; non-executive directors should meet as a group at least once a year without executive directors being present, and the annual report should indicate whether such meetings have occurred; chairmen and chief executives should consider implementing executive development programmes to train and develop suitable individuals in their companies for future director roles; the board should inform shareholders as to why they believe a certain individual should be appointed to a non-executive director, and resources should be available for ongoing development of directors; the performance of the board, its committees and its individual members, should be evaluated at least once a year, 'the annual report should state whether these reviews are being held and how they are conducted; a full-time executive director should not hold more than one non-executive directorship or become chairman of a major company; no one non-executive director should sit on all three principal board committees (audit, remuneration, nomination). There was substantial opposition to some of the recommendations but they nonetheless helped to inform the Combined Code. Good practice suggestions from the Higgs Report were published in 2006.

Smith (2003)

The Smith Review of audit committees, a group appointed by the Financial Reporting Council, reported in January 2003. The review made clear the important role of the audit committee: 'While all directors have a duty to act in the interests of the company, the audit committee has a particular role, acting independently from the executive, to ensure that the interests of shareholders are properly protected in relation to financial reporting and internal control'. The review defined the audit committee's role in terms of a high-level overview it needs to satisfy itself that there is an appropriate system of controls in place but it does not undertake the monitoring itself.

Combined Code (2003)

The revised Combined Code, published in July 2003, incorporated the substance of the Higgs and Smith reviews. However, rather than stating that no one non-executive director should sit on all three board committees, the Combined Code stated that 'undue reliance' should not be placed on particular individuals. The Combined Code also clarified the roles of the chairman and the senior independent director (SID), emphasizing the chairman's role in providing leadership to the non-executive directors and in communicating shareholders' views to the board; it also provided for a 'formal and rigorous annual evaluation' of the board's, the committees', and the individual directors' performance. At least half the board in larger listed companies were to be independent non-executive directors.

Cadbury Report (1992)

The Cadbury Report recommended a Code of Best Practice with which the boards of all listed companies registered in the UK should comply, and utilized a 'comply or explain' mechanism. Whilst the Code of Best Practice is aimed at the directors of listed companies registered in the UK, the Committee also exhorted other companies to meet its requirements. The main recommendations were as follows:

1. Title Board of Directors

1. The board should meet regularly, retain full and effective control over the company, and monitor the executive management.

2. There should be a clearly accepted division of responsibilities at the head of a company, which will ensure a balance of power and authority, such that no one individual has unfettered powers of decision. Where the chairman is also the chief executive, it is essential that there should be a strong and independent element on the board, with a recognized senior member.

3. The board should include non-executive directors of sufficient calibre and number for their views to carry significant weight in the board's decisions.

4. The board should have a "formal schedule" of matters specifically reserved to it for decision to ensure that the direction and control of the company is firmly in its hands.

5. There should be an agreed procedure for directors in the furtherance of their duties to take independent professional advice if necessary, at the company's expense.

6. All directors should have access to the advice and services of the company secretary, who is responsible to the board for ensuring that board procedures are followed and

that applicable rules and regulations are complied with. Any question of the removal of the company secretary should be a matter for the board as a whole.

7. Non-executive directors should bring an independent judgement to bear on issues of strategy, performance, resources, including key appointments, and standards of conduct.

8. The majority should be independent of management and free from any business or other relationship which could materially interfere with the exercise of their independent Judgement, apart from their fees and shareholding. Their fees should reflect the time which they commit to the company.

9. Non-executive directors should be appointed for specified terms and reappointment should not be automatic.

10. Non-executive directors should be selected through a formal process and both this process and their appointment should be a matter for the board as a whole.

11. Directors' 'service contracts should not exceed three years without shareholders' approval.

12. There should be full and clear disclosure of directors' total emoluments and those of the chairman and highest-paid UK director, including pension contributions and stock options. Separate figures should be given for salary and performance-related elements and the basis on which performance is measured should be explained.

13. Executive directors' pay should be subject to the recommendations of a remuneration committee made up wholly or mainly of non-executive directors.

2. Reporting and Controls

1. It is the board's duty to present a balanced and understandable assessment of the company's position.

2. The board should ensure that an objective and professional relationship is maintained with the auditors.

3. The board should establish an audit committee of at least three non-executive directors with written terms of reference which deal clearly with its authority and duties.

4. The directors should explain their responsibility for preparing the accounts next to a statement by the auditors about their reporting responsibilities.

5. The directors should report on the effectiveness of the company's system of internal control.

6. The directors should report that the business is a going concern, with supporting assumptions or qualifications as necessary.

Source: Cadbury Code (1992)

Today, the recommendations of the Cadbury Report and subsequent UK reports on corporate governance are embodied in the UK's Combined Code (2006). Various sections of the Combined Code are referred to in appropriate chapters and the full text of the Combined Code (2006) is included as an appendix.

Basle Committee

The Basle Committee (1999) guidelines related to enhancing corporate governance in banking organizations. The guidelines have been influential in the development of corporate governance practices in banks across the world. Sound governance can be practised regardless of the form of a banking organization.

In 2006, the Basle Committee issued new guidance comprising eight sound corporate governance principles:

Principle 1 board members should be qualified for their positions, have a clear understanding of their role in corporate governance, and be able to exercise sound judgement about the affairs of the bank;

Principle 2 the board of directors should approve and oversee the bank's strategic objectives and corporate values that are communicated throughout the banking organization;

Principle 3 the board of directors should set and enforce clear lines of responsibility and accountability throughout the organization;

Principle 4 the board should ensure that there is appropriate oversight by senior management consistent with board policy;

Principle 5 the board and senior management should effectively utilize the work conducted by the internal audit function, external auditors, and internal control functions;

Principle 6 the board should ensure that compensation policies and practices are consistent with the bank's corporate culture, long-term objectives, and strategy, and control environment;

Principle 7 the bank should be governed in a transparent manner;

Principle 8 the board and senior management should understand the bank's operational structure, including where the bank operates in jurisdictions, or through structures, that impede transparency (i.e. 'know your structure').

Source: Basle Committee on Banking Supervision (2006) Enhancing Corporate Governance for Banking Organisations.

Sarbanes-Oxley Act 2002

More recently, and following directly from the financial scandals of Enron, Worldcom, and Global Crossing, in which it was perceived that the close relationship between companies and their external auditors was largely to blame, the US Congress agreed reforms together with changes to the NYSE Listing Rules that have had a significant impact not just in the USA but around the world. The changes are embodied in the Accounting Industry Reform Act 2002, widely known as the 'Sarbanes-Oxley Act'.

Initially, one of the most publicized aspects of the Sarbanes-Oxley Act, was the requirement for CEOs and CFOs to certify that quarterly and annual reports filed on forms 10-Q, 10-K, and 20-F are fully compliant with applicable securities laws and present a fair picture of the financial situation of the company. The penalties for making this certification when aware that the information does not comply with the requirements are severe: up to $1m fine or fitness or imprisonment of up to ten years or both.

The Sarbanes-Oxley Act seeks to strengthen (external) auditor independence and also to strengthen the company's audit committee listed companies, for example, must have an audit committee comprised only of independent members, and must also disclose whether they have at least one 'audit committee financial expert' on their audit committee. The 'audit committee financial expert' should be named and the company should state whether the expert is independent of management (for listed companies, the audit committee should comprise only independent members).

The Act establishes a new regulatory body for auditors of US listed firms the Public Company Accounting Oversight Board (PCAOB) with which all auditors of US listed companies have to register, including non-US audit firms. Correspondingly, the Securities Exchange Commission (SEC) has issued separate rules that encompass the prohibition of some non-audit services to audit clients, mandatory rotation of audit partners, and auditors' reports on the effectiveness of internal controls. The SEC implementation of the Sarbanes-Oxley Act prohibits nine non-audit services that might impair auditor independence. In many cases, these effectively prohibit the audit firm from either auditing accounting services provided by the audit firm's staff or providing help with systems that will then be audited by the audit firm. These nine areas cover :

    (a) book-keeping or other services related to the accounting records or financial statements of the audited company;

    (b) financial information systems design and implementation;

    (c) appraisal or valuation services, fairness, opinions, or contribution-in-kind reports (where the firm provides its opinion on the adequacy of consideration in a transaction);

(d) actuarial services;

(e) internal audit outsourcing services;

(f) management functions/human resources (an auditor should not be a director, officer, or employee of an audit client, nor perform any executive role for the audit client such as supervisory, decision-making, or monitoring);

(g) broker or dealer, investment adviser, or investment banking services;

(h) legal services or expert services unrelated to the audit;

(i) any other service that the PCAOB decides is not permitted.

There are also requirements relating to the rotation of audit partners such that the lead audit partner should rotate every five years, and is then subject to a five-year period during which he/she cannot be the audit partner for that company. Similarly, other partners involved with the audit, but not acting as the lead partner, are subject to legal attrition followed by a two-years bar. Any member of the audit team is barred for one year from accepting employment in certain specified positions in a company that he/she has audited.

The auditor is required to report to the audit committee various information, which includes all critical accounting policies and practices, and alternative accounting treatments.

The Sarbanes-Oxley Act provides for far-reaching reform and has caused much disquiet outside the USA because the Act applies equally to US and non-US firms with a US listing. However, some of the provisions of the Sarbanes-Oxley Act are in direct conflict with provisions in the law/practice of other countries. In reality, this has led to some companies delisting from the NYSE and has deterred other non-US firms from applying to be listed on the NYSE.

1.6. CORPORATE GOVERNANCE MECHANISM AND OVERVIEW

4 P's of Corporate Governance

Corporate Governance can also be explained on the basis of 4 P's (People, Purpose, Process and Performance). Business cannot run with only profits, but there must be recognition for human aspects too. This is made possible by corporate governance. Corporate Governance has the integrated framework, where the people are formally either trained or helped to develop to work for a definite and defined purpose in applying the systematic processes consistently to give constant growth by better performance.

The above defined 4P's play a vital role in modern management. The purpose and the process are well defined and explained to achieve better performance by the people. Corporate Governance highlights the foci of the entire business in a corporation. This is centralized to people. People work for the people. Corporate governance also emphasizes the structure that should be formally defined and organized. With the consistent process having a common goal, the performance that is expected from the people is planned and oriented in order to get the overall insight of development of the business and the company respectively.

People

People are the heart of any organisation in general and Corporate Governance in particular. People associated with any organisation include Investors, Employees, Partners, Customers, Suppliers, Lenders, Government and Society. Equity, Ethics and Relationship

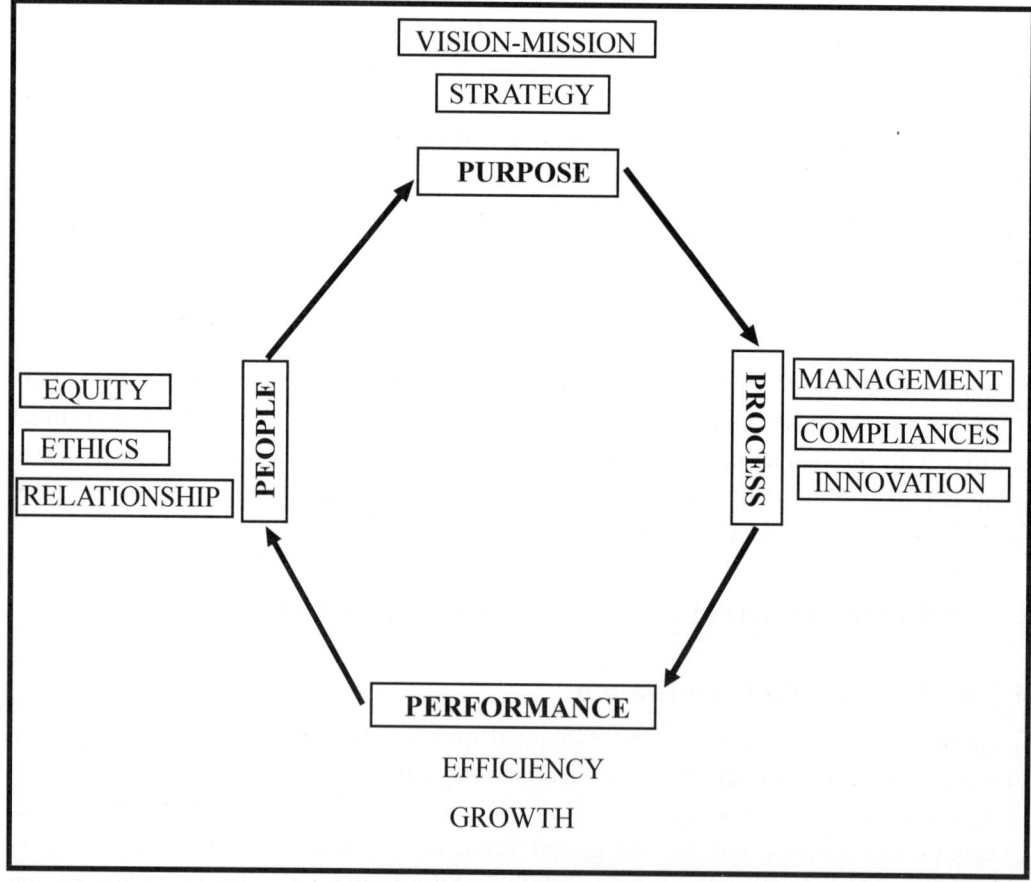

Fig. 1.2: 4 P's of Corporate Governance

are the parameters on which People orientation of Corporate Governance can be measured. Equity means fair and equitable treatment to all. Equity has two aspects positive and negative. In positive sense, equity means same behaviour, award, appreciation etc to all on achievement of the objectives. In the negative sense, equity talks about same extent and kind of punishment on commitment of an act of commission or omission of any misconduct or contravention of any rule or regulation. Equity also means equitable distribution of wealth among the people. Protection of human rights of people is also an important aspect of it.

Ethical practices of organisations include integrity, propriety and independence, transparency in operations and disclosure of information, anticorruption practices, non money-laundering, protection of intellectual property rights, non counterfeiting, avoidance of harmful products, proper insolvency or closure process.

Relationship includes stakeholders' empowerment and engagement in management and decision making process, harmonious culture resolution of conflict between employee and employee, between employee and employer, environment, health, safety and confidentiality.

Purpose

Another 'P' of corporate governance is "Purpose". The Purpose should be Established, Measurable, Actionable and Communicated. Different aspects of Purpose are Vision-Mission and Strategy. The Vision and Mission are decided based on Unified and Shared Values of the Organisation, Stakeholders' policies and Organisation Commitment. The Strategy leads to Strategic Action Plan, Perform Metrics, Capacity Building Planning and Strategic Management Team.

Process

This 'P' of Corporate Governance includes Process Management. Process Compliances and Process Innovation. The Process should be Established, Integrated, Documented, Automated, Implemented and Maintained. Process Management has different aspects such as Organisation Management, Resource Management, Supply Chain Management, Marketing and Brand Promotion, Outsourced Process Management, Environment and Energy Management, Relationship Management. Information System Management, Risk and Crisis Management. The Plant/Unit/Organisation has to comply with various rules, regulations, statutes and laws enforced by state and central government which includes compliance management, Independent Assurance Mechanism and Whistle Blowing. When certain things are going wrong or violating any norm, rule regulation or any act, the employees should have freedom to sound it off to the top management to take preventive action.

Performance

Performance should be measured, analysed and communicated in order to achieve Growth through Efficiency, Efficiency can be segregated into different types as Operational Efficiency, Asset or Infrastructure Efficiency and Management Process Efficiency. Growth in any Organisation brings in increase in income of organisation and stakeholder, increase in net worth or market share, expansion and diversification and increase in opportunities in stakeholders.

Mechanism of Corporate Governance

1. Companies Act

The Indian companies are generally regulated by the companies act 1956. It is one of the biggest legislation with 658 sections and 14 schedules. The aim of companies act are quite long and touch each and every aspect of the companies existence in relation to companies act. Corporate Governance provides some legal rights to the shareholders, they are as follows:

(A) To elect Directors who are responsible for specifying objectives and laying down policies.

(B) To determine remuneration of director and CEO.

(C) For removal of directors.

(D) To take active part in the annual general meetings.

Thus companies act aimed at strengthening corporate democracy, protecting the interest of minority shareholders and providing maximum flexibility to the companies in responding to the market needs.

2. Security Law

The primary security law in our country is SEBI Act.

Since its inception in 1992, the board had taken a number of initiatives towards investor's protection. One such initiative is important information discloser, both in prospective and in annual account. e.g. like in T.V. we have several ad companies regarding the issue of shares they are offering for investment, in those ads the whole information is disclosed.

3. Capital Market

Capital market itself has considerable impact on Corporate Governance. What makes capital market discipline so much more attractive than regulatory market?

(A) Unlike regulatory market is very good at micro level judgment and decision.

(B) Unlike regulatory, market is not bound by board rules and can exercise business judgment.

4. Company Boards

Developments banks hold large blocks of shares in companies. They are equally big shareholders too being equity holders; these investors have their nominees in the boards of companies. The nominees can effectively block resolutions which may be detrimental to their interest. Unfortunately the role of nominees directors has been passive as has been pointed by several companies.

5. Statutory Audit

It is one of the most important and necessary mechanism for good Corporate Governance. Auditors are the actual keepers of shareholders, lenders and others who have financial stakes in companies. The auditing process basically insures whether the financial statement and accounts and complete creditable financial statement an essential for business enterprise to raise capital and for society to have trust on companies.

6. Code of Conduct

Code of conduct means set of rules and regulations. The code is based on checks and balances, especially at the level of board of directors and the chief executive to guide against under concentration of power and adequate disclosure to enable those entitled to have the information they need, in order to exercise their rights. It has four sectors :

1. Role of board of directors.
2. Role of non executive directors.
3. Executive directors.
4. Financial reporting and control.

"The famous 'Code of best practice' was advocated by the Cadbury committee in the UK."

Factors Influencing Corporate Governance

1. The ownership structure.
2. The structure of company boards.
3. The financial structure.
4. The institutional environment.

1. *The Ownership Structure*

The structure of ownership of a company determines, to a considerable context, how a corporation is managed and controlled. The ownership of the company can be either dispersed among individual and institutional shareholder or can be concentrated in the hands of a few large shareholders.

Our corporate sector is characterized by the coexistence of state owned, private and multinational enterprise. The share of these enterprise are held by institutional as well as small investors like the team lending institutions, institutional investors, directors and their relatives, foreign investors.

2. *The Structure of Company Boards*

Along with the structure of ownership, the structure of company boards has considerable influence on the way the Company is managed and controlled. The board of directors is responsible for establishing corporate objectives, developing board policies and selecting top level executive to carry those objectives and policies. Company's boards are permitted to vary in size, composition and structure so as to best serve the interest of the corporation and the shareholder.

3. *The Financial Structure*

In Corporate Governance the notion that the financial structure of the company i.e. Proportion between debt and equity has implication for the quality of Governance. It is no secret that exercise significant influence on the way, it is managed and controlled, banks, for an example, as a creditor can perform the important function of screening and monitoring companies as they are better informed than other investors.

4. *The Institutional Environment*

The legal document and political environment within which company operates determines in large measure the quality of Corporate Governance, for example, the extent to which shareholder can control the management depends on their voting right as defined in their company law, and the extent to which the creditor will be able to exercise financial claims on a bankrupt will depend on bankrupt laws and procedures.

The Present

The Corporate Governance in India picked up momentum after the debate of big companies such as Enron world and BCCI Bank. Those were times when the confidence of the financial communities, shareholders and investors look a beating the world over. It was around that time that foreign institutions started invested money in Indian companies which also triggered the need for great accountability. Today, fund managers view firms such as Tata Motors, ITC, Ranbaxy, Infosys, and hero Honda motors are having higher governing standards. Luckily many companies are expecting good governance standards.

The Future

As we go into the future, Corporate Governance will become more relevant and a more acceptable practice. Seeds are already sown towards honest business practices, more and more progressive companies are drawing and enforcing code of conduct, are accepting tough accounting standards and are following more stringent disillusive norms than are mandated by law. These tendencies would be further strengthened by a variety of forces that are acting today and would become stronger in years to come.

1.7. MODELS OF CORPORATE GOVERNANCE

The Anglo-US Model

The Anglo-US Model is characterized by share ownership of individual, and increasingly institutional, investors not affiliated with the corporation (known as outside shareholders or "outsiders"); a well-developed legal framework defining the rights and responsibilities of three key players, namely management, directors and shareholders; and a comparatively uncomplicated procedure for interaction between shareholder and corporation as well as among shareholders during or outside the AGM.

Key Players in the Anglo-US Model

Players in the Anglo-US model include management, directors, shareholders (especially institutional investors), government agencies, stock exchanges, self-regulatory organizations and consulting firms which advise corporations and/or shareholders on corporate governance and proxy voting.

Of these, the three major players are management, directors and shareholders. They form what is commonly referred to as the "corporate governance triangle." The interests and interaction of these players may be diagrammed as follows :

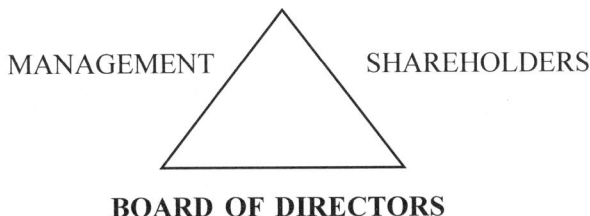

MANAGEMENT   SHAREHOLDERS

BOARD OF DIRECTORS

The Anglo-US model, developed within the context of the free market economy, assumes the separation of ownership and control in most publicly held corporations. This important legal distinction serves a valuable business and social purpose: investors contribute capital and maintain ownership in the enterprise, while generally avoiding legal liability for the acts of the corporation. Investors avoid legal liability by ceding to management control of the corporation, and paying management for acting as their agent by undertaking the affairs of the corporation. The cost of this separation of ownership and control is defined as "**agency costs**".

The interests of shareholders and management may not always coincide. Laws governing corporations in countries using the Anglo-US model attempt to reconcile this conflict in several ways. Most importantly, they prescribe the election of a board of directors by shareholders and require that boards act as fiduciaries for shareholders' interests by overseeing management on behalf of shareholders.

Share Ownership Pattern in the Anglo-US Model

In both the UK and the US, there has been a marked shift of stock ownership during the postwar period from individual shareholders to institutional shareholders. In 1990, institutional investors held approximately 61 percent of the shares of UK corporations, and individuals held approximately 21 percent. (In 1981, individuals held 38 percent.) In 1990, institutions held 53.3 percent of the shares of US corporations.

The increase in ownership by institutions has resulted in their increasing influence. In turn, this has triggered regulatory changes designed to facilitate their interests and interaction in the corporate governance process.

Composition of the Board of Directors in the Anglo-US Model

The board of directors of most corporations that follow the Anglo-US model includes both "insiders" and "outsiders". An "insider" is as a person who is either employed by the corporation (an executive, manager or employee) or who has significant personal or business relationships with corporate management. An "outsider" is a person or institution which has no direct relationship with the corporation or corporate management.

A synonym for insider is executive director; a synonym for outsider is non-executive director or independent director.

Traditionally, the same person has served as both chairman of the board of directors and chief executive officer (CEO) of the corporation. In many instances, this practice led to abuses, including: concentration of power in the hands of one person (for example, a board of directors firmly controlled by one person serving both as chairman of the board of directors and CEO); concentration of power in a small group of persons (for example, a board of directors composed solely of "insiders"; management and/or the board of directors' attempts to retain power over a long period of time, without regard for the interests of other players (entrenchment); and the board of directors' flagrant disregard for the interests of outside shareholders.

In this model, the board appoints and supervises the managers who manage the day-to-day affairs of the corporation. While the legal system provides the structural framework, the stakeholders in the company will be suppliers, employees and creditors. However, creditors exercise their lien over the assets of the company. The policies are framed by the board of directors and implemented by the management. The board oversees the implementation through a well-designed information system. The board of directors, being responsible to their appointers – the shareholders – commits to them certain returns within the board contours of the market framework.

It will ensure an efficient organization for production, exchange and performance monitoring. However, there is no agreement on the cost effectiveness or efficiency of the model (Macey, 1998). Wile Fischel and Easterbrook (1991) and (Romano, 1993) made a very optimistic assessment of the U.S. flawed. It will not be costless for the market to provide a greater supply of institutional investor monitoring.

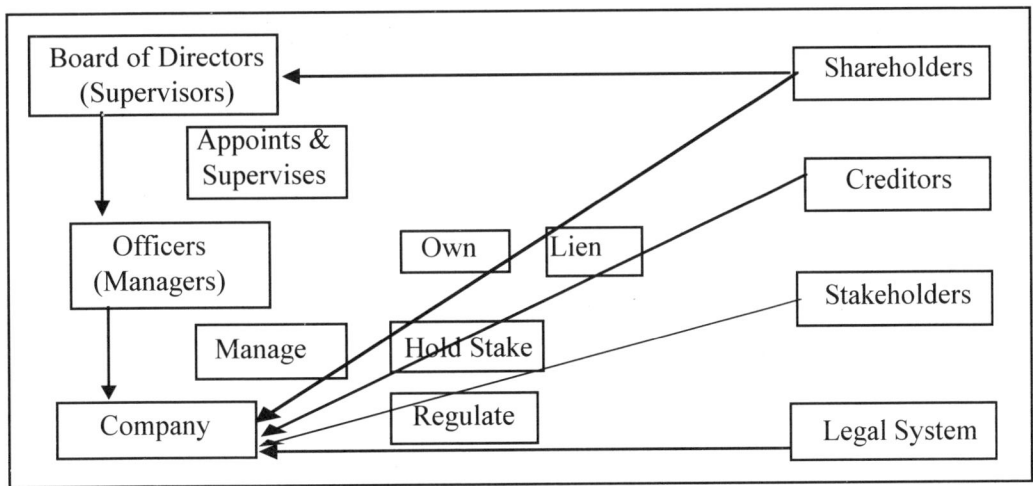

Fig. 1.3: The Anglo-US Model

The distinctive features are :

(i) Clear separation of ownership and management, which minimizes conflict of interests.

(ii) Companies are run by professional managers who have negligible ownership stakes linked to performance. CEO has a major role to play.

Regulatory Framework in the Anglo-US Model

In the UK and US, a wide range of laws and regulatory codes define relationships among management, directors and shareholders.

In the US, a federal agency, the Securities and Exchange Commission (SEC), regulates the securities industry, establishes disclosure requirements for corporations and regulates communication between corporations and shareholders as well as among shareholders.

The regulatory framework of corporate governance in the UK is established in parliamentary acts and rules established by self-regulatory organizations, such as the Securities and Investment Board, which is responsible for oversight of the securities market. Note that it is not a government agency like the US SEC. Although the framework for disclosure and shareholder communication is well-developed, some observers claim that self-regulation in the UK is inadequate, and suggest that a government agency similar to the US SEC would be more effective.

Stock exchanges also play an important role in the Anglo-US model by establishing listing, disclosure and other requirements.

Disclosure Requirements in the Anglo-US Model

As noted above, the US has the most comprehensive disclosure requirements of any jurisdiction. While disclosure requirements are high in other jurisdictions where the Anglo-US model is followed, none are as stringent as those in the US.

The Japanese Model

The Japanese model is characterized by a high level of stock ownership by affiliated banks and companies; a banking system characterized by strong, long-term links between bank and corporation; a legal, public policy and industrial policy framework designed to support and promote "keiretsu" (industrial groups linked by trading relationships as well as cross-shareholdings of debt and equity); boards of directors composed almost solely of insiders; and a comparatively low (in some corporations, non-existent) level of input of outside shareholders, caused and exacerbated by complicated procedures for exercising shareholders' votes.

Equity financing is important for Japanese corporations. However, insiders and their affiliates are the major shareholders in most Japanese corporations. Consequently, they play a major role in individual corporations and in the system as a whole. Conversely, the interests of outside shareholders are marginal. The percentage of foreign ownership of Japanese stocks is small, but it may become an important factor in making the model more responsive to outside shareholders.

Key Players in the Japanese Model

The Japanese system of corporate governance is many-sided, centering around a main bank and a financial/industrial network or keiretsu.

The main bank system and the keiretsu are two different, yet overlapping and complementary, elements of the Japanese model. Almost all Japanese corporations have a close relationship with a main bank. The bank provides its corporate client with loans as well as services related to bond issues, equity issues, settlement accounts, and related consulting services. The main bank is generally a major shareholder in the corporation.

In the Japanese model, the four key players are: main bank (a major inside shareholder), affiliated company or keiretsu (a major inside shareholder), management and the government. Note that the interaction among these players serves to link relationships rather than balance powers, as in the case in the Anglo-US model.

In contrast with the Anglo-US model, non-affiliated shareholders have little or no voice in Japanese governance. As a result, there are few truly independent directors, that is, directors representing outside shareholders.

The Japanese model may be diagrammed as an open-ended hexagon:

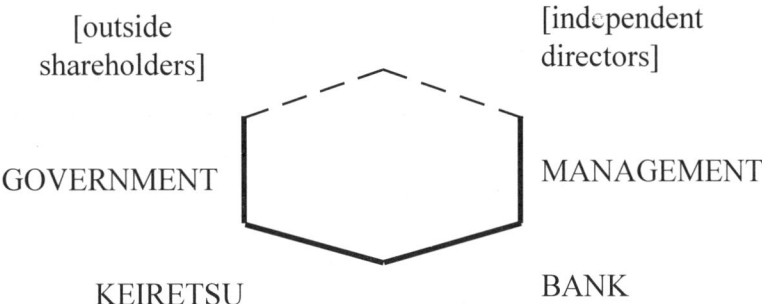

[outside shareholders]

[independent directors]

GOVERNMENT

MANAGEMENT

KEIRETSU

BANK

The base of the figure, with four connecting lines, represents the linked interests of the four key players: government, management, bank and keiretsu. The open lines at the top represent the non-linked interests of non-affiliated shareholders and outside directors, because these play an insignificant role.

Regulatory Framework in the Japanese Model

In Japan, government ministries have traditionally been extremely influential in developing industrial policy. The ministries also wield enormous regulatory control. However, in recent years, several factors have weakened the development and implementation of a comprehensive industrial policy. First, due to the growing role of Japanese corporations at home and abroad, policy formation became fragmented due to the involvement of numerous ministries, most importantly, the Ministry of Finance and the Ministry of

International Trade and Industry. Second, the increasing internationalization of Japanese corporations made them less dependent on their domestic market and therefore somewhat less dependent on industrial policy. Third, the growth of Japanese capital markets led to their partial liberalization and an opening, albeit small, to global standards. While these and other factors have limited the cohesion of Japanese industrial policy in recent years, it is still an important regulatory factor, especially in comparison with the Anglo-US model.

Disclosure Requirements in the Japanese Model

Disclosure requirements in Japan are relatively stringent, but not as stringent as in the US. Corporations are required to disclose a wide range of information in the annual report and/or agenda for the AGM, including: financial data on the corporation (required on a semi-annual basis); data on the corporation's capital structure; background information on each nominee to the board of directors (including name, occupation, relationship with the corporation, and ownership of stock in the corporation); aggregate data on compensation, namely the maximum amount of compensation payable to all executive officers and the board of directors; information on proposed mergers and restructurings; proposed amendments to the articles of association; and names of individuals and/or companies proposed as auditors.

Japan's disclosure regime differs from the US regime (generally considered the world's strictest) in several notable ways. These include: semi annual disclosure of financial data, compared with quarterly disclosure in the US; aggregate disclosure of executive and board compensation, compared with individual data on the executive compensation in the US; disclosure of the corporation's ten largest shareholders, compared with the US requirement to disclose all shareholders holding more than five percent of the corporation's total share capital; and significant differences between Japanese accounting standards and US Generally Accepted Accounting Practices (US GAAP).

German Model

The German corporate governance model differs significantly from both the Anglo-US and the Japanese model, although some of its elements resemble the Japanese model.

Banks hold long-term stakes in German corporations, and, as in Japan, bank representatives are elected to German boards. However, this representation is constant, unlike the situation in Japan where bank representatives were elected to a corporate board only in times of financial distress. Germany's three largest universal banks (banks that provide a multiplicity of services) play a major role; in some parts of the country, public-sector banks are also key shareholders.

There are three unique elements of the German model that distinguish it from the other models outlined in this article. Two of these elements pertain to **board composition** and one concerns **shareholders' rights :**

First, the German model prescribes two boards with separate members. German corporations have a two-tiered board structure consisting of a **management board** (composed entirely of insiders, that is, executives of the corporation) and a **supervisory board** (composed of labour/employee representatives and shareholder representatives). The two boards are completely distinct; no one may serve simultaneously on a corporation's management board and supervisory board. Second, the size of the supervisory board is set by law and cannot be changed by shareholders.

Third, in Germany and other countries following this model, **voting right restrictions** are legal; these limit a shareholder to voting a certain percentage of the corporation's total share capital, regardless of share ownership position.

Most German corporations have traditionally preferred bank financing over equity financing. As a result, German stock market capitalization is small in relation to the size of the German economy. Furthermore, the level of individual stock ownership in Germany is low, reflecting Germans' conservative investment strategy. It is not surprising therefore, that the corporate governance structure is geared towards preserving relationships between the key players, notably, banks and corporations.

The system is somewhat ambivalent towards minority shareholders, allowing them scope for interaction by permitting shareholder proposals, but also permitting companies to impose voting rights restrictions.

The percentage of foreign ownership of German equity is significant; in 1990, it was 19 percent. This factor is slowly beginning to affect the German model, as foreign investors from inside and outside the European Union begin to advocate for their interests. The globalization of capital markets is also forcing German corporations to change their ways. When Daimler-Benz AG decided to list its shares on the NYSE in 1993, it was forced to adopt US GAAP. These accounting principles provide much greater financial transparency than German accounting standards. Specifically, Daimler-Benz AG was forced to account for huge losses that it could have "hidden" under German accounting rules.

In this model, although the shareholders own the company, they do not entirely dictate the governance mechanism. As shown shareholders elect 50 percent of members of supervisory board and the other half is appointed by labour unions. This ensures that employees and labourers also enjoy a share in the governance. The supervisory board appoints and monitors the management board. There is a reporting relationship between them, although the management board independently conducts the day-to-day operations of the company.

The distinctive features are :

(i) Banks and financial institutions have substantial stake in equity capital of companies.

(ii) Labour Relations Officer is represented in the management board. Worker participation in management is practised.

(iii) Both shareholders and employees have equal say in selecting the members of the supervisory board.

Key Players in the German Model

German banks, and to a lesser extent, corporate shareholders, are the key players in the German corporate governance system. Similar to the Japanese system described above, banks usually play a multi-faceted role as shareholder, lender, issuer of both equity and debt, depository (custodian bank) and voting agent at AGMs. In 1990, the three largest German banks (Deutsche Bank AG, Dresdner Bank AG and Commerzbank AG) held seats on the supervisory boards of 85 of the 100 largest German corporations.

In Germany, corporations are also shareholders, sometimes holding long-term stakes in other corporations, even where there is no industrial or commercial affiliation between the two. This is somewhat similar, but not parallel, to the Japanese model, yet very different from the Anglo-US model where neither banks nor corporations are key institutional investors.

The mandatory inclusion of labour/employee representatives on larger German supervisory boards further distinguishes the German model from both the Anglo-US and Japanese models.

Regulatory Framework in the German Model

Germany has a strong federal tradition; both federal and state (Laender) laws influence corporate governance. Federal laws include: the Stock Corporation Law, Stock Exchange Law and Commercial Law, as well as the above mentioned laws governing the composition of the supervisory board are all federal laws. Regulation of Germany's stock exchanges is, however, the mandate of the states.

A federal regulatory agency for the securities industry was established in 1995. It fills a former void in the German regulatory environment.

Disclosure Requirements in the German Model

Disclosure requirements in Germany are relatively stringent, but not as stringent as in the US. Corporations are required to disclose a wide range of information in the annual report and/or agenda for the AGM, including: corporate financial data (required on a semi-annual basis); data on the capital structure; limited information on each supervisory board nominee (including name, hometown and occupation/affiliation); aggregate data for compensation of the management board and supervisory board; any substantial shareholder holding more than 5 percent of the corporation's total share capital; information on proposed mergers and restructurings; proposed amendments to the articles of association; and names of individuals and/or companies proposed as auditors.

The disclosure regime in Germany differs from the US regime, generally considered the world's strictest, in several notable ways. These include: semi-annual disclosure of financial data, compared with quarterly disclosure in the US; aggregate disclosure of executive compensation and supervisory board compensation, compared with individual data on executive and board compensation in the US; no disclosure of share ownership of members of the supervisory board, compared with disclosure of executive and director's stock ownership in the US; and significant differences between German accounting standards and US GAAP.

1.8. DEFINING GOOD GOVERNANCE

Good governance refers to the ability to deliver goods to the stakeholders. When we talk of good governance, we are referring to the elimination of mall governance and the establishment of good governance through democratic processes and rule of law, so that citizens of a country and members of a society do not suffer. To talk of good governance is to make the various agents of political system work for the betterment of all the citizens especially the marginalized and the vulnerable communities. Good governance is enhancing the ability of the people to gain better and dignified life, greater options to choose from and ensuring transparency in administration etc.

At one time, the notion of 'governability' was also used to discuss the facets of good governance. The concept of governability directs attention to a state's capacity to govern. For the Indian situation, the issues of its growing crisis of governability will refer to three types of problems; 1) the absence of enduring coalition; 2) policy ineffectiveness, an incapacity to accommodate political conflict without violence (Kohli, 1990). One way of measuring governability is to set up standards, whereby some objective definition of a society's problems would be sought and against which the capacity of a government to solve problems would be assessed. Most governments would fall short of these goals. A political establishment that repeatedly fails to fulfil their stated objective is deemed to be a government with a low capacity to govern. It is at this juncture that crisis engulfs the government which, in turn, affects not only the state but also the society.

In many of the political discourses, good governance is equated to political convenience, that is, performing well so as to continue to remain in power. This is the political compulsion before the political parties, due to which they engage in political manipulations. It is also not weighing things on the cost benefit analysis, as if people do not matter. Good governance is ensuring a better today and a brighter future for all the citizens. In the Asia-Pacific region, in a special way, local governance has generated a great deal of interest among the people, since it is the localised communities that represent the aspirations and hopes, culture and society of people.

Bureaucracy and Good Governance

It was also taken for granted that in most of the nations bureaucracy stands committed to a set of values enshrined in the constitution. The administrative structure in many countries could be relied upon for ensuring good governance, people oriented governance. But, the whole set of values for which the bureaucracy stands committed is under the threat of displacement in the name of liberalization. In other countries, where other primordial identifiers are surfacing either in terms of racism, casteism, and regionalism, the bureaucracy itself is undergoing major change. In a modern state where traditional communitarian forms of administrative structures are sacrificed for the sake of state administration, it becomes all the more essential to insist upon good governance.

The ever sharpening understanding of good governance not only recognizes the plurality of actors involved in the process of governance, they also address themselves to the substance of governance. This means, governance is no longer simply equated with civil service reform, or with the application of management strategies devised in the private sector to public organisations. Instead, now there is a greater emphasis on participation, decentralisation, accountability, and governmental responsiveness, and even broader concerns such as those of social equality and justice. This new emphasis has been facilitated by a parallel process: the discrediting of the conventional definition of development as economic growth, and the adoption by international agencies of the human development perspective associated with the writings of Amartya Sen and Mahbub-ul-Haq, most recently linked also with the agenda of human rights.

In the recent past, many political establishments have been caught in scams and scandals, with leaders amassing wealth, resources and power. The magnitude of the systematic fraud, deceit, chicanery, embezzlement and theft is shocking. Moreover, this appropriation of public funds directly affects the social sector investments in general and the poor in particular. In this regard, good governance also refers to accountability and transparency. Thus the basic principles of good governance are equality, justice, prosperity and democracy. These principles are ensured through participation, decentralisation, accountability and transparency.

In a developing country like India, governance concerns necessarily have a wide ambit. The recognition that governance takes place in domains other than that of exclusively formal institutionalized political and administrative structures means that governance concerns encompass a variety of spheres. These include the political (equal application of rule of law, accountability and transparency, the right to information, and corruption in public life); the economic (corporate governance, the regulation of private sector, and financial markets); the civil society (in its various manifestations, not-excluding uncivil associations) (Jalal & Pai; 2001). It is these aspects that would lead to long term sustainability of communities, societies, nations and the universe at large.

Good Governance and Sustainable Development

The Draft Plan of Implementation of the World Summit on Sustainable Development considers the following as the three basic components of sustainable development :

    (a) economic development,

    (b) social development

    (c) environmental protection

The Draft Plan further states that these three components are interdependent and mutually reinforcing. Poverty eradication, changing unsustainable patterns of production and consumption, and protecting and managing the natural resource base of economic and social development–are overarching objectives and essential requirements for sustainable development.

Challenges faced by Good Governance

Since the political establishment is habituated to be the central source of power, wants to control the fate of all the citizens, it is not easy for the politicians to give up power. In the past, participation in electoral politics was considered to be the central reality of good governance. While this is true, good governance today involves participation in electoral democracy, ensuring the benefit of development to all the citizens and making the rulers accountable to the lives and hopes of the public. The citizens of a country also could get so used to be ruled by the powerful elite, that they might not like to take the responsibility of governing themselves. More than ever the local situation is extremely interconnected with the national and international milieu. Hence, today many nation states also cannot exercise their sovereignty. At this juncture it is a tall claim that good governance can be ensured everywhere.

Many scholars emphasize on the social indicators for measuring the quality of governance. Social inequality in India retards both balanced development and distorts the logic of democracy. It is precisely this distorting logic of democracy in an unequal society that necessitates state-welfare for the protection of the vulnerable, for the concerns of distributive justice can not be fulfilled by governance alone. The answer, therefore, is not to look towards the state, but at different ways of approaching and defining both democracy and development: a view of democracy, for instance, that goes beyond the procedural to seek the substantive democratization of not only the state, but also society and social relations; and a view of development that possibly departs from the conventional ways of measuring this goal by focusing not on GDP and GNP, but on the enlargement of human capabilities and the enhancement of the quality of life for all citizens. The multiple meanings these concepts have acquired in particular societies have emerged out of rich histories of political practice and discourse.

Current Status of Good Governance in India: Empirical Evidences

The quality of governance is an issue of increasing concern in countries around the world, both developed and developing. The UN Secretary General has stated, "Good governance is perhaps the single most important factor in eradicating poverty and promoting development." However, a lack of systematic data, both over time within countries as well as between countries around the world, ensures that fundamental questions remain to be answered adequately. How can we best measure governance? How does governance performance differ across time and space? Which are the most critical issues of governance?

By undertaking a set of systematic, comprehensive assessments of governance at the national level, the goal of the World Governance Survey (WGS) was to provide some further insight into these issues. Using a cohesive framework and questionnaire, governance assessments were undertaken in 16 developing and transitional societies, representing 51% of the world's population. We believe that experience of the project indicates the ability to generate valid and valuable data despite the contested nature of the governance concept and the considerable methodological problems in collecting data on this set of issues. The project had three main achievements. First, it developed a comprehensive framework and process oriented set of indicators for assessing governance that were acceptable across

Good Governance

Governance for the ruled and governance for the rulers become different for complex societies. Corporate governance is a way of life and not a set of rules. A way of life that necessitates taking into account the shareholders interests in every business decision.

A key element of good governance, is transparency projected through a Code of good governance which incorporates a system of checks and balances between key players boards, management, auditors and shareholders.

"Governance has three major components, that of process, content, and deliverables." (as suggested by Harsh Mander, Mohammed Asif).

The process of governance includes such factors as transparency and accountability. Content includes values such as justice and equity. Deliverables indicate the ways the rules meet the basic needs of citizens; go by Rule of Law i.e. transparent.

Good governance is always responsive to the needs of the people and copes with emerging challenges in society through its accountability. Driven by the spirit of vision, it adds value to life and explore the unknown and thereby stretching towards tomorrow, today. Good governance strives to inspire confidence and blend together certainty and competitive technology not only to satisfy the local customers but also will meet global standards through:

Customer service focus

Empowered employees

Innovative services

Cost efficiency.

Some of the definitions of good governance are as follows:

The World Bank in the report (1994) described "Good governance is epitomized by predictable, open, and enlightened policy making (that is, transparent processes), Executive imbued with professional ethos, is accountable for its actions".

Further United Nations Development Programme (UNDP) describes: Governance is the exercise of economic, political and administrative authority to manage a country's affairs at all levels. It comprises mechanisms, processes, and institutions, through which citizens and groups articulate their interests, exercise their legal rights, meet their legal obligations and mediate their difference.

Governance can be good only if efforts are made to minimize corruption, takes minority interests, and interests of the society at large.

Hon'ble Justice Sri M.N.Venkatachalia, former Chief Justice of India vouchsafes: the subject has assumed sharper focus with the unpleasant experiences of the more recent corporate collapses, of the human factor in financial, managerial and audit supervision. These developments shook the foundation of the Corporate credibility. Actually Corporate governance adds considerable value to operational performance of companies.

M. Damodaran, Chairman, SEBI, says "Companies should see cost of compliance with corporate governance norms as an investment in protecting the interests of stakeholders and not as an expenditure".

Rahul Bajaj, Chairman, Bajaj Auto Ltd., says "Good Corporate Governance is truly the need of the hour. Shareholder confidence in company management is essential for the sound development of the capital market and hence of companies. It is the competence and integrity of the management of a company, including the Director, which determines the quality of its governance".

Excellent Corporate Governance practices enhance the individual company's performance. "Good governance supports the objectives of regulators responsible for protecting the confidence in financial systems and markets and encourages investment flows, increasing overall economic prosperity", says, A.K. Agarwal, Executive Director (Corporate Affairs) Bharat Petroleum Corp. Ltd.)

"Practicing Good Corporate Governance refers to leadership and support within the limits of an organisation's resources of publicly important purposes says, S. Ramadorai, Chairman, CMC Ltd. Such purposes might include improving education and health care in our community, environmental excellence, resource conservation, community service, and improving industry and business practices. Good Corporate Governance entails influencing other organizations, private and public. Good Corporate Governance is the primary requirement for the organization to survive, grow and flourish at all teems to come."

Corporate Governance is about protecting shareholder value but good corporate governance is about enhancing shareholder value and the corporate brand.

In the present scenario of globalization and the onslaught of W.T.O. coupled with the recession the world over the Indian Business Community will have no way but to seriously and continuously strike for Excellence in Corporate Governance" to the maximum possible extent. In this direction, it is incumbent on corporations to adopt business ethics at all levels.

Good Corporate Governance not only helps in building trust with all stakeholders including customs, suppliers, creditor and diverse investors, Brijmohan Lall Munjal , Chairman, Hero Honda Motors Ltd. says, corporate ingenuity will be determined by the extent of commitment to the practice of key governance principles of trusteeship, transparency, control and ethical corporate citizenship."

Essential elements of good governance:

Good governance is concerned primarily with the utilization of authority to promoting national development and improving public welfare. Legitimacy of the government and human rights guarantees it provides, are at the core of good governance.

Basic elements of good governance are rule of law, accountability, transparency, participation and people's control. All these are based on equity justice and rights. The ultimate form of equity in governance is seen as people's control over governance and its goals.

Rule of law: It is one of the responsibilities of the state to build a democratic and just society. Rule of law is an established legal framework that should be fair and enforced impartially. The formal discussion of Rule of Law, namely justice, fairness and liberty focus as a more basic level: the processes of formulating and applying rules. Moreover the rule of law encompasses well-defined rights and duties, as well as mechanisms for enforcing them and settling disputes in an impartial manner.

Five elements are critical: A set of rules known in advance rules that are actually in force, ensuring application of the rules, conflict resolution, amendment procedures.

Effective Corporate Governance

1. Ensuring the Basis for an Effective Corporate Governance Framework

The corporate governance framework should promote transparent and efficient markets, be consistent with the rule of law and clearly articulate the division of responsibilities among different supervisory, regulatory and enforcement authorities.

To ensure an effective corporate governance framework, it is necessary that an appropriate and effective legal, regulatory and institutional foundation is established upon which all market participants can rely in establishing their private contractual relations. This corporate governance framework typically comprises elements of legislation, regulation, self regulatory arrangements, voluntary commitments and business practices that are the result of a country's specific circumstances, history and tradition. The desirable mix between legislation, regulation, self-regulation, voluntary standards, etc. in this area will therefore vary from country to country. As new experiences accrue and business circumstances change, the content and structure of this framework might need to be adjusted.

Countries seeking to implement the Principles should monitor their corporate governance framework, including regulatory and listing requirements and business practices, with the objective of maintaining and strengthening its contribution to market integrity and economic performance. As part of this, it is important to take into account the interactions and complementarily between different elements of the corporate governance framework and its overall ability to promote ethical, responsible and transparent corporate governance practices. Such analysis should be viewed as an important tool in the process of developing an effective corporate governance framework. To this end, effective and continuous consultation with the public is an essential element that is widely regarded as good practice. Moreover, in developing a corporate governance framework in each jurisdiction, national legislators and regulators should duly consider the need for, and the results from, effective international dialogue and cooperation. If these conditions are met, the governance system is more likely to avoid over-regulation, support the exercise of entrepreneurship and limit the risks of damaging conflicts of interest in both the private sector and in public institutions.

A. The corporate governance framework should be developed with a view to its impact on overall economic performance, market integrity and the incentives it creates for market participants and the promotion of transparent and efficient markets.

The corporate form of organisation of economic activity is a powerful force for growth. The regulatory and legal environment within which corporations operate is therefore of key importance to overall economic outcomes. Policy makers have a responsibility to put in place a framework that is flexible enough to meet the needs of corporations operating in widely different circumstances, facilitating their development of new opportunities to create value and to determine the most efficient deployment of resources. To achieve this goal, policy makers should remain focused on ultimate economic outcomes and when considering policy options, they will need to undertake an analysis of the impact on key variables that affect the functioning of markets, such as incentive structures, the efficiency of self-regulatory systems and dealing with systemic conflicts of interest. Transparent and efficient markets serve to discipline market participants and to promote accountability.

B. The legal and regulatory requirements that affect corporate governance practices in a jurisdiction should be consistent with the rule of law, transparent and enforceable.

If new laws and regulations are needed, such as to deal with clear cases of market imperfections, they should be designed in a way that makes them possible to implement and enforce in an efficient and even handed manner covering all parties. Consultation by government and other regulatory authorities with corporations, their representative organisations and other stakeholders, is an effective way of doing this. Mechanisms should also be established for parties to protect their rights. In order to avoid over-regulation, unenforceable laws, and unintended consequences that may impede or distort business dynamics, policy measures should be designed with a view to their overall costs and benefits. Such assessments should take into account the need for effective enforcement, including the ability of authorities to deter dishonest behaviour and to impose effective sanctions for violations.

Corporate governance objectives are also formulated in voluntary codes and standards that do not have the status of law or regulation. While such codes play an important role in improving corporate governance arrangements, they might leave shareholders and other stakeholders with uncertainty concerning their status and implementation. When codes and principles are used as a national standard or as an explicit substitute for legal or regulatory provisions, market credibility requires that their status in terms of coverage, implementation, compliance and sanctions is clearly specified.

C. The division of responsibilities among different authorities in a jurisdiction should be clearly articulated and ensure that the public interest is served.

Corporate governance requirements and practices are typically influenced by an array of legal domains, such as company law, securities regulation, accounting and auditing standards, insolvency law, contract law, labour law and tax law. Under these circumstances, there is a risk that the variety of legal influences may cause unintentional overlaps and even conflicts, which may frustrate the ability to pursue key corporate governance objectives. It is important that policy makers are aware of this risk and take measures to limit it. Effective enforcement also requires that the allocation of responsibilities for supervision, implementation and enforcement among different authorities is clearly defined so that the competencies of complementary bodies and agencies are respected and used most effectively. Overlapping and perhaps contradictory regulations between national jurisdictions is also an issue that should be monitored so that no regulatory vacuum is allowed to develop (i.e. issues slipping through in which no authority has explicit responsibility) and to minimise the cost of compliance with multiple systems by corporations.

When regulatory responsibilities or oversight are delegated to non-public bodies, it is desirable to explicitly assess why, and under what circumstances, such delegation is desirable. It is also essential that the governance structure of any such delegated institution be transparent and encompass the public interest.

D. Supervisory, regulatory and enforcement authorities should have the authority, integrity and resources to fulfil their duties in a professional and objective manner. Moreover, their rulings should be timely, transparent and fully explained.

Regulatory responsibilities should be vested with bodies that can pursue their functions without conflicts of interest and that are subject to judicial review. As the number of public companies, corporate events and the volume of disclosures increase, the resources of supervisory, regulatory and enforcement authorities may come under strain. As a result, in order to follow developments, they will have as significant demand for fully qualified staff to provide effective oversight and investigative capacity which will need to be appropriately funded. The ability to attract staff on competitive terms will enhance the quality and independence of supervision and enforcement.

Looking at him dressed in indigo blue jeans and a casual half-sleeved shirt, it's hard to tell that Jamshed J. Irani has slept for all of three hours the previous night. His face, chubby and calm, doesn't betray the slightest hint of fatigue, or turmoil, that the managing director of Tata Steel must be experiencing. Barely five hours earlier, at three in the morning, he's flown into Mumbai from the steel major's sprawling works at Jamshedpur. But it's not the long hours that must be stressing out the workaholic Irani. Rather, it's the unpleasantness of the task carried out over the previous 16 hours that must weigh on the soft-spoken 65-year-old's mind.

India's oldest steel manufacturer, you see, is going through a painful phase of restructuring. Last year, it decided to do something it has never done in its 94-year history: sack its officer level employees. Under a Performance Ethic Programme (PEP), Tata Steel has identified 1,200 officers it wants to either repurpose or let go. And Irani had spent the previous day at Jamshedpur, personally explaining to the employees most of whom have never dreamt of a life outside the corporate cocoon in Jamshedpur why the move is an issue of organisational survival. With that, he also signaled the formal rollout of the dreaded programme "Tough times call for tough decisions," says Irani, poker faced, sitting on a sofa in his upmarket apartment on Mumbai's Marine Drive.

If there's consternation inside and outside the industrial township, it's for good reason. Even before the first stake was driven into the soil of Sakchi (later renamed Jamshedpur) on February 27, 1908, the idea was to create an industrial island, one that none of its tribal workers would want to leave. In fact, in a letter to his son Dorabji, the founder Jamsetji Nusserwanji Tata actually listed what all the new steel factory must have (right from wide roads to quick governing trees to places of religious worship) for those who toiled for the new mill. The ensuing managements more than delivered on the founder's vision, even setting up poultry and dairy farms, besides schools and hospitals.

The Utopia called Jamshedpur worked fine as long as its benefactor could work on a cost-plus basis. But liberalization of the sector in the early 90s suddenly made supporting a workforce of 78,000 and a township of several lakhs well nigh impossible. Coupled with the downturn in steel, profits slid rapidly. Between 1996 and 1999, net profits crashed from Rs. 56 crore to Rs. 28 crore. It wasn't until last year that manic cost cuttings and improvements in efficiencies helped profits rebound to Rs 43 crore. This year, profits are expected to be significantly higher than the previous year's.

Why is Good Corporate Goveranance Important?

Policy makers, practitioners and theorists have adopted the general stance that corporate governance reform is worth pursuing, supporting such initiatives as splitting the role of chairman/chief executive, introducing non-executive directors to boards, curbing excessive executive performance-related remuneration, improving institutional investor relations, increasing the quality and quantity of corporate disclosure, inter alia. However, is there really evidence to support these initiatives? Do they really improve the effectiveness of corporations and their accountability? There are certainly those who are opposed to the ongoing process of corporate governance reform. Many company directors oppose the loss of individual decision making power, which comes from the presence of non-executive directors and independent directors on their boards. They refute the growing pressure to communicate their strategies and policies to their primary institutional investors. They consider that the many initiatives aimed at 'improving' corporate governance in UK have simply slowed down decision-making and added an unnecessary level of the bureaucracy and red tape {refer to summary Richard Branson's experiment with the stock market : Appendix A.4}. The Cadbury Report emphasized the importance of avoiding excessive control and recognized that no system of control can completely eliminate the risk of fraud (as in the case of Maxwell) without hindering companies' ability to compete in a free market (Cadbury Report, 1992, p. 12, para 1.9). This is an important point, because human nature cannot be altered through regulation, checks and balances. Nevertheless, there is growing perception in the financial markets that good corporate governance is associated with prosperous companies. The research of Solomon and Solomon A. showed some evidence to support the agenda for corporate governance reform. The findings indicated that the institutional investment community considered both company directors and institutional investors in corporate governance reform, viewing the reform process as a 'help rather than a hindrance'.

Their findings endorsed many of the issues relating to the agenda for corporate governance reform in UK. For example, they show, that institutional investors agreed strongly with the Hampel view that corporate governance is as important for small companies as for larger ones. The results also indicated significant support from the institutional investment community for the continuance of a voluntary environment for corporate governance. The respondents' agreement that there should be flintier reform in their investee companies also added support to the ongoing reform process. Lastly, the institutional investors perceived a role for themselves in corporate governance reform, as they agreed that the institutional investment community should adopt a more activist stance.

1.9. GOVERNING ACTION

In order for a social-political instrument, regardless of design or application, to work, the action element of governing interactions relies upon convincing and socially penetrating images and sufficient social political will or support.

In the literature there is a tendency towards overcoming the dichotomy of voluntarism and determinism in action theories, and to interpret both action and structure as non-opposed but interdependent variables of a unique process. Although accounts vary, action and structure are now considered as inextricably grounded in practical interaction.

Governing action occurs at the intentional level of social political interactions, while simultaneously societal circumstances limit or enhance such action.

Literature on social-political action predominantly deals with collective action through participation in political activities, the action of interest groups, protest movements and other social groups.

Therefore much political governing can be seen as a reaction to actions of other societal actors. However, one has to be careful here: in diverse, dynamic and complex societies, or parts thereof, it is often not easy to ascertain whose action calls for whose reactions Governing action and reaction are best seen as continuous and recurrent processes, where in a given circumstance one governing entity takes the initiative and in a following circumstance another does: governing as pattern of chains of action-reaction.

Coordination is a major governing mechanism at the international level of governing interventions in handling complex societal issues; whatever its shortcomings, bureaucracy serves as a relatively controllable hierarchical structure for those coordinating interactions.

J.J. Irani Committee Report on Company Law

Background

Legal framework for corporate entities is essential to enable sustainable economic reform. Such framework has to be in tune with emerging economic scenario, encourage good corporate governance and enable protection of the interests of the stakeholders, including investors.

It is appropriate that comprehensive reviews of the Companies Act, 1956 has been taken up through a consultative process initiated by Ministry of Company Affairs by exposing a Concept Paper on Company Law through electronic media. Such broad based consultation would enable working out an appropriate legislative proposal to meet the requirements of India's growing economy and should form an integral part of the law making exercise.

Approach towards new Company Law

Company Law should comprise of the basic principles guiding the operation and governance of different kinds of corporate entities in India and be available in a single, comprehensive, centrally administered framework. Such legal framework should provide a smooth and seamless transition from one form of business entity to another and be amenable to adaptation to new business models as they emerge.

Company law should be compact. While essential principles should be retained in the substantive Law, procedural and quantitative aspects should be addressed in the Rules.

Classification and Registration of Companies

The law should take into account the requirements of different kinds of companies prescribing the essential requirements of their corporate governance structure.

Small and Private Companies should be provided greater flexibility and freedom of operation while enabling compliance at low cost. To unleash the entrepreneurial talent of the people in the information and technology driven environment, law should recognize One Person Company (OPC). Such companies should be provided with a simpler legal regime through exemptions.

Government Companies should be treated at par with other companies and be subject to a similar compliance standards.

Process of registration should be speedy, optimally priced and compatible with e-Governance initiatives. Companies should be required to make necessary declarations and disclosures about promoters and directors at the time of incorporation. Stringent consequences should follow if incorporation is done under false or misleading information.

Strong action should be taken under law against companies that vanished with the investors' funds. Preventive action in respect of such "vanishing companies." should begin with registration itself and should be sustained through a regime that requires regular and mandatory filing of statutory documents. This should be followed up with clearly provided legal process for tracking and causing disgorgement of ill-gotten gains. Corporate veil should be lifted to enable access to the individuals responsible.

Regular filing should be made easy efficient and cost effective. Non-filing of documents or incorrect disclosures should be dealt with seriously. Delays in filing should be penalized through non-discretionary late fee relatable to the period of default. There should be a system of random scrutiny of filings of corporates to be carried out by the registration authorities.

Limited liability partnerships should be facilitated through a separate enactment.

Companies Act need not prescribe limitations on the number of members.

Law should require transparency in functioning of charitable and licensed companies.

Management and Board Governance

Law should provide an appropriate framework of governance that should be complied with by all companies without sacrificing the basic requirement of exercise of discretion and business judgment in the interests of company and its stakeholders.

There should be an obligation on the part of a company to maintain a Board of Directors as per the provisions of Law and to disclose particulars of the directors through statutory filings of information.

Law should provide for only the minimum number of directors necessary for various classes of companies. There need not be any limit to maximum number of directors. Government should not intervene in the process of appointment and removal of directors in non-Government companies. No age limit for directors need be specified in the Act other than procedures for appointments to be followed by prescribed companies for appointment of directors above a particular age.

Every company to have at least one director resident in India. Requirement of obtaining approval of Central Govt. under Companies Act for appointment of non-resident managerial persons should be done away with. Duty to inform the Registrar of particulars regarding appointment, resignation, death etc. of directors should be that of the company.

Presence of independent director on the boards of companies having significant public interest would improve corporate governance. Law should recognize the principle of independent directors and spell out their attributes, role, qualifications, liability and manner of appointment along with the criteria of independence. However, prescription of the number and proportion of such directors in the Board may vary depending on size and type of company and may be prescribed through Rules.

Decision on remuneration of directors should not be based on a "Government approval based system" but should be left to the company. However, this should be transparent, based on principles that ensure fairness, reasonableness and accountability and should be properly disclosed. No limits need be prescribed. In case of inadequacy of profits also the company may be allowed to pay remuneration recommended by remuneration committee (wherever applicable) and with the approval of shareholders.

Certain committees to be constituted with participation of independent directors should be mandated for certain categories of companies where the requirement of independent directors is mandated. In other cases constitution of such committees should be at the option of the company. Law should specify the manner and composition of various committees of the board like (i) Audit Committee (ii) Stakeholder's Relationship Committee and (iii) Remuneration Committee along with obligation on the part of the company to consult them in certain matters.

Board meetings to be held every three months with a minimum of four meetings to be held in a year. The gap between two meetings not to exceed four months. Meetings by electronic means to be allowed. In the case of companies where Independent Directors are prescribed, notice period of 7 days has been recommended for Board Meetings with provisions for holding emergency meetings at a shorter notice. Consent of shareholders by way of special resolution should be mandatory for certain important matters.

Every company should be required to appoint a Chief Executive Officer, Chief Finance Officer and Company Secretary as its Key Managerial Personnel whose appointment and removal shall be by the Board of Directors. Special exemption may be provided for small companies, who may obtain such services, as may be required from qualified professionals in practice.

Minority Interests

'Minority' and 'Minority Interest' should be defined in the substantive law. Law must balance the need for effective decision making on corporate matters through consensus without permitting persons in control to stifle action for redressal arising out of their own wrong doing.

Accounts and Audit

Accounting Standards should be notified under the Companies Act early.

Consolidation of financial statements should be made mandatory. Requirement of attaching financial statements of subsidiary company(ies) with the holding company to be done away with.

Format of financial statements should be prescribed in the Act/Rules. Cash Flow Statement to be made part of mandatory financial statements. Financial year should be aligned to uniformly end on 31st March. Option to maintain books in electronic form should be given to companies. Books of account should be preserved by a company for a period of seven years.

Shareholder Democracy

◆ The shareholders are the real owners who provide the risk capital base on which the entire corporate edifice rests.

♦ The general body of shareholders constitutes the fountain head of all powers which are entrusted to different organs of the company.

♦ The shareholders are both the cause and end of a corporate enterprise. A corporation derives sustenance from them and exists for them.

♦ The company law is based on the concept of ownership vesting the ultimate control of the company with the shareholders.

Shareholder Democracy and Protection of Minority interests – Legal Position in India

Voluntary efforts to promote investors' relations, though laudable, have not been sufficient. Legal provisions are needed to compel corporates to look after the interests of investors and to secure their approval in respect of important corporate decisions. Corporations are, at least, legally owned by shareholders who subscribe to their equity capital on the basis of a public offer or private placement. Such subscriptions are made on the faith of stated objectives in the offer documents. The investors exercise their rights in the general meetings. Currently, the company law requirements mandate a 7596 majority in certain matters and a simple majority in other cases, of those present and voting (personally or through recorded proxies) at the meeting. A show of hands is usually enough for the chairman to determine if a resolution has the requisite majority. There is of course, a provision for poll in case of any doubts or when demanded by eligible number of shareholders.

Because of their continued dependence on the information provided by the company and those responsible for its governance, investors seek and are entitled to some protection from being deceived or unfairly treated by those in operational control. Reporting and disclosure requirements and best practices are developed to meet this need. More importance is also attached to protecting the interests of minority shareholders because they may lack the resources to protect themselves. But no protection can be provided or should be expected by any investor including minority shareholder in respect of equity risk, which is taken while investing in risky instruments like company shares. SEBI requirements for highlighting risk factors in equity offers is an example of how potential investors should be made aware of the nature and extent of risks involved in investing. Protection of investors should be in the matters relating to transparency in accounting and reporting, majority oppression, biased management, non-conformance to obligatory requirements and so on.

Rights of Shareholders

Equity investors have certain inherent property rights by virtue of their equity holding.

For example, an equity share can be bought, sold or transferred. An equity share also entitles the investor to participate in the profits of the company. The liability of the equity holder is limited to the amount of investment. The ownership of an equity share gives the holder the right to information about the corporation and a right to influence the corporation by participating in the general meetings and by exercising his right of voting. The corporate governance framework should protect the rights of shareholders. Below is given a description of the rights of shareholders which are basic from the point of view of better corporate governance :

(1) Basic Shareholders Rights : These include the following :

    (a) To secure methods of ownership registration;

    (b) To convey or transfer shares;

    (c) To obtain relevant information about the corporation on a timely and regular basis;

    (d) To participate and vote in general shareholder meetings;

    (e) To elect members of the Board;

    (f) To share the profits of the corporation; and

    (g) To share the company's residual profits.

(2) Right to participate and to be sufficiently informed about decisions concerning fundamental corporate changes.

(3) Opportunity to participate in general meeting.

(4) Efficient and transparent functioning of markets for corportate control :

    (a) The rules and procedures governing the acquisition of corporate control in the capital markets, and extraordinary transactions such as mergers and sales of substantial portions of corporate assets should be clearly articulated and disclosed so that investors understand their rights and recourse. Transactions should occur at transparent prices and under fair conditions that protect the rights of all shareholders.

    (b) Anti-takeover devices should not be used to shield management from accountability.

(5) Shareholders, including institutional investors, should consider the costs and benefits of exercising their voting rights.

(6) Equitable treatment of shareholders including minority and foreign shareholders.

    Investors must have a confidence that their investments will be protected against misuse or misappropriation by the management, board or controlling shareholders.

One of the methods by which shareholders can enforce this right is the legal provisions which enable them to initiate legal and administrative proceedings against management and board members. In other words, effective, cheaper and expeditious grievance redressal methods must exist to reinforce the confidence of minority investors.

At the same time, appropriate provisions must exist to protect management and board members against litigation abuse. This may be ensured by introducing 'safe harbours' for management and board members such as the business management rule. Therefore, there must be a balance between the right of investors to seek remedies against infringement of ownership rights and abuse of this right.

(7) Equal treatment of shareholders of the same class.

(a) Within any class, all shareholders should have the same voting rights.

(b) All investors should be able to obtain information about the voting rights attached to all classes of shares before making the purchase.

(c) Any change in voting rights should be submitted for approval of the general meeting by a specified majority.

(d) The custodians or nominees should cast their votes in a manner agreed upon with the beneficial owner of shares.

(e) Processes and procedures for general meeting should provide for equitable treatment of all shareholders. These should not make it unduly difficult or expensive to cast votes.

(8) Prohibition of insider trading and abusive self-dealing: The above mentioned transaction takes place when persons having close relationship with the company exploit their relationships to the detriment of the company and the investors. Since insider trading involves manipulation of the capital markets, this has been declared illegal in most of the countries. These practices constitute a breach of good corporate governance inasmuch as they violate the principle of equitable treatment of shareholders.

(9) Members of the Board and management should be required to disclose any material interest in transactions or matters affecting the corporation.

(10) The use of voting rights by institutional investors should not be opposed to the interests of small private investors.

(11) The use of voting rights should take into account increasing internationalization of corporation's shareholder base and not be confined to domestic shareholders.

1.10. WEALTH CREATION IN GOVERNANCE

Wealth Creation, Management and Distribution

Wealth management is also a very important entity which precedes wealth creation. Wealth management is well achieved by improving the functional performance which is insured with risk management.

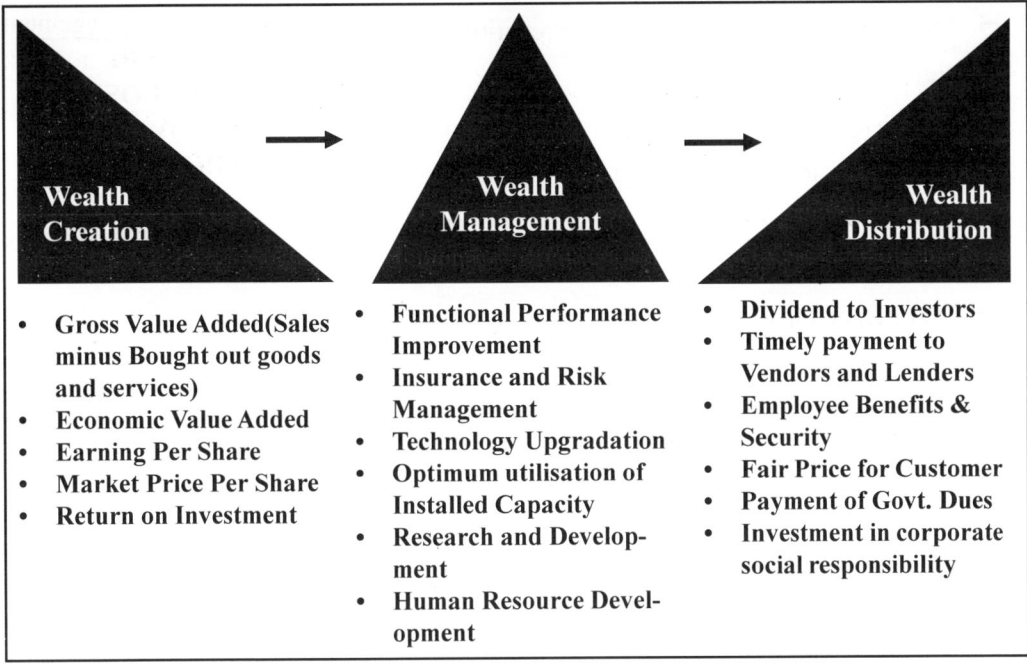

Fig. 1.4: Wealth Creation, Management and Distribution

The competent world needs fast actions and better quality in goods and services. This can only be maintained with the upgradation of the latest and advanced technology.

With all the above, corporate governance weighs more with the optimum utilization of both capital and manpower with their respective growth. No company can survive without meeting the expectations of the consumers and the other stakeholders. One of the ways to be a consistent player of the game is to be with innovations. With the market competition and people's choice, there needs to be improved and advanced products which are given only by research and development. In present day world, good research and development is the key to sustain in the market.

Wealth distribution is another challenging and vital part of the process. Distribution should meet the expectations of all the stakeholders, beginning with the equity in providing the dividend to investors, timely payment to vendor and lenders, securing the employees with the benefits and fair prices for customers, this will help in keeping up the trust and creating the goodwill. Any company should be responsible in developing the nation; this can be done by payment of taxes, dues and taking up the social responsibility.

```
☐  Fixation and justification of Issue Price

☐  Risk Factors and Management Perception
                                                    ┌─────────────────┐
☐  Industry Analysis Report                         │      SEBI       │
                                                    │   (DISCLOSURE   │
☐  Installed Capacity and Capacity Utilisation      │   & INVESTOR    │
                                                    │   PROTECTION)   │
☐  Past Track Record and Projected Financial Analysis│   GUIDELINES   │
                                                    └─────────────────┘
☐  Comparison of Financial Data with Industry Averages

☐  Stock Market Data-Analysis

☐  Significant Financial Ratios

☐  Profitability Ratios

☐  Earning Per Share (EPS)

☐  NAV Per Share

☐  Return on Net Worth

☐  P/E Ratio
```

Fig. 1.5: Disclosure in Offer Documents

Disclosure in Offer Documents

Disclosure and transparency of operations of an organisation is the most desirable thing

```
■  Management Discussion and Analysis of Financial Condition

■  Industry Structure and Developments                    ┌──────────────┐
                                                          │   LISTING    │
■  Segment-wise and Product-wise Performance              │  AGREEMENT   │
                                                          │  CONDITIONS  │
■  Market Price Data - Monthly High and Low               └──────────────┘

■  Performance in relation to broad based indices like BSE Sensex

■  Matters to be placed before the Board Annual Operating Plans and Budgets

■  Foreign Exchange Exposure Material default in Financial Obligation

■  Report on Risk Management - SWOT Analysis
```

Fig. 4.6: Clause 49 of Companies Act, 1956 - Corporate Governance

in the stakeholder's world. SEBI has given some guidelines for Disclosure and Investor Protection. Few of the aspects included in it are as follows: Issue price should be fixed and the justification for the same should be provided, Industry Analysis Report. Installed capacity and Capacity Utilisation, Past Track Record and Projected Financial Analysis, Comparison of Financial data with Industry Averages and, Stock Market Data Analysis.

A Brief about Clause 49

In India the Companies Act, 1956 lays the responsibility of financial statements on the board of directors. The board is also responsible for maintaining proper books of account which would give a true and fair view of the financial position and comply with the Accounting Standards. The principles of 'truth and fairness' of the books of account embody the robustness of internal controls, the responsibility of which also lies with the board. The auditors under the Indian laws, since 1975 nearly 30 years before SOX (Sarbanes-Oxley Act. US) were issuing an opinion on the internal controls over critical processes of a company.

The Effort Outlay

Internal controls are systems that act as checks and balances over the operation of an enterprise to preserve the sanctity of the transactions the company has with others and itself.

The controls need to be in operation for an entire year to be able to deliver robust year-end financial statements. To establish such controls for 90 days and test them for efficacy is inadequate.

Independent Directors

The mythical angels called independent directors now have the sole responsibility of keeping Corporate India in check. The CII and Prime Database have created a large database of individuals who qualify to be independent directors. Needless to say such a database gets populated by ail retired, retiring, out of work and aged professionals who probably did not make it to the board position of the companies they worked for. We seem to forget that a director holds a fiduciary responsibility and needs to be carefully chosen. You do not go through a public database to find a director for a listed company.

An independent director is an oxymoron in a country where owning/promoting families have substantial ownership and control. Under the Companies Act, all directors need to be appointed and remunerated through resolutions approved by the shareholders. The family usually owns a majority or a substantial stake and wields overwhelming influence over such enterprises. No independent director would be appointed or remunerated sans the family's nod. Where would his interest lie? Rarely have we seen independent directors taking a vastly different view than the promoters. If they do they usually end up voting with their feet. The recent Escorts case is an example when some independent directors stood up for what they believed was just and equitable and resigned which reinforces the fait accompli faced by independent directors either you are with us or against us.

Our insurance against mismanagement would not lie in definitions and rules of appointing independent directors but in the ethical calling and reputational risks some of these directors will have to protect themselves from.

Investment Versus Speculation

While it is difficult to draw the line of distinction between investment and speculation, it is possible to broadly distinguish the characteristics of an investor from those of a speculator as follows :

	Investor	**Speculator**
Planning horizon	An investor has a relatively longer planning horizon. His holding period is usually at least one year.	A speculator has a very short planning horizon. His holding period may be a few days to a few months.
Risk disposition	An investor is normally not willing to assume more than moderate risk. Rarely does he knowingly assume high risk.	A speculator is ordinarily willing to assume high risk.
Return expectation	An investor usually seeks a modest rate of return which is commensurate with the limited risk assumed by him.	A speculator looks for a high rate of return in exchange for the high risk borne by him.
Basis for decisions	An investor attaches greater significance to fundamental factors and attempts a careful evaluation of the prospects of the firm.	A speculator relies more on hearsay, technical charts, and market psychology.
Leverage	Typically an investor uses his own funds and eschews borrowed funds.	A speculator normally resorts to borrowings, which can be very substantial, to supplement his personal resources.

Gambling: Gambling is fundamentally different from speculation and investment in the following respects:

- Compared to investment and speculation, the result of gambling is known more quickly. The outcome of a roll of dice or the turn of a card is known almost immediately.

- Rational people gamble for fun, not for income.

- Gambling does not involve a bet on an economic activity. It is based on risk that is created artificially.

- Gambling creates risk without providing any commensurate economic return.

Approaches to Investment Decision Making

The stock market is thronged by investors pursuing diverse investment approaches which may be subsumed under four broad approaches :

- Fundamental approach
- Psychological approach
- Academic approach
- Eclectic approach

Fundamental Approach: The basic tenets of the fundamental approach, which is perhaps most commonly advocated by investment professionals, are as follows :

There is an intrinsic value of a security and this depends upon underlying economic (fundamental) factors. The intrinsic value can be established by a penetrating analysis of the fundamental factors relating to the company, industry, and economy.

At any given point of time, there are some securities for which the prevailing market price will differ from the intrinsic value. Sooner or later, of course, the market price will fall in line with the intrinsic value.

Superior returns can be earned by buying under-valued securities (securities whose intrinsic values exceed the market prices) and selling over-valued securities (securities whose intrinsic values are less than the market prices).

Psychological Approach: The psychological approach is based on the premise that stock prices are guided by emotion, rather than reason. Stock prices are believed to be influenced by the psychological mood of the investors. When greed and euphoria sweep the market, prices rise to dizzy heights. On the other hand, when fear and despair envelop the market, prices fall to abysmally low levels. I.M.Keynes described this phenomenon in eloquent terms: "A conventional valuation which is established as the outcome of the mass psychology of a large number of ignorant individuals is liable to change violently as the result of a sudden fluctuation of opinion due to factors which do not really make much difference to the prospective yield."

Since psychic values appear to be more important than intrinsic values, the psychological approach suggests that it is more profitable to analyze how investors tend to behave as the market is swept by waves of optimism and pessimism which seem to alternate. The psychological approach has been described vividly as the 'castles-in-the-air' theory by Burton G. Malkiel.

Those who subscribe to the psychological approach or the 'castles-in-the-air' theory generally use some form of technical analysis which is concerned with a study of internal market data, with a view to developing trading rules aimed at profit-making. The basic premise of technical analysis is that there are certain persistent and recurring patterns of price movements which can be discerned by analysing market data. Technical analysts use a variety of tools like bar chart, point and figure chart, moving average analysis, breadth of market analysis, and so on.

Academic Approach: Over the last five decades or so, the academic community has studied various aspects of the capital market, particularly in the advanced countries, with the help of fairly sophisticated methods of investigation. While there are many unresolved issues and controversies stemming from studies pointing in different directions, there appears to be substantial support for the following tenets.

Stock markets are reasonably efficient in reacting quickly and rationally to the flow of information. Hence, stock prices reflect intrinsic value fairly well. Put differently :

Market price@ Intrinsic value

- Stock price behaviour corresponds to a random walk. This means that successive price changes are independent. As a result, past price behaviour cannot be used to predict future price behaviour.

- In the capital market, there is a positive relationship between risk and return. More specifically, the expected return from a security is linearly related to its systematic risk (also referred to as its market risk or non-diversifiable risk).

Eclectic Approach: The eclectic approach draws on all the three different approaches discussed above. The basic premises of the eclectic approach are as follows :

- Fundamental analysis is helpful in establishing basic standards and benchmarks. However, since there are uncertainties associated with fundamental analysis, exclusive reliance on fundamental analysis should be avoided. Equally important, excessive refinement and complexity in fundamental analysis must be viewed with caution.

- Technical analysis is useful in broadly gauging the prevailing mood of investors and the relative strengths of supply and demand forces. However, since the mood of investors can vary unpredictably excessive reliance on technical indicators can be hazardous. More important, complicated technical systems should ordinarily be regarded as suspect because they often represent figments of imagination rather than tools of proven usefulness.

- The market is neither as well-ordered as the academic approach suggests, nor as speculative as the psychological approach indicates. While it is characterized by some inefficiencies and imperfections, it seems to react reasonably efficiently and rationally to the flow of information. Likewise, despite many instances of mispriced securities, there appears to be a fairly strong correlation between risk and return.

The operational implications of the eclectic approach are as follows :

- Conduct fundamentla analysis to establish certain value.

- Do technical analysis to assess the state of the market psychology.

- Combine fundamental and technical analyses to determine which securities are worth buying, worth holding, and worth disposing off.

- Respect market prices and do not show excessive zeal in 'beating the market'. Accept the fact that the search for a higher level of return often necessitates the assumption of a higher level of risk.

Common Errors in Investment Management

Investors appear to be prone to the following errors in managing their investments :

- Inadequate comprehension of return and risk
- Vaguely formulated investment policy
- Naive extrapolation of the past
- Cursory decision making
- Untimely entries and exits
- High costs
- Over-diversification and under-diversification
- Wrong attitude toward losses and profits.

DIRECTORS AND THEIR ROLE

THE BOARD OF DIRECTORS

Let us first explain what is a board of directors, what it is supposed to do and what its roles are?

A group of top executives and very senior managers in a company constitute its board of directors. The board is a bridge which links the persons who are shareholders (equity or money providers) with those who manage or create value (management and managers) for the company. The board, therefore, is an overlap between the diffused and widely spread shareholders, sometimes across the globe, and the powerful experts who manage the business operations. Managers are small in number whereas the equity providers are innumerable but powerless. In reality, the providers delegate the responsibilities of managing the operations of the company to the management/managers. A board is thus like a group of middlemen.

The law of a country defines the provisions for stringent and absolute fiduciary duties to be performed by the board of directors. These duties include ensuring that the organization is run effectively in the best interests of the shareholders, viz. the owners of the organization. So, the board's prime duties entail monitoring the management's or managers' performance, on behalf of the shareholders.

Growth as well as survival of any organization is contingent upon keeping a balance between the two sets of power that is, the power of equity holders (money power) and the power of the managers who manage a given corporate body. So, success of a corporation depends upon the flow of capital from the equity owners and the effective management which runs the enterprise. The organization provides ample opportunities to people who manage the enterprise, to explore ways and means, make systemic changes, etc., so that, it results into higher bottom line, increased organizational image. A better, professionally managed organization also commands respect of the investors. All these are also reflected in higher earning capacity of the enterprise shares in the capital market. Karl Marx and Adam Smith did not believe in corporate type of organization. They were opposed to the idea of creation of a structure in a corporation because of the apparent difference between the ownership and control by two different bodies. Adam Smith was critical of joint stock companies; his criticism was focused on both the owners and the managers. His logic was that the equity owners quite often show that they have no knowledge or understanding of the business of the company. And, as for the directors of an enterprise, he said that they cannot be expected to oversee the business activities with the same vigil and interest as the partners of a private organization. In other words, their interest will be low since the equity invested in the business does not belong to them.

Whereas the views of Karl Marx and Adam Smith are quite valid but in the backdrop of the emerging global market and explosion of expertise and, knowledge there has got to be two sets of bodies one that provides the capital and the one which manages. Of course, it would have been ideal if the two sets of bodies were just one. We have seen in the case on Warren Buffett where Buffett was considered an ideal modern investor. If investors take the role model of Buffett or Pierre Du Pont, corporations will not only grow and prosper but the property will reflect value addition to the corporation in the capital market.

With rapid growth and expansion of businesses round the world, a new class of professional managers is born and the managers are ready to take responsibility for managing the business. Management is a profession, a discipline like medicine or law, and the senior management of the business needs the expertise of a professional manager to manage the business. The fact that management is a profession is borne out by the number of MBAs in demand in business, commerce and industry and other organizations and there has been a mushroom growth of management institutions round the world. An entrepreneur cannot spend money on resources which do not add value to the business. To rephrase it, uneconomic expenditures are not attractive to entrepreneurs.

It is pertinent to note that for the body of shareholders who are so widespread and diversified, often across the globe, it just may not be physically possible to oversee the management activities, let alone the question that management requires the skill, knowledge and competence. Managers are hired and shareholders delegate the responsibilities to manage the business. The investors are empowered to elect their representatives who on their behalf oversee the management activities and business operations. These representatives are the directors of the board of the enterprise. In smaller businesses, the directors may also be the investors. The directors are assigned the duty under the law of the land to safeguard the assets of the enterprise. But, the investors do need assurances that the power to manage is not misused or used for wrong purposes. At the same time, a pertinent point arises as to how to allow the management-wide discretionary power in managing the business and also ascribe to the management, accountability for use of the authority. This is a moot point today in the area of corporate governance. In any case, managers need to be given the power they need for managing. And, the authority and power delegated without accountability, will not achieve the results and objectives of a corporation.

Duties of the Board of Directors

The US laws related to corporate governance at board level have been undergoing changes. However, as per the law of the land, a director has to perform two sets of duties:

- duty of care
- duty of loyalty

A director's conduct will be assessed by the "rule on business judgment". Let us explain these two duties.

Duty of care: Duty of care implies that the director is obliged to exercise adequate diligence in decision making. While making a decision, he must have in possession, at command, all the facts, figures and relevant information of the issues and concerns, and he should be able to explain while making a decision on relevant topics that he has considered all the alternative ways and means.

Duty of loyalty: Duty of loyalty means a director must have uncompromising loyalty to the organization which he must demonstrate through his actions. A director sitting on the boards of two different corporations having conflicting interests, will be required to resign from one of the boards. He could not demonstrate loyalty to both the firms by his decisions. Such a conflict may arise in case both the companies are interested to buy another business of same nature and goals.

The American Law Institute has outlined the following duties/functions of the directors.

- Selection, evaluation and dismissal of CEOs, as and when necessary.
- Carrying out the functions defined by the legal framework and/or ascribed to the board of directors as per the norms of the corporation. Approving corporate plans/ activities and reviewing the same, which the board and the senior management consider as significant. Further, to effect changes in the accounting systems/ procedures which in the eyes of the board are very important.

- Keeping an eye on the way the business is run and evaluating the conduct of the same on regular basis. The objective being to ensure that the company resources are utilized in furtherance of value addition to the shareholders stocks, following the legal provisions and all ethical values. Further, to ensure that a reasonable portion of the resources are being put to humanitarian and public welfare usage.

- Defining the management compensation scheme and review of the company succession plan system. Regulation, evaluation and appointment/ selection or replacement of CEOs.

- Planning, reviewing and approving finance linked objectives and the prime strategies and game plan of the company.

- Counseling and advising the top management.

- Checking adequacy of systemic procedures/methods in compliance of the legal provisions.

Selecting and giving recommendations to the investors, the list of suitable persons for electing them for the board. And also evaluating the organizational performance and the board linked procedures/methods.

The board of directors have the responsibility to oversee the operations of the business being carried out by the management. They have to ensure that the business enterprise is managed effectively. But, the problem is that in practical or legal terms they are not allowed to closely involve themselves in running the operations of the enterprise. The board exists to evaluate the performance of the company and also to initiate prompt corrective actions, once the level of performance is poor.

To reiterate, directors of a board do not perform the job of the executives, that is running the day-to-day business. As such, the job of the directors is pretty difficult. They are required to select the right talents to run the business and in case of poor results, their services are liable to be terminated. As directors, they are held accountable for the overall organizational performance and not for just making decisions on day-to-day basis.

Business ambience is full of challenges created by the changing conditions. Business, commerce and other organizations need professionals and experts at managerial levels for value addition through use of new technologies, methods and systemic processes as well as organizational mission and vision. It is needless to say that organizations demand high professionalism from the incumbents to the post of director.

Organizations go out of the bounds of the corporations to look for professional directors. In the changing environment, corporations do not need clones. In particular, at the top level, they need professionals who are innovative and have the potential to take organizations to greater height.

Board of Directors Lays the Ethical Foundation

In the wake of the mass corporate scandals involving forgery, dishonesty, financial scams, both at corporate levels and at the state levels in the past couple of years, the need for practising ethical values and integrity has come to the fore. The Telgi stamp scam (2004) involving several hundred crores of rupees is the latest of its nature which has shaken the public conscience. The case is still under investigation. One well-known management consultant says values in corporates have crashed and so the need exists for corporates to redefine their ways and means of managing the business in the corrupt social and political environment. For Azim Premji, the Chairman of Wipro, value-based management is the only way to run an enterprise successfully.

In fact, the society in our country strongly feels that the internals and those coming from the same business family should be forbidden from taking positions of directors in their parental organization. This kind of legal provision is expected to be included soon so as to stop the family feuds that are going on in business houses in India in the last couple of years. In the year 2005, Reliance Group of Industries, has caught the attention of the public. Their stocks crashed due to the ongoing family feud. The disputes within the family of controllers of the business empire, adversely impacted the capital market. Business and industry are part and parcel of the society and as such the activities and health of the organizations directly affect the economy of the country. There have also been other cases of family run business houses in this country. Such instances spoil the mood of the investors, create anxiety in the minds of the small investors and uncertainties all over the business environment. Post-Enron case at Dhabol, a major change was seen in the public outlook. The three values on corporate governance, namely, probity, transparency and accountability are being watched more carefully.

Ethics refer to values and principles which are manifested in one's behaviour to another person. Values refer to beliefs translated into actions or thoughts. Acceptance of the values implies acceptance of what is right or wrong, bad or good in the society.

Bob Garratt says in his book 'Thin on Top' that the West lives in a 'post ethical society'... Everything is relative, with no firm values except self-interest. He also prays to God that the West does not head for decline. In view of the declining values, he expresses his apprehension that if the basic fabric that binds people to people ceases to exist, besides the concern for the other, civility and obedience to law goes down the drain. Such an ambience will be analogical to what the Australians term as the 'white outing' of the social system and the society. But, this may lead to a collapse of the total system.

The period of 1980s and 1990s has witnessed continuous decline in the value system in the Indian context. Large scale scams, whether it is the capital market scam triggered by Harshad Mehta, the case of ITC in the early nineties and the subsequent incidents are just a few instances of the fraud and corrupt practices that the society has witnessed. But, the growing level of unemployment, deteriorating law and order, the lack of will to punish the guilty in the states and the country as a whole as well as the deterioration in the educational standards, triggered by mushrooming growth of universities and institutions, have resulted in negative multiplier effect on the morale of the people. The situation is getting worse when the young people are looking for shortcuts in all that they do, besides, having ambition for quick money.

In the field of governance, there is an urgent need for improvement in its quality. Those systems, namely, the political, judiciary, bureaucracy as well as the police, which are the pillars of governance, are under pressure for improvement. But, cracks are visible in many areas. Structure of governance has already come under heavy fire publicly. The media is active and the organizations are under public vigil. Capital market is suffering from uncertainties. Small investors, especially, those retired from service have lost faith in the system. In this case B.D. is to refurbish the sullied image by highlighting the expectations gaining adequate employment, earnings and savings. But, in case aspirations of the majority of youth of getting adequate opportunities for employment is not achieved, they have the potential to become a strong pressure group, leading to a great social as well as political turmoil for the country.

If the above challenges are successfully dealt, economic growth of the country can accelerate rapidly. But, in all these challenges, the key word is 'we' – that is internal cohesion in the country is a significant factor. Internal cohesion has been the most significant factor in the rise of Japan, Soviet Union and China. It is like a prerequisite for any concerted effort for improvement. Unity is strength. Loss of internal cohesion was the main reason for fall of the once mighty Soviet Union.

Supremacy of the Board of Directors

The Cadbury Report (Adrian Cadbury, the Governor of Bank of England) has established supremacy of the board of directors of an enterprise. And, supremacy of the board can be maintained by the professional directors who have to act within the bounds of the legal and ethical standards. And, the standards have to be reflected in the performance of their duties. Their duties are outlined below :

1. Duty of legitimacy

2. Duty of care

3. To maintain independent views and give critical review

4. Duty of trust

5. To uphold primary loyalty of a director

6. To uphold three values of corporate governance

7. To own social responsibility

8. To keep at heart, protection on the interests of the minority owners.

9. To pay attention to task performance and delivery of primary roles

10. To learn, develop and communicate.

1. Duty of legitimacy: A director needs to be fully conversant with the basic legal and other rules/regulations governing corporations since he bears personal liability for the offences of a company. The offences could pertain to breaking health and safety rules or be related to corporate manslaughter, fraud or environmental pollution. There are two fundamental documents which a director has to fully understand. These are the Memorandum/ Articles of Association and Shareholders Agreement. Let us explain what these documents are about. The Memorandum of association inter-alia deals with purpose of the organization and the Articles of association is an addendum to the memorandum. It lays down the guiding principles of the organization.

(a) **Memorandum/Articles of Association:** This document deals with the legal process which gives birth to a company and the company is maintained. The memorandum covers such issues as the legal ceiling on a company's ability to operate, besides the procedure on voting in annual general and extraordinary general meetings, issues of shares, declaration of dividends, selection of the board members as well as their dismissal. Any violation of these provisions is an indication of dereliction of the duties and infringement of the law. The regulators and the owners can sue the company for civil wrong. The CEO and the Secretary of the company have to ensure, under the provisions of the memorandum, to follow the entire processes in boardroom activities.

(b) **Shareholders Agreement is a contractual document.** This is meant for the owners of the company, i.e., the shareholders. For the directors of the board, the agreement is like a license to operate and so the directors must know the broad action plan/strategies of the owners. The agreement is also a crucial document because it outlines the basic expectations of the owners. The agreement has to be followed vis-a-vis the Memorandum and Articles of Association for observing the legal provisions.

2. Duty of care: Directors have got to perform their roles/functions and tasks with full commitment and competence, and are accountable for the same. They should take care that what they do contributes to the value of the company. In the UK, acceptance of the TurnBull Report has brought to light new practices, e.g., selection of information by the board, strategic thinking, assessing risks, critical review, decision making, testing values on processes of the board, etc. The duty of care involves :

- directors must earmark enough time to the board matters/papers prior to the meeting.
- directors must budget enough time to discharge their duty of care.
- duty of care is not limited to just attending meetings of the board and saying 'yes' or 'no' to proposals furnished by the executives.
- directors must be properly trained in competence building and be subject to regular appraisal.
- it is not possible for any director to hold more than four posts of directorship.

Holding more than four posts of directorship will be at the cost of their duty on the other boards. Directors ought to be paid higher, but not in the form of stock options. High compensation is essential to motivate them for greater attention to duty.

3. Independent views and critical thinking: Directors are expected to have the awareness and maturity to make own judgments on giving direction and should select prudent control procedures in the best interests of the company.

An adequate induction program for the directors under the guidance of the Chairman will be helpful to the new incumbents to fully understand their duties. In the US, a director who is not able to act independently is liable to get twenty years of imprisonment. He must say 'no' to the CEO's proposals in case he finds the same unfit/incongruent to his judgment and information is scarce. All the directors are equal and have equal rights. Good companies have a practice to have free and open relationship between the CEO and the Chairman. Duty of independent critical review enhances the duty of the directors. Their decisions are implemented by the executives.

4. Duty of trust: The law requires the directors to hold their trust all the time. The key role of the directors is related to their long-term fiduciary duty. In turbulent periods, shareholders may take their equity and run away. In case the number of the shareholders is large, it can create tensions between them (shareholders) and the directors. In such a case, the directors may feel that demands of the shareholders will most likely pose a threat to the future of the company and so they may take appropriate stand. Of course, the owners are empowered to vote out the directors at the AGM (Annual General Meeting). The legal position in such a case is not clear.

5. To uphold the primary loyalty of a director: Right from the time of his appointment, a director's primary duty is to be loyal to the company which has its legal entity. The loyalty is not to the person(s) who appointed him. The immediate needs of the owners do not assume any importance to the directors since they pose a threat to the future of the company. A competent chairman guides the board successfully in discharge of his duty to the company.

6. To uphold three values of governance: Accountability, probity and openness are the accredited values in all effective/good governance.

These values can be further explained to include :

(a) Transparency to owners in assessment of risks and the decision process

(b) Honesty in all dealings within and without the board

(c) Be accountable to the shareholders

Let us explain the meaning of the above three values.

(a) **Transparency to owners:** The shareholders should know that there exists an agreed transparent process through which directors make decisions. The process includes appraisal by the board and the directors. As a result of the decisions made by the board the share owners are able to finally arrive at a decision as to which director is to be voted back in the AGM.

Honesty in all dealings within and without the board: This is another significant value to be observed for effective corporate governance. This value is the basis for competence of the board and a safeguard against corrupt practices. The dealings are based on mutual trust based on confidence amongst the colleagues, which can be tested but not taken for granted.

(b) **Honesty in dealings within and without:** The basic principles remain the same as explained in honest dealings within the board. In business dealings with overseas clients/organizations, the highest rule in practice in the parent country should be exercised, rather than going by the local customs and rituals. Ethical values of the board and the shareholders is of paramount significance which must be upheld. In the backdrop of ethical investments going up, it is easier to go by principles of high ethical values in all dealings overseas. Countries infested with corruption, bribery and dishonest practices, lawlessness, stand to lose foreign investments. Countries such as Nigeria, Cambodia and North Korea are the worst case illustrations.

(c) **Be accountable to the shareholders:** The owners appoint the board of directors on the premise that on their behalf, the directors will be totally accountable individually and collectively for their activities. In olden days, there were few environmental changes in the reporting system at the AGM (Annual General Meeting) profits and dividend accounts used to be discussed and disposed. Only a few queries used to be raised. But, with the emergence of the environmental turbulence in the backdrop of changing economic, political, social, technological, environmental and design development conditions; and the world trade being globalised, the annual reporting in the AGMs is proving to be inadequate. This has resulted in supremacy of the board and not the CEOs, as the prime source of regular information on business performance for the owners/ shareholders and the regulators. In the circumstances, the quarterly reports are proving not of much value. Half yearly reports are sufficient.

The monthly internal report and audit system need to be more accurate and have to be linked to an information system which shows the trend and not just figures in isolation. Directors need to exert their prudent control on the operations. Since the liability of the company is limited in relation to the paid up shareholders, legally the board and the directors owe unlimited liability. Therefore, the directors and the board must spend enough time in discharge of their accountability.

7. To own social responsibility: This is one subject which is controversial and viewed differently by different agencies. But, it is a significant responsibility abroad in the US and European countries if the company has to survive. There is one view on the subject which underscores supremacy of the free market and in absolute terms, the right of the board to do what it plans within the ambit of the legal provisions. Advocates of free market have put forward their thoughts that in case there are no laws at present, the board of directors is empowered to harness the market potential and maximize profitability. The board can seek monopoly in the market in the best interests of the shareholders.

There is another viewpoint put forward by the anti-capitalist group. The group views international trade essentially as an evil of the society. International organizations are considered as agents for repressing the labour, in complete disregard to the development of the local economy and human rights. Developed countries are looked upon as making deliberate efforts to suppress the natural bondages of love, cooperation, and stable state. They are accused of developing purely materialistic economies. However, there are also others who perceive the usefulness in striking a balance creation of wealth along with direct or indirect prosperity for everyone in the society. This thought is closer to Adam Smith's concept on creation and distribution of wealth.

8. To keep the interests of the minority owners at heart: The duty is taken as an extension of the duty to uphold the three values of corporate governance. In those countries where the law is not strong, the minority shareholders encounter difficult times. This is more so in those countries where privatization is the key word for efficiency/effectiveness and the concerned government is alluring substantial funds from FIIs (foreign institutional investors) as a part of the strategic economic development plan of the nation. In places where small investors are not treated well, the concerned government has begun to take in all earnest, the thoughts of effective corporate governance and tightening laws on personal property and also using ombudsmen concept/and in dispute cases going to international arbitration. In this context, it is only pertinent to say that codes of conduct on corporate governance from World Bank, CACG, UN, OECD as well as the European Union would be helpful on the subject. Garratt hints that if the concept of "shareholder sum-age" becomes a law, it will be a boon on this subject.

9. To pay attention to task performance and delivering primary roles: Directors come across many dilemmas while making decisions. The following points will be helpful in performing their roles/making correct decisions :

- while considering short-term local demands, take decisions which balance with the regional, international and national trends.
- take appropriate, long-term, well considered decisions to take the company forward but keep it under prudent control of the board.
- should be thoroughly aware of the functioning of the company so as to be responsible for its operations, but apportion enough time for designing long-term fact based views on developments outside/ external to the business.
- All out efforts ought to be focused on commercial demands of the business objective, but also act responsibly to take care of the interests of the other (external) stakeholders. The above four roles of the board of directors may be further divided into,
- strategic thinking
- ensuring accountability
- policy formulation
- supervising the management

10. To learn, develop and communicate: Learning continuously from own actions/ decisions in discharge of the duty, developing and appraising the members of the organization rigorously on ongoing basis and communicating with the stakeholders, both internal and external, constitutes a major duty of the directors. Garratt observes, from his long years of experience in top board positions round the world that directors, by and large, are averse to learning on assuming the role as a director. They feel that acquisition of the post of directorship is a reward of a successful executive. And so, they can have a cozy retirement. Long years of experience makes them successful to hold top post as a director. But, such a thought is nothing but a myth. Since a director directs, he does not manage and directing requires training just as an executive needs training to excel in his executive position. Directorship is the beginning of a new, challenging chapter in one's career which demands systematic induction, training and appraisal. It is the chairman's duty to ensure that his company has a systematic induction and training system for the directors and appraisal at the end of the year. The Institute of Directors, London, has developed a process for testing the competence of directors. The process launched in 1999 includes oral as well as written examination which assesses both experience and knowledge of the directors. Directors are required to sign a declaration on code of conduct and vouch for continuing with professional development programs. This is the world's first international testing process for the directors.

Board-management Intersay Relationship

There is an intersay relationship between the board and the management which is inevitable for overseeing the management and select the management cadre of executives who are expected to give the best performance. In case the executives do not perform and meet the defined results, the board has the right to replace them. This is the role that the board is expected to perform in terms of the legal provisions. However, theoretically, the management works at the pleasure of the board of directors. But, in practice, the position is just the other way. The directors of the board are obliged to the management for information, compensation as well as nomination. Further, in many cases most directors are not willing to invest time and energy for overseeing the corporation activities. They are also not able/willing to give a financial commitment for the success of the corporation. In companies where this kind of situation prevails, there are different factors of the board which do not work in the interests of the representation of the shareholders.

Directors depend upon the management for accurate, timely and relevant information. The management controls full, accurate and relevant information. The dealing officials can withhold information. The source of all the information may be the team leader. Outsiders may detect the problems on their own. The board is expected to solve such problems. But, when the information is not fed to the board, how can the board detect, analyze or solve the issues. The decisions based on partial data/facts can only be shortsighted/invalid or meaningless. In these circumstances, no independent director or constitution of a committee or a board committee or a change in the procedure of nomination of the directors, can do much. So, the directors must get the total, accurate and timely information in order to analyze and take decisions appropriate to the needs of the company.

2.1. BOARD STRUCTURES, PROCESSES AND EVALUATION

Scholars and practitioners have often wrestled with questions relating to board composition, size, demographics, and so on, and their influence on corporate performance and behaviour. The best that one can possibly do is to work towards an appropriate board, in terms of its size, profile, competencies and other criteria in the context of the company's perceived needs. Also, an ideal board so arrived at is not a constant for all time and all circumstances. The dynamics of business, its risk profile, performance of the executive, ownership and other stakeholder patterns, all change from time to time. These factors impact upon the demands on the board and will have to be addressed proactively, in terms of its size, balance, competencies, independence, conflict potential, deliberation processes, and so on. This chapter and the next discuss boards and their committees from this perspective. The following chapters will address issues relating to independence and conflicts.

Board Composition

Every incorporated company must have directors, who collectively are referred to as the board of directors, or just the board. Directors fall under different categories, such as executive or whole time directors, including managing directors, non-executive directors who may be independent or not, who may be nominees or appointees to represent discrete constituencies, and so on. These are described below:

Executive Directors

Company managing directors, functional directors, and other such persons, who hold a full-time appointment in their company fall into this category. It is useful to recall at this stage the legal connotation of the terms, managing director and whole time director. Companies Act, 1956, defines managing director as, "a director who, by virtue of an agreement with the company or of a resolution passed by the company in general meeting or by its Board of directors or, by virtue of its memorandum or articles of association, is entrusted with substantial powers of management which would otherwise be exercisable by him, and includes a director occupying the position of a managing director, by whatever name called."

There is a very important proviso to this sub-section. It states:

"Provided further that a managing director of a company shall exercise his powers subject to the superintendence, control and direction of its Board of directors."

This proviso incorporates a fundamental tenet of good corporate governance, namely, board primacy and supremacy in terms of its suzerainty in all matters concerning the corporation, and reflects the board's overall accountability and concomitant authority, circumscribed only by the legislative and regulatory regimes in the country. It also offers a humbling and sombre recognition of the gap that exists around the world between precept and practice. It brings out the fact that despite the overwhelming larger-than-life image of the all-powerful corporate CEO, that position is in fact subordinate to the even higher authority of the board, notwithstanding the often limited demonstration of that superior authority in practice.

A whole time director has not been specifically defined in the Companies Act, and is only referred to, as "whole-time director" includes a director in the whole-time employment of the company.

In many ways, the position of a whole time director under Indian company law is not very dissimilar to that of a managing director except in hierarchical terms, and also in the context of delegated authority. The managing director has "substantial" powers of "management", while a whole time director's authority is usually limited to some part of the total operations, dealing with a function or a business segment or a geographical area, and so on. Also, it is not necessary for a whole time director to have any significant corporate level management content in the job except to the extent of managing his or her function or operation and its team of subordinates. This could be purely a specialisation in areas such as technology, research, sales, production, operations, finance, human resources, and so on.

In theory, a company could have as many managing directors and/ or whole time directors as necessary and can be justified. Indian company law, even while limiting the number of managing directorships that can be held by an individual to two, subject to prescribed approval processes, places no limit on the number of managing directors a company may have. The position under English Law also is similar as articulated by the following commentary :

"..Many people assume that a company will have one, and only one, person entitled to be called its 'managing director', and that such an officer can be expected to have plenary powers. But company law has never supported this view. Act provides that the board may appoint one or several managing (or 'executive') directors and that they may have varying degrees of authority and responsibility. In modern business practice, any large company will almost invariably have one figure at the top of the hierarchy who can be identified as its 'chief executive officer'. He or she will, of course, be in company law terms a 'managing director'; but not necessarily the only one."

Courts have held managing directors not to be employees. Under certain circumstances and with appropriate approvals, company law at present permits an individual to be a managing director of not more than two companies. However, there is no bar against them and other whole time directors being on other boards as non-executive directors, within the statutory ceilings on total number of permitted directorships. It would be legitimate and responsible for the shareholders of a listed company to expect their whole time directors, including managing directors, to devote all or substantially all of their time to the affairs of the company.

There may however, be a case for managing and even other whole time directors to be permitted to join other company boards one or two years before their scheduled retirement, or the expiry of their contracts. This would enable them to continue to be active after their "executive employment" ceases, while at the same time also allowing for a gradual process of orderly succession within the company. In any event, the board should unanimously agree to such external directorships and the members in general meeting should annually approve their continuing directorships on other company boards. The exception to this general principle would of course be such directorships in subsidiary companies and controlled affiliates, whether incorporated or not, as part of their managerial supervisory role in such organisations. But, good practice would demand that even such appointments be approved by the employing board and annually endorsed by members in general meeting.

Non-executive Directors

Non-executive directors by definition are largely also independent. However, in some countries, especially those, like India, with a preponderance of concentrated ownership structures, provisions do exist for recognising non-executive directors.

As a result, this group includes all members on company boards other than those employed whole time by the companies themselves, and covers board members whether they are independent or non- independent, including all nominee, ex-officio and constituency directors. Potential non-executive directors may have to seek adequate information from inviting companies, to help them assess the level of time commitment expected, prior to accepting such positions. Their holding fewer directorships would naturally help the company get adequate time allocation and attention from them.

The influence of outside directors on corporate governance is a matter of board structure and process, but ultimately it depends on the persons so appointed. Non-executive directors are as responsible for the company's progress and success as their executive colleagues. They can play an important role in the formulation of company's strategies and in monitoring their implementation. In fact, the more they play their full and active role in the board's functions, the more effective is their influence likely to be on issues bearing upon its corporate governance.

The Cadbury Committee Report specially referred to the role of outside directors in terms of board structure as well as the conduct of the board. The code of best practices as recommended by the committee, for instance, suggests that where the chairman is also the chief executive, it is essential to have a strong and independent element on the board, with a recognised senior member. (In India, the listing agreements mandate that half the board should be non-executive and independent but do not prescribe a senior or "lead" outside director in such cases). Since all directors have broadly equal legal responsibilities, it is their independence of judgement that distinguishes the contribution of outside directors in board deliberations and decisions. Some of the key points that have been made in relation to the role of non-executive directors in corporate governance are as follows.

1. Non-executive directors should bring an independent judgement to bear on issues of strategy, performance, key appointments and standards of conduct.

2. The majority on the board should be independent of management and free from any business, or other relationship (apart from their directors' fees and shareholding) that could materially interfere with the exercise of their independent judgement.

3. Non-executive directors should be appointed for specified terms and reappointment should not be automatic.

4. Non-executive directors should be selected through a formal process and both, this process and their appointment, should be a matter for the board as a whole, operating through its nominations or governance committee, where one is in place.

Most of these are now standard requirements in many countries. In India, many of these recommendations have been fully or partially addressed by corporate legislation and SEBI's regulatory provisions.

Nominee and Ex-Officio Directors

Nominee and -directors form a distinct sub-group of the non-executive director category. Besides the contentious issues relating to the system of institutional nominee directors, there are other practical difficulties associated with this practice. More often than not, nominees are senior full time employees of the nominating organizations and have little free time (unless being such nominee directors is itself their whole time job, or is suitably factored into their job time scheduling) to devote to the affairs of the companies of which they are nominee directors. Equally, senior bureaucrats with enormously responsible jobs in government or other public institutions will have difficulty attending to the affairs of the companies on whose boards they sit by virtue of their office. As a result, their contribution could often be limited. Nominating organisations and potential incumbents would do well to weigh these additional demands before nominating or accepting non-executive positions on company boards. Requirements of, and contributions by the nominating institutions or government agencies could be better achieved respectively by appropriate monitoring procedures and invitations to specific sessions rather than inflicting onerous responsibilities associated with such directorships on their executives.

Constituency Directors

Constituency directors are those who are appointed to the board of the company from a particular constituency, such as workmen, or small shareholders of the company, and so on. It is important to note that even though the person may be appointed from a particular constituency, once he becomes a director, his fiduciary responsibility extends to all the shareholders of the company and is not limited to the particular constituency he represents.

The board chair may be executive or non-executive. A senior or lead director is usually appointed where the position of board chair and CEO are occupied by the same person. Independent directors are non-executive directors who fulfil conditions intended to assure their objectivity in judgement. These topics are dealt with in some detail later, in this and the following chapters.

Rights, Responsibilities and Default Liability of Directors

As has been mentioned earlier, the role and responsibility of the directors collectively as a board is to supervise, control and direct. It is the executive's prerogative and obligation to manage and ensure that policies approved or, prescribed by the board are efficiently, effectively arid ethically implemented in pursuit of the goal to maximise shareholder wealth. Any consequential liability of directors (executive and non-executive) should be limited to the appropriate discharge or otherwise of these responsibilities. Directors have to be judged on whether they have acted in good faith, in the interests of the corporation and all its shareholders, prescribed appropriate systems and processes; ensured they were put in place and monitored from time to time, and were being followed and complied with (by regular compliance, certification by management, supplemented by validation checks by internal audit or other surveillance systems), and whether they have reacted, researched and responded to instances of repetitive systems breaches as any normal prudent businessman would do in similar circumstances.

Board structures

As an umbrella expression, structure covers various attributes and dimensions of boards ranging from their size and diversity to their balance and independence. Some key parameters that apply to board structures are :

Board Structures : An Analytical Framework

Board structures have been fascinating field of scholarly study for several decades and yet no definitive consensus has been possible on the determinants that dictate optimal board structures. Much of the divergence in findings can be attributed to the kind of governance approach the researchers adopt. Arguably, agency and stewardship theories are mentioned in earlier chapter.

Agency Theory Approach

Board structures in such circumstances are overly loaded in favour of monitoring and assurance mechanisms, a majority or significant presence of independent, objective directors free from any domineering influence of the company and its executive, whistle blowing processes, and so on. Major decisions are taken away from the executive domain, and reserved for board exercise. As Fama and Jensen point out, one way for the board to achieve independent control is by separating the initiation and implementation of decisions, that is, "decision management", from the ratification and monitoring of decisions, that is, "decision control."

Stewardship Theory Approach

In this approach, the manager is more than the exclusively self-interested, rational, economic individual focused on personal pecuniary benefits. The stewardship theory recognizes that non-pecuniary objectives and aspirations may motivate people in management, for example, "the need for achievement and recognition, the intrinsic satisfaction of successful performance, respect for authority and the work ethic." Galbraith puts this succinctly in the context of a generally theory of motivation :

"Pecuniary compensation need not be the main motivation of members of the technostructure. Identification and adaptation may be the driving forces. Above a certain level, these may operate independently of income. Maximization of income for the techno-structure is neither needed, nor sought. The question of what goals members of the techno-structure identify themselves with, and to what personal goals they seek adaptation, remains. But it will be clear that there is no necessary conflict with the stockholders as there would be if both were seeking to maximize pecuniary return".

This recognition negates the agency theory assumption of goal incongruence inherent in the separation of ownership and control. On the contrary, the stewardship approach finds virtue in the assumption of control by the executive, empowering them to maximize corporate wealth creation, endowed as it is with technical and managerial expertise, superior information access, and most importantly, a firm commitment to the corporate objective and well being. Such an approach would strongly favour insider dominated boards, fewer, if any, outside, non-aligned, independent directors, and reduced focus on control with a concomitant increased emphasis on contribution to the firm's wealth creation efforts.

Resource Dependence Approach

This perspective, advocated by Pfeffer and Salancik, postulates that firms exert control over their environment by coopting the resources needed to survive and grow. As they state, "the most direct method for controlling dependence is to control the source of that dependence. One is not always in a position to achieve control over dependence through acquisition and ownership, however." But linkages could be forged except where they are prescribed. They also describe four categories of benefits that companies look for in such linkages :

- First, information exchange about the activities of that organization which may impinge on or affect the focal organisation. For example, interlocking directors among competitors may provide each with information about the other's costs and pricing and market strategy plans.

- Second, opening up a communication channel between organisations to convey information. For example, a banker on a board, learning of its requirements, may convey a funding business opportunity to his bank.

- Third, a support commitment from important elements of the environment. For example, a coopted board member, exposed to the perspectives of the company, tends to align his views and communications accordingly.

- Fourth, association of prestigious co-opted directors legitimises the company and adds to its reputational value.

In this approach, company boards would prefer to co-opt individuals who can provide them with necessary linkages to the external environment necessary to subserve their objectives. A related cooptation initiative would be to induct people whose presence on the board would help to blunt to some extent any opposition they may have while being outside the board.

"When an organization appoints an individual to a board, it expects the individual will come to support the organization, will concern himself with its problems: will favorably present it to others, and will try to aid it. A board member is publicly identified with the organization, and thus may be expected to accept some responsibility for its actions. The feeling of participating in setting organizational policy makes the individual both more identified with, and more committed to that policy."

This category would include persons with name and fame in their chosen field of activity and those who have networking connections with external organisations, or authorities who have the power and potential to help or hinder the corporation in achieving its goals, and opinion makers who adopt apparently adversarial positions. Interlocking boards (a subject that will be discussed later in this chapter) is a common mechanism in this approach, aligning people with access to resources the company is dependent upon with the interests of the firm.

Structure as Indicator of Board Independence

A key characteristic of a good board is its collective independence, especially in the context of its overarching responsibility of oversight and monitoring of executive management's performance. This calls for an unbiased and objective board that can tackle its responsibility with an all-shareholders perspective. Board independence rests on three pillars: first, the balance between executive and non-executive members including their diversity; second, the independence of the board chair with the concomitant separation of chair/ CEO duality, and the concept of lead directors; and third, independence and objectivity of individual directors.

Governance guidelines and codes around the world highlight the imperatives of board independence in the context of the distancing of ownership from control. In Canada for example, the Joint Committee on Corporate Governance states that, "Boards must have the capacity, independent of management, to fulfil [these] responsibilities, and to engage in constructive and mature relationship with management. This requires a culture that provides opportunity for both directors and management to feel comfortable when management positions are challenged."

It should be emphasised that an independent board need not necessarily be a confrontational board. Constructive critiquing and articulation of contrarian and complementary viewpoints can often help to bring out the finer nuances and implications of various decisions from totally different perspectives. Such comprehensive consideration of matters of critical policy and importance to the corporation can only render the decision-making process stronger and more robust. The key to successful board performance is for its members to be assertive but not aggressive, firm but polite, and have mutual respect for each other's expertise and viewpoints. Convince or be convinced should be the watchwords. Constructive dissent must be encouraged and viewed as the board's strength, not its weakness.

While constructive engagement is appropriate in most circumstances, it should not be allowed to become a divisive factor in board cohesion and develop into an "us vs them" situation. Fortunately, real life experience in the United Kingdom seems to suggest that, "while there might be a tension, there was no essential contradiction between the monitoring and strategic aspects of the role of the non-executive director. Polarized conceptions of the role bear little relation to the actual conditions for non-executive effectiveness."

Our discussion on the component elements of board structures should be viewed in the larger context of their role both as determinants and indicators of board independence.

Board Size

There is no legislative mandate in India on board size except for certain minimum requirements two in the case of private limited companies, and three in case of public limited companies. The Companies (Amendment) Bill, 2003 had proposed a minimum of seven directors in the case of listed companies. This bill also sought, for the first time in Indian corporate legislative history, to limit the board size to a maximum of 15. The Companies Bill, 2009, has further reduced this number to 12. While such maxima numbers are perhaps quite adequate under most circumstances and for a vast majority of companies, and also appears in governance guidelines in certain countries and well-regarded companies, the wisdom or even the compulsions for legislating such a maximum are open to debate. It takes away a legitimate right, and even worse, it relieves company boards of their responsibility to determine the most appropriate size for their needs and circumstances.

Several studies have examined the relationships between board size, and board effectiveness, company performance, and so on. Lipton Lorsch and Jenson in 1993 were arguably the first to raise this question academically, and came to the conclusion that smaller boards were associated with better board effectiveness. They argued that when boards get beyond seven or eight people they are less likely to function effectively and are less easy for the CEOs to control. In a more recent work though, Carter and Lorsch have extended these numbers: "If pushed to offer a number, we would suggest a maximum of ten directors. We believe eight to ten are appropriate for some companies, and even fewer perhaps six to eight are sufficient for smaller or less complex companies, although with these smaller boards, scheduling committee meetings may be a problem." Modern day regulatory requirements mandating certain committees of the board for listed companies, and the desire not to overlap members too much on different committees, do dictate additional board strength. In the Indian context, smaller boards often translate into inadequate time allocations to committee meetings. With directors serving on more than one committee, and the desire to compress board and committee meetings within a short span of time, usually a single day or even less, the effectiveness of board and committee meetings may well suffer & the Higgs' Report in the United Kingdom cautiously ventured: "An effective board should not be so large as to become unwieldy. It should be of sufficient size that the balance of skills and experience is appropriate for the requirement of the business and that changes in the board's composition can be managed without undue disruption."

Indian Experience

Corporate boards in India have traditionally been compact in size, given the dominant ownership patterns widely prevalent in the country. Boards were largely seen as legal necessities and their usefulness was limited to fulfilling the compliance requirements of the day. In other spheres, there was little that executive management expected their boards to deal with. This position began to change gradually in the 1990s when companies found themselves in global competition, thanks to the economic liberalisation initiatives taken by the government in 1991. Regulatory requirements, especially for listed corporations, got tougher and in any case companies with overseas listings had to comply with stricter requirements in their host countries. In terms of board size therefore, Indian practice is increasingly falling in line with international experience. Some empirical data on board size from three recent studies is provided.

In a six-year (1998-2003) timeframe study of 164 listed companies included in the Bombay Stock Exchange's BSE 200 Index (excluding banks and companies not listed for all the six years), Ajay Garg found the median size of their boards to be constant at ten directors. Within this sample, state owned enterprises registered an increased median from nine in 1998 to eleven in 2003; there was no change in the median of the domestic private sector and foreign multinational sector companies, which remained constant at ten and nine members, respectively.

A study focusing on the top corporations included in the National Stock Exchange's Nifty index found that as of 31 March 2006, their board size median was, 11, with some 40 per cent of the companies in the sample clustering in the range of nine to eleven directors. This sample was further analysed by different category groups such as industry, parentage, sales revenue, net worth, and market capitalisation.

Director Independence

"The welfare of a nation is dependent upon the state, and the welfare of the state is dependent upon the direction that a philosopher or advisor provides."

Independence of thought and expression are key attributes of free people, namely, those who are equipped and willing to think for themselves on the basis of what is right under the circumstances before them. A dictionary meaning of independence is being "free from influence, guidance or control of another or others, not [being] dependent on or affiliated with a larger or controlling group or system." In effect, independence is the antithesis of subjugation, or being subject to the exercise of power by others. Almost invariably it is also a state of the mind and strength of character, not entirely dictated by material influences. To judge objectively the fairness and equity of a proposal or action in a given situation, is to be ethical and therefore independent. Independence cannot be legislated; it has to come from within and very often it tends to be contrarian. History is replete with independent thinkers who could and did envision ideas and concepts not in conformity with the generally accepted norms of the day.

A key ingredient of independence of thought is therefore an ability to evaluate situations and decision alternatives objectively, without being influenced by the consequences of their impact on those likely to be affected by such decisions. Objectivity in such decision-making may be eroded by various extraneous considerations often dictated by implicit or explicit power sources. A good example of an explicit power source, in a corporate context, is the authority that nominates a director on the board of the company; the individual then feels obliged to consider the interests of the nominating authority when making decisions. Similarly, an example of implicit power source is a CEO who "invites" an individual to join his board; the person then feels obliged towards his "benefactor" and may support his plans and actions even if not entirely convinced about the propriety of such actions. Power as a concept and practical reality is therefore an integral ingredient of individual independence.

Relevance of Independence in Corporate Governance

Why is such independence necessary in corporate boards and in individual directors? John Stuart Mill's description in the context of civil and social liberty, of the struggle between liberty and authority, offers a sound justification for the institution of independent directors in a corporate governance context.

"By liberty, was meant protection against the tyranny of the political leaders. The rulers were conceived as in a necessarily antagonistic position to the people they ruled. Their power was regarded as necessary, but also as highly dangerous; as a weapon which they would attempt to use against their subjects, no less than against external enemies. To prevent the weaker members of the community being preyed upon by innumerable vultures, it was needful that there should be an animal of prey stronger than the rest, commissioned to keep them down. But as the king of vultures would be no less bent upon preying on the flock than any of the minor harpies, it was indispensable to be in a perpetual attitude of defense against his beak and claws. The aim, therefore, of the patriots was to set limits to the power which the ruler should be suffered to exercise over the community; and this limitation was what they meant by liberty."

The analogous situation in corporations is quite striking. The executive management has the skills and expertise required to run the business of the entity, not unlike the strong animal portrayed by Mill. There is both theoretical and empirical support for fearing that such a powerful executive may well "prey" not only on the competition as is reasonably intended but also upon the community of shareholders and other stakeholders, by seeking to appropriate for itself a more than just and equitable share of the created wealth and the associated wealth-creating assets. There is a need to have a countervailing force to contain and if possible pre-empt such excesses; and the independent directors on the board of the companies are the "defence" that Mill refers to in containing the potential excesses of the executive through its "beak and claws."

In similar vein, Indian tradition also highlights the need for the protection of the weak against the strong, while also cautioning against abuse of authority and power. The purpose of public policy lies in "protecting the small fish from the big fish. In that process, the state must not turn itself into the biggest fish of all. The king is best in whose realm the people live without fear, just as sons live without fear in the house of their father."

It all goes back to the paradigm of mistrust and surveillance. We have already noted Adam Smith's conviction, expressed more than two centuries ago, that managers cannot be expected to take care of "other peoples' monies" with the same dispatch and alacrity as they would their own. The Berle and Means postulates of separation of ownership and control and its consequences of potential misappropriation of shareholders' wealth, and La Porta's reconfirmation that the interests of minority or external shareholders need protection both from managers and dominant owners in control, underline the need for some institutionalised mechanism to protect absentee investors.

The concept of independent directors who would act, as trustees and arbiters on behalf of such minority or absentee shareholders against the combined power of management and dominant owners in control. Donald Clarke identifies in the US context at least three good reasons why executive managements may not only tolerate but also welcome independent directors on their boards: first, "This is a bonding device whereby management signals to potential investors that it is willing to be monitored effectively, and thereby reduce the firm's cost of capital."; second, "In order to protect itself from liability in shareholder suits. In such case transactions that either are or look very much like self dealing or in some other way implicate a conflict of interest, the blessing of directors who are both disinterested in the transaction in question and independent of management can be invaluable"; and third, "Independent and disinterested directors are of vital importance in shareholder litigation." At least the first two of these three drivers are quite relevant to India.

Recent examples of corporate scams such as at Enron, Parmalot and Satyam, corporate misdemeanours, executive greed and expropriatory tendencies of the last two decades are potent pointers to the scope and probability of abuse that need curbing! Whether the institution of independent directors has the wherewithal to do this, and more importantly, whether it has been able to do so, are questions that need to be explored. If the institution has not succeeded to the extent desired, the solution may lie in strengthening its capabilities through capacity building among the independent directors, and promoting an enabling market and legal environment that would facilitate its functioning. One could also argue that the institution of independent directors is perhaps an appropriate "risk management" measure, a kind of insurance against unfettered exposure to the risk of executive or other kinds of abuse; and accept that all such measures can only provide a "reasonable assurance", and cannot guarantee absolute protection. To administer a reality check, such countervailing measures even in civic society, have never been able to guarantee total protection against political or bureaucratic abuse of power, anywhere in the world.

Non-executive Directors

In corporate governance literature, and in legislative and regulatory documentation around the world, several terms such as independent, unrelated, outside, disinterested, and non-executive directors, are somewhat interchangeably used to denote broadly independent directors. In fact, the connotations of these terms may vary. Academic attention has been devoted to discerning these precise meanings and the practical applications in which their specific attributes are relevant. An umbrella description, non-management director, has been suggested to cover all such directors on the basis that their common attribute in most jurisdictions requires them not to be part of the executive management of their companies.

The Companies Act in India does not refer to independence anywhere in relation to directors. The Companies Bill 2009 is set to change this position, with provision requiring independent directors in case of listed and other specified unlisted companies. It only recognises executive (namely, managing and other whole time) directors and non-executive (comprising all other) directors.

The Kumar Mangalam Birla Report (and the listing agreements) recognises three distinct groups of directors, namely, executive, non-executive non-independent, and non-executive independent. The category of non-executive non-independent directors is presumably a recognition of ground realities that requires accommodation, in case of several family-run and other closely controlled companies, of family and friends on company boards, besides the general objective of bringing a broader view to the company's activities. The listing mandates however, draw a line when it comes to unbiased and objective decision-making, by insisting upon the proportion of independent directors on a board and also in constituting key committees of the board and their chairs.

Disinterested Directors is an expression that seems to be well established, despite its negative connotations, in the state corporation laws in US, which do not specifically incorporate the concept of independence or independent directors in their statutes. Their emphasis is more on whether any fiduciary is involved in a conflict situation in any particular transaction. Hence the derived expression 'disinterested'. In the event of any litigation it is up to the courts to determine whether a director was interested or not, and as noted earlier, in doing so they have the freedom to look beyond financial or economic criteria alone, impairing such independence. This amounts to a transactions based approach to director independence; while it has its costs, the claimed advantage is that it does not impose the costs of independent directors on company boards based on a set of straight jacketed, universal criteria of independence. Of course, this does not mean that independent directors on boards of companies in the US are not mandatorily required; in fact they are, but by virtue of federal legislation and market regulation in respect of listed corporations.

Finally, what is required on boards is a set of people who can stand back and look objectively at actions and inactions of executive management to ensure that overall shareholder interests are protected. While all the other expressions discussed have specific application in appropriate circumstances, the fact remains that such non-aligned directors must at all times have the required level of independence to discharge their duties effectively.

While independence, as noted earlier, may reflect the state of the mind and strength of individual character, in practice, some articulation is required for purposes of denotation. To quote Nietzsche, the 19th century German philosopher, "Independence is for the very few; it is a privilege of the strong." Indeed, there are (and will continue to be) people of independent spirit who bring to bear upon their work the seal of integrity and loyalty to the causes they espouse or the constituencies they represent, without fear or favour, brooking no interference. More apparent measures however are required for the vast majority of others to ascertain their independent status. One should not overlook the age-old maxim that "Caesar's wife must be above suspicion." In the case of large listed companies such independence should not only be ensured but also be so perceived.

Concept of Power

Max Weber famously defined power as "the probability that one actor within a social relationship will be in a position to carry out one's own will despite resistance, regardless of the basis on which this probability rests," and went on to state that "all conceivable qualities of a person and all conceivable combinations of circumstances may put oneself in a position to perform one's will in a given situation." Power is also closely related to dominance, defined as the probability that a command with a given specific content will be obeyed by a given group of persons. Dominance itself could be the function of economic, political, traditional or intellectual superiority, actual or perceived and is often the result of probable adverse consequences of the commands express or implied not being obeyed. In the context of exercise of power (and its corollary, subjugation to such exercise), it is useful to note the three categories of power that Galbraith identified: condign power that involves exercise of authority with the threat of punishment in case of disobedience, compensatory power that promises rewards in return for compliance, and conditioned power that envisages psychological brainwashing of the subject to an extent that he or she is led to believe that what is being done is out of his or her own volition. It is not difficult to envision the alarming potential, in the context of corporate boards and directors, for exercise of any or all of these three types of power, in a subtle and often not so subtle manner. A high profile example is of the American tobacco and food giant, RJR Nabisco, where the independence of non-executive directors was systematically eroded over a period of time by use of an astounding range of apparently courteous and hospitable measures. Independent directors, as men and women of conscience and objectivity, should constantly be aware of the dangerous pitfalls and threats to their independence that may come in many pleasant and otherwise acceptable forms.

2.2. THE ROLE OF INDEPENDENT DIRECTORS

Introduction

The corporate governance process focuses on the role of the board of directors. Directors have been considered as the "brain and soul of the organization" (Pearce and Zahra, 1991). Fama and Jenson (1983) describe the board as the "apex of the firm's decision control system". Forbes and Milliken (1999) characterize boards of directors as providing the "formal" link between the shareholders of the firm and its managers. However, not much is known about their contributions to the board process and strategy. Since the 1990's more and more professional non-executive directors (NEDs) have come into existence, Chief executives are beginning to realise the importance of the role of these highly experienced individuals.

Non-executive Director

The Council of Investors (USA) views independence as the most critical to a properly functioning board. This is even more so since the effects of conflict of interest cannot be detected either by the board or the shareholders. An independent director is a person whose directorship is the only connection to the corporation; who should not be employed by the corporation for the past five years or employed by the director of an affiliate; affiliates includes predecessor companies; employee, owner or director of a firm that is one of the corporation's or its affiliates' paid advisor or consultant that receives a revenue of $50,000.

According to the American Law Institute a majority of directors should be free of any significant relationship with company or its senior executives.

National Association of Corporate Directors (NACD)

The NACD's Blue Ribbon Commission on Director Professionalism (1998) recommended that a substantial majority of directors should be independent Directors (NACD).

NACD BRC 1999 on improving the effectiveness of the Audit Committee recommended that companies with market capitalization of over $200 million should have an audit committee composed of independent directors.

Business Round Table (BRC) statement on corporate governance (1997) suggests that a substantial majority of directors should be independent.

National Association of Securities Dealers (NASD) recommends for the corporations listed in NASDAQ that boards maintain a minimum of two independent directors and that the audit committee should be comprised of independent directors.

The New York Stock Exchange (NYSE) recommends for the listed companies on the NYSI that the audit committee must be maintained and comprised entirely of independent NYSE revised definition 2003: a director or whose immediate, family receives $ 100,000 per year in direct compensation is presumed not to be independent director until five years after he ceases to receive more than $100,000 in such compensation. A board overrides this presumption in case the compensatory relationship is not material.

According to the International Finance Corporation, an independent director should not have been employed by the company or its related parties in the past five years, and is not affiliated with the company, i.e., as an advisor or consultant to the company or its related parties.

Non-executive directors (NEDs), who are also referred to as "part time," "independent" or "outside" directors, sometimes form the majority on corporate boards. Their role is to act as advisors to the management of the organisation and to ensure that the organisation is being run in the interests of its owners. The role of the NED in the UK is enshrined within prescribed governance practices. The collapse of the Maxwell

Empire, BCCI, Polly Peck and the Guinness scandal reveal negligent practices. The Cadbury Report suggests that the importance of corporate governance lies in the contribution that it can make to both the accountability and the prosperity of an organisation (Cadbury, 1992).

The role and contribution of NEDs remains largely ambiguous, and the available research presents a mixed evidence of the role and performance of independent directors.

It is suggested that much of NEDs work is probably non-routine; or it is of a strategic nature, involving the "acquisition and exchange" of information regarding changing environments and organisational response to it. However there may be role conflicts between the CEO and independent directors as also other categories of directors which may constrain the efficacy of the role of independent directors.

It is the whole time directors who are responsible for the day-to-day supervision and management of the company's affairs and their involvement is deeper and greater than that of non-whole time independent directors. They are not involved in the day-to-day affairs of the companies and their role is limited to attending and effectively participating in the board and other meetings with the object of safeguarding the interests of the shareholders.

Having non-executive directors on a company's board is a universal practice. This is needed for the integrity and accountability of the companies. It is in the interest of shareholders who invest in the company. A non-executive director can bring valuable external business, which provides strategic success to the company. Some independent directors are unable to perform duties as they are influenced by the chief executive, specially in CEO dominated board, the role of non- executive is needed for the purpose of transparency and accountability of the companies.

The independence of the director is interpreted in terms of independence of thoughts and action as also that such directors are not involved in a conflict of interest.

Non-executives are important for the board functioning and achievement of corporate governance standard. The influence of the non-executive director depends on the board structure, process and the person concerned.

Independent directors participate in strategy formulation and are partially responsible for the organization success. Effectiveness of the independent directors is linked with the effectiveness of corporate governance process. It is the independence of judgment which enables the non-executive directors to play a role in the board deliberation and decisions.

The Cadbury Committee also recommended that it is preferable to have an independent director in case the chairman is working as the chief executive.

CII India (1998) suggested that effective corporate governance depends upon proper functioning of the board of directors and core group of the professional independent directors. Non-executive directors should participate actively in the board's activities and should have clearly defined responsibilities within the board. CII also suggested that an independent director should not hold directorship in more than 10 companies in order to ensure properly carrying out the proper responsibilities. Similarly, the Birla Committee (2000) and Narayan Murthy Committee 2003 also make significant recommendations on the role of the independent directors.

According to the Naresh Chandra Committee :

An Independent Director of the company is a non-executive director who :

1. Apart from receiving director's remuneration, does not have any material pecuniary relationships or transactions with the company, its promoters, its senior management or its holding company, its subsidiaries, and associated companies.

2. Is not related to promoters or management at the board level, or one level below the board. (Spouse and dependent parents, children and siblings).

3. Has not been an executive of the company in the last three years.

4. Is not a partner or an executive of the statutory auditing firm, the internal audit firm that is associated with the company, and has not been a partner or an executive of any such firm for the last three years. This will also apply to legal firms and consulting firms that have a material association with the entity.

5. Is not a significant supplier, vendor or customer of the company.

6. Is not a substantial shareholder of the company, i.e., owning 2 percent or more of the block of the voting shares.

7. Has not been a director, independent or otherwise, of the company for more than three terms of three years each (not exceeding nine years in any case).

8. An employee, executive director or nominee of any bank, financial institution, corporations or trustees of debentures and bond holders, who is normally called the nominee director will be excluded from the pool of directors in the determination of the number of independent directors, in other words, such a director will not feature either in the numerator or the denominator.

9. Moreover, if an executive in, say, company X, becomes an non-executive director in another company Y, while another executive of the company Y becomes a non-executive director in company X, then neither will be treated as an independent director.

10. The committee recommends that the above criteria be made applicable for all the listed companies, as well as unlisted public limited companies with a paid up share capital and free reserves of Rs. 10 crores and above or turnover of Rs. 50 crores and above with effect from the financial year beginning 2003.

"Independent Directors are those who, apart from receiving director's remuneration, do not have any material pecuniary relationship or transactions with the company, its promoters, its management or its subsidiaries, which in the judgment of the board may affect their independence of judgment."

The Role of Independent Directors :

- To monitor and control the chairman/chief executive,
- To serve as a link with external environments,
- To provide an international perspective.
- To improve board processes,
- To bring in specialist knowledge,
- To provide continuity,
- To help identify alliances and acquisition, to help maintain an ethical climate, among others.

Who is an Independent Director?

An independent director is an independent person appointed to the board to ensure that his view is not internally focused. The actual role varies among the most common roles such as: part time chairman, confidant of the chief executive, expert with specialist knowledge, a community conscience, a contact maker, conferrer of organization status.

The actual role performed by the independent director depends upon :

- Background and experience
- Company situation
- Current composition of the board
- Relations between the chairman and independent directors
- Recruitment process
- Training and Development

What should an Effective Independent Director do?

- Actively monitor the company's progress and identify problem and potential areas for growth.

- To participate and give advice in regard to strategic planning at the board level.
- To contribute to the quality of the board process.
- To advise the board on the issues related to compensation to the directors and senior executives, share option scheme and pension scheme.
- To act as a chairperson of the board and subcommittee meeting as and when required.

Why do people become Independent Directors?

1. The challenge of a new business area
2. Broadening their experience
3. Gaining reputation as a professional adviser
4. Developing a Network
5. Developing leadership qualities
6. Learning from other Companies
7. Engaging in creative pursuits like strategy development and institution building.

The role of the independent directors is constrained a great deal in the case of CEO-duality, specially when the legally required number of independent directors is not appointed. There are prospects of contribution by the independent director in case of CEO non-duality.

Competencies: It is very essential that the independent directors have competencies to develop strategies of problem solving, team building, lateral leadership skills, and creativity and negotiation skills, among others.

Percentage of independent directors: According to the Birla Committee, (2000) at least 50% of the board of a listed company should consist of non-executive directors. In the case of CEO duality, and a third of the board should be comprised of independent directors in the case of non CEO duality.

Ensuring independence of judgment: Defining independence is not sufficient to ensure the independence of judgment. The choice of the directors and the skills that the independent directors bring to the board room, the conduct of the board meetings, the quality and quantity of the financial, operational and strategic information supplied by the management to the board, independent evaluation of the management, etc. are important factors.

Duration and the Conduct of the Board Meetings

According to the Ganguly Committee Report (RBI) (2002), the Independent /non-executive executive directors should raise in a meeting, questions relating to business strategy including loans and recovery policy, housekeeping and internal control systems, record of exposure to various factors/industries by way of credit and investment, risk management systems, internal audit, accounting policy, senior management development and other important aspects of the banks and investor relations.

Most of the organization's specialist departments spend a lot of their time in discussing the legal and regulatory requirements, but the total time spent for the board meeting is on an average about 2-3 hours. According to the Committee, this much of time is not sufficient for discussing the legal and regulatory issues.

Financial and Non-financial Information at the Board Level

The Committee suggested that the Clause 49 of the listing agreement mandate the information that must be given to the board of directors. The Committee believes that this list of mandated disclosure is adequate to properly inform independent directors about the basic financial and non-financial performance of the company. In addition to above disclosure, the company should make press release and analyse presentation. This disclosure would help the company in projecting itself to the general public as well as to investors.

Audit Committee and its Independence

Audit committee of all the listed companies as well as UNLISTED public limited companies with a paid up share capital and free reserves of Rs 10 crore and above, or turnover of Rs. 50 crore and above should consist exclusively of independent directors. This will not apply to: 1) unlisted public companies, which have not more than 50 shareholders and which are without debt of any kind from the public, banks or financial institutions, as long as they don't change their character; 2) Unlisted subsidiaries of the listed companies.

In addition to disclosing the names of members of the audit committee and the dates and frequency of meetings, the chairman of the audit committee must annually certify whether and to what extent each of the functions listed in the committee charter were discharged in the course of the year. This will serve as the committee "action taken report" to the shareholders.

Regarding three key issues like remuneration of independent directors, legal liabilities of non-executive and independent directors, and the training of directors, the Committee wishes to emphasize the importance of sustainable and long-term reforms in the corporate governance.

Selection of the Independent Director

According to Ganguly Committee Report (2002) the appointment and nomination of independent/non executive directors to the boards of banks for both public and private sector should be from a group of professional people to be trained and maintained by RBI. In case of any deviation in this procedure, prior permission of RBI is required.

Identification of people requires extensive and time consuming networking as most of the appointments are done on the basis of networking. The management consultants, business journalists and public relations specialists can provide the suggestions for such vacancies. Other networks can be industry federations, charities, training and enterprise councils and so on.

Remuneration of Independent Directors

The maximum sitting fee in India is now revised upto Rs. 20,000 per board meeting.

The sitting fees of the independent directors are not very lucrative. Profit making companies are permitted to pay up to 1% of their net profits as commission to the independent directors which can be considered attractive for the independent directors. The Indian banks and public sector enterprises can't pay any commission to independent directors.

The need is to get independent directors of the highest standards of skill, to perform the functions necessary to turn around loss making companies. The non-profit making companies cannot get the services of the best independent directors. With the level of present remuneration, it is difficult to attract the best talent to the boardroom of corporate India.

Legal Responsibilities of an Independent Director

According to the law, the independent director has the same responsibilities and liabilities as any other director.

Civil Liability : The duties of a director are to act honestly and in good faith in the best interests of the company. These liabilities apply to independent directors as well as to the executive director.

Criminal Liability : The criminal liability depends on the nature of the offence. Some of the requirements under the law constitute, in their non-performance or performance of a criminal offence, and attract the liability. Proof of any knowledge and or complicity is not required. The offence basically requires proof of failure to exercise the due care (negligence) or of dishonesty.

The liability of the independent director depends upon the level of involvement and knowledge. Thus the independent director is more liable when the necessary steps to avoid a breach of the criminal code has not been taken.

Protecting the Independent Director from the Liabilities

1. Assess how well the existing directors are informed about the conduct of the organization.
2. Examine whether the ethical issues are discussed at the board level or not.
3. Assure whether the board minutes are properly recorded or not.
4. Participate in the board meetings and attending the meeting regularly.
5. Contribute to due diligence in acquisition and joint venture because the legal liability increases in such deals.
6. The independent director should duly consider moral conduct as it involves the policy implications.

Liabilities of Non-executive and Independent Directors

Wrongful disclosure by the chairman and members of the audit committee in company's annual report should attract disqualification and penalties. If the non-executive director had the knowledge of unlawful acts by the management or the board and fails to act according to the law, then the said director should be made legally liable for such ignorance.

The different liabilities of the executive directors and non-executive or independent counterpart should be considered. The persons considered responsible for the contravention committed by the company are; (i) The managing director; (ii) Executive or wholetime director; (iii) Managers; (iv) The company secretary; (v) any person in accordance with whose instructions the board is accustomed to act; (vi) any person who has been entrusted and charged by the board to be an officer in default subject to his or her consent.

Non-executive directors are far less liable for the ignorance of the provisions in the Companies Act than their executive counterparts.

Exempting Non executive Directors from Certain Liabilities

To attract high quality managers on the boards of the Indian companies would be difficult if they have to constantly worry about the criminal liabilities under the different acts. The Birla Committee recommends that non-executive directors be exempt from certain liabilities. The different Acts in question are :

Act, Negotiable Instruments Act, Provident Fund Act, ESI Act, Factories Act, Industrial Act and Electricity Supplies Act.

Independent directors should be free from litigation and other related costs. There is one possible legal issue that has concerned members of the committee. The three strata of directors are executive directors, non-executive/independent directors who are members of the chairmen of audit committee, and other non-executive/independent directors. Executive directors ought to be more liable than non-executive or independent director.

The companies are expected to purchase a reasonable amount of insurance for directors and officers (D&O). This should cover independent directors even when they cease to be directors. If the offences relate to the period when they were directors, then this insurance cover would pay for the litigation and pecuniary penalties, if any, and mitigate the corporate and individual risk of being an independent director.

Status of Independent Director

The difference between the independent director and his duties is far from the real issues of the business. The managing director, or chairman of the board has the power to take decisions. Directors collect their fees for attending the board meetings and enjoying a good lunch.

An independent director adds value to the board process by his expertise and strategic business insights.

The independent director represents the larger shareholders within the company. Now, shareholders want to approve the board decisions before they are taken.

The importance has been given to the independent director by the regulator as well. The audit committee and remuneration committee consists of independent director as chairman. Independent director needs to "Whistle Blow" or resign when companies are not willing to address the concerns raised by shareholders, Independent director should help the board in this regard. The shareholder's interest is to be seen by all the directors not just by the part-time directors. Independent directors, are being considered as a peer group and changes are recommended to enable them to play a dominant role. So it is suggested that the workload of independent director is expanded to make the board effective. Board reforms are taking place in the fast pace in that direction. Independent directors are considered as peer group to control the management.

Role of Independent Director in India

1. In India, the board can delegate powers to the whole-time or executive director. The obligations of the board are diligence, care, loyalty, avoidance of conflicts and skills in performing the duties. There should be same standards of care for executive and

independent directors, except where executive directors act in a management function delegated to them by the board and is separated from the board functions. Directors should have access to training, to fully understand their rights, responsibilities, duties and liabilities.

2. Board members have an obligation to treat all shareholders to exercise the right of appeal to SEBI if they feel treated unfairly.

At least two-thirds of the board of directors should be rotational. One-third consists of permanent directors, which include promoters, executive directors and nominee directors. Section 53, IA, Clause 49, requires issuers to have at least one-third independent directors, if the functions of chairman of the board and CEO are decoupled. An independent director is defined as a non-executive director who, inter alia, has no material pecuniary relationships or transactions with the company, its promoters, senior management or its holding company, its subsidiaries and associated companies, which in the judgment of the board may affect the independence of judgment of the director, and is not related to promoters or management at the board level, or at one level below the board, their relatives, lawyers, consultants, employees of associated companies, etc.

Policy recommendations: It has been argued that the institutional nominee directors representing DFIs do not bring specialised knowledge and hence, contribute little to the deliberation of the boards. An alternative would be for DFIs to nominate expert independent directors on their behalf. This would make them more independent. Such directors would not face the same conflicts of interest in situations where the repayment of loans is discussed as do current and former DFI employees. The maximum term of independent directors should be capped.

3. The board should ensure compliance with applicable law and take into account the interests of stakeholders.

The company secretary ensures that the board complies with its statutory duties and obligations. The board reports annually on company activities, including company performance on environmental issues, labour issues, tax compliance and provisions of the Competition Act.

4. The board should be able to exercise objective judgment on corporate affairs independent, in particular, from management: (1) Boards should consider assigning a sufficient number of non-executive board members capable of exercising independent judgment to tasks where there is a potential for conflict of interest. Examples of such key responsibilities are financial reporting, nomination, and executive and board remuneration. (2) Board members should devote sufficient time to their responsibilities.

Audit, nomination and remuneration/compensation committees are common. The audit committee should have at least three members, all non-executive, with a majority being independent and at least one director having financial and accounting expertise. Its chairman should be independent. The audit committee's role, composition, functions, powers and attendance requirements are detailed in Clause 49 (2000), Section II. The audit committee's recommendations are binding on the board. Reportedly, in some companies, audit committee meetings take place hurriedly before the full board meeting.

A director may be a member of up to 15 company boards. Clause 49 (2000) caps the number of committee chairmanships to five and the number of committee memberships to ten. Independent director compensation has two components: a small sitting fee and a commission of up to 1 percent of net profits. Loss-making companies, banks and public sector companies cannot pay commissions except with the express authorization of the pertinent regulatory authority.

Policy recommendations: Given that multiple board membership held by the same person can interfere with the performance of directors, companies and shareholders should consider whether such a situation is desirable. Audit committee members have sufficient financial and accounting knowledge to understand financial information, ask informed questions to the internal and external auditors and conduct meaningful meetings. Special training courses should be developed, including possibly a certification programme. Adequate across-the-board compensation for independent directors will help ensure that they devote sufficient time to their responsibilities and will increase the supply of high quality candidates. Compliance with the audit committee requirements should be monitored closely by regulators.

5. In order to fulfil their responsibilities, board members should have access to accurate, relevant and timely information.

In the past, management has shared little substantive information with outside directors, who were often selected with the tacit understanding that they would not ask for such information. Clause 49 mandates the information that must be placed before the board. This information is sufficient to inform directors about the financial and non-financial situation of the firm. Non-executive and independent board members also have access to expert's advice at company expense. The audit committee has the right to: (i) investigate any activity within its terms of reference; (ii) seek information from any employee; (iii) obtain outside legal or other professional advice, and (iv) Secure attendance of outsiders with relevant expertise.

Role of Independent Directors in Indian Public Enterprises

Several measures have been initiated to professionalize the management of Public Enterprises. Induction of professionals on the Boards of PSEs as non-official part-time Directors is being done. As per the guidelines issued by Department of Public Enterprises (DPE) in March 1992, the number of such non-official part-time Directors should be at least 1/3rd of the actual strength of the Board. The guidelines also envisage that the number of Government Directors on the Boards should not be more than one-sixth of the actual strength of the Board and in any case should not exceed two. Apart from this, there should be some functional Directors on each Board whose number could be up to 50% of the actual strength of the Board. As per SEBI's guidelines on corporate governance, in the cases of the listed companies headed by non-executive Chairman at least 1/3rd of the Board should comprise Independent Directors and in the cases of companies headed by an executive Chairman, at least half of the Board should comprise Independent Directors.

Appointment of non-official part-time Directors on the Boards of PSEs is made by the administrative Ministries/ Departments from the panel prepared in consultation with the Department of Public Enterprises. In so far as Navratna and Miniratna PSEs are concerned, the panel of non-official part-time Directors is prepared by a Search Committee consisting of Chairman (PESB), Secretary (OPE), Secretary of the administrative Ministry/Department of the concerned PSE, and four non-official Members. According to the Navratna and Miniratna schemes, the Boards of these companies should be professionalised by inducting a minimum of four non-official Directors in the case of Navratnas and three non-official Directors in the case of Miniratnas before the Boards exercise the enhanced powers.

Non-official part-time Directors have been appointed on the Boards of all the nine Navratna PSEs. In July, 1997 the Government had identified nine Public Sector Enterprises that had comparative advantages and potential to emerge as global giants as Navratnas. These PSEs are given enhanced autonomy and delegation of powers to incur capital expenditure, to enter into technology Joint Ventures/Strategic Alliances, to effect organizational restructuring, to create and wind up below Board level posts and to raise capital from domestic and international market. Restructuring of Board by inducting at least four non-official Directors is a pre-condition for exercise of the enhanced powers. The nine Navratna PSEs are BHEL, BPCL, GAIL, HPCL, LOC, MTNL, NTPC, ONGC and SAIL.

The committee has identified 42 Miniratnas. The criteria for conferring the status of Miniratna are (i) the PSE should be profit making for the last three years continuously and should have positive net worth, (ii) should not have defaulted in repayment of loans/ interest payment on loans due to government, (iii) should not depend upon budgetary support or government guarantee and (iv) its Board is restructured by inducting at least three non-official Directors. PSEs which have made pre-tax profit of Rs.30 crore or more in at least one of the three years, will be given Category I, while others are given Category II status. The administrative Ministries are empowered to declare a PSE as a Miniratna if it fulfils the eligibility conditions.

The enhanced powers given to Miniratna PSEs include the power to incur capital expenditure, enter into joint ventures, set up technological and strategic alliances and formulate schemes of human resources management. Presently, there are 42 Miniratna PSEs (29 Category I and 13 Category II). The names of Miniratna PSEs are given in Annexure-51. Exercise of enhanced powers by these PSEs is subject to the condition that adequate number of non-official Directors are inducted on their Boards. The Search Committee has made selections in another 17 cases, which are under process in the concerned Ministries/Departments.

Role of Independent director :

Some of the leading roles of Independent Directors are to:

1. Monitor and control the chairman/chief executive, to provide an international perspective;

2. Steer the company through a difficult or sensitive transition: sort out the conflicts for the managerial position in acquisition cases;

3. To bring specialist knowledge;

4. To provide continuity: facilitate changes that encourage new ideas;

5. To help identify alliances and acquisition: to help maintain an ethical climate, and improve the status of the company among others.

2.3 DISCLOSURE AND TRANSPARENCY

The corporate governance framework should ensure that timely and accurate disclosure is made on all material matters regarding the corporation, including the financial situation, performance, ownership, and governance of the company.

Public disclosure is typically required, at a minimum, on an annual basis though some countries require periodic disclosure on a semi-annual or quarterly basis, or even more frequently in the case of material developments affecting the company. Companies often make voluntary disclosure that goes beyond minimum disclosure requirements in response to market demand.

Disclosure promotes real transparency monitoring of companies and is central to shareholders' ability to exercise their ownership rights on an informed basis. Experience shows that disclosure can also be a powerful tool for influencing the behaviour of companies and for protecting investors. A strong disclosure regime can help to attract capital and maintain confidence in the capital markets. By contrast, weak disclosure and non-transparent practices can contribute to unethical behaviour and to a loss of market integrity at great cost, not just to the company and its shareholders but also to the economy as a whole. Shareholders and potential investors require access to regular, reliable and comparable information in sufficient detail for them to assess the stewardship of management, and make informed decisions about the valuation, ownership and voting of shares. Insufficient or unclear information may hamper the ability of the markets to function, increase the cost of capital and result in a poor allocation of resources.

Disclosure also helps improve public understanding of the structure and activities of enterprises, corporate policies and performance with respect to environmental and ethical standards, and companies' relationships with the communities in which they operate. The OECD Guidelines for Multinational Enterprises are relevant in this context.

Disclosure requirements are not expected to place unreasonable administrative, cost burdens on enterprises. Nor are companies expected to disclose information that may endanger their competitive position unless disclosure is necessary to fully inform the investment decision and to avoid misleading the investor. Material information can be defined as information whose omission or misstatement could influence the economic decisions taken by users of information.

The Principles support timely disclosure of all material developments that arise between regular reports. They also support simultaneous reporting of information to all shareholders in order to ensure their equitable treatment. In maintaining close relations with investors and market participants, companies must be careful not to violate this fundamental principle of equitable treatment.

A. Disclosure should include, but not be limited to, material information on :

1. The financial and operating results of the company

Audited financial statements showing the financial performance and the financial situation of the company (most typically including the balance sheet, the profit and loss statement, the cash flow statement and notes to the financial statements) are the most widely used source of information on companies. In their current form, the two principal goals of financial statements are to enable appropriate monitoring to take place and to provide the basis to value securities. Management's discussion and analysis of operations is typically included in annual reports. This discussion is most useful when read in conjunction with the accompanying financial statements. Investors are particularly interested in information that may shed light on the future performance of the enterprise.

Arguably, failures of governance can often be linked to the failure to disclose the "whole picture". It is therefore important that transactions relating to an entire group of companies be disclosed in line with high quality internationally recognised standards and include information about contingent liabilities and off-balance sheet transactions, as well as special purpose entities.

2. Company objectives

In addition to their commercial objectives, companies are encouraged to disclose policies relating to business ethics, the environment and other public policy commitments. Such information may be important for investors and other users of information to better evaluate the relationship between companies and the communities in which they operate and the steps that companies have taken to implement their objectives.

3. Major share ownership and voting rights

One of the basic rights of investors is to be informed about the ownership structure of the enterprise and their rights vis-a-vis the rights of other owners. The right to such information should also extend to information about the structure of a group of companies and intragroup relations. Such disclosures should make transparent the objectives, nature and structure of the group. Such disclosure might include data on major shareholders and others that, directly or indirectly, control or may control the company through special voting rights, shareholder agreements, the ownership of controlling or large blocks of shares, significant cross shareholding relationships and cross guarantees.

Particularly for enforcement purposes, and to identify potential conflicts of interest, related party transactions and insider trading, information about record ownership may have to be complemented with information about beneficial ownership. The OECD template Options for Obtaining Beneficial Ownership and Control Information can serve as a useful self-assessment tool for countries that wish to ensure necessary access to information about beneficial ownership.

4. Remuneration policy

Remuneration policy for members of the board and key executives, information about board members, including their qualifications, selection process, other company directorships and whether they are regarded as independent by the board.

Investors require information on individual board members and key executives in order to evaluate their experience and qualifications and assess any potential conflicts of interest that might affect their judgement. For board members, the information should include their qualifications, share ownership in the company, membership of other boards and whether they are considered by the board to be an independent member. It is important to disclose membership of other boards not only because it is an indication of experience and possible time pressures facing a member of the board, but also because it may reveal potential conflicts of interest and makes transparent the degree to which there are inter-locking boards.

Information about board and executive remuneration is also of concern to shareholders. Of particular interest is the link between remuneration and company performance. Companies are generally expected to disclose information on the remuneration of board members and key executives so that investors can assess the costs and benefits of remuneration plans and the contribution of incentive schemes, such as stock option schemes, to company performance. Disclosure on an individual basis (including termination and retirement provisions) is increasingly regarded as good practice and is now mandated in several countries. In these cases, some jurisdictions call for remuneration of a certain number of the highest paid executives to be disclosed, while in others it is confined to specified positions.

5. Related party transactions

It is important for the market to know whether the company is being run with due regard to the interests of all its investors. To this end, it is essential for the company to fully disclose material related party transactions to the market, either individually, or on a grouped basis, including whether they have been executed at arms-length and on normal market terms. In a number of jurisdictions this is indeed already a legal requirement. Related parties can include entities that control or are under common control with the company, significant shareholders including members of their families and key management personnel.

Given the inherent opaqueness of many transactions, the obligation may need to be placed on the beneficiary to inform the board about the transaction, which in turn should make a disclosure to the market. This should not absolve the firm from maintaining its own monitoring, which is an important task for the board.

6. Foreseeable risk factors

Users of financial information and market participants need information on reasonably foreseeable material risks that may include: risks that are specific to the industry or the geographical areas in which the company operates; dependence on commodities; financial market risks including interest rate or currency risk; risk related to derivatives and off-balance sheet transactions; and risks related to environmental liabilities.

7. Issues regarding employees and other stakeholders

Companies are encouraged, and in some countries even obliged, to provide information on key issues relevant to employees and other stakeholders that may materially affect the performance of the company. Disclosure may include management/employee relations, and relations with other stakeholders such as creditors, suppliers, and local communities.

Some countries require extensive disclosure of information on human resources. Human resource policies, such as programmes for human resource development and training, retention rates of employees and employee share ownership plans, can communicate important information on the competitive strengths of companies to market participants.

8. Governance structures and policies

Governance structures and policies, in particular, the content of any corporate governance code or policy and the process by which it is implemented.

Companies should report their corporate governance practices, and in a number of countries such disclosure is now mandated as part of the regular reporting. In several countries, companies must implement corporate governance principles set, or endorsed, by the listing authority with mandatory reporting on a "comply or explain" basis. Disclosure of the governance structures and policies of the company, in particular the division of authority between shareholders, management and board members is important for the assessment of a company's governance.

As a matter of transparency, procedures for shareholders meetings should ensure that votes are properly counted and recorded, and that a timely announcement of the outcome is made.

B. Information should be prepared and disclosed in accordance with high quality standards of accounting and financial and non-financial disclosure.

The application of high quality standards is expected to significantly improve the ability of investors to monitor the company by providing increased reliability and comparability of reporting, and improved insight into company performance. The quality of information substantially depends on the standards under which it is compiled and disclosed. The Principles support the development of high quality internationally recognised standards, which can serve to improve transparency and the comparability of financial statements and other financial reporting between countries. Such standards should be developed through open, independent, and public processes involving the private sector and other interested parties such as professional associations and independent experts. High quality domestic standards can be achieved by making them consistent with one of the internationally recognised accounting standards. In many countries, listed companies are required to use these standards.

C. An annual audit should be conducted by an independent, competent and qualified, auditor in order to provide an external and objective assurance to the board and shareholders that the financial statements fairly represent the financial position and performance of the company in all material respects.

In addition to certifying that the financial statements represent fairly the financial position of a company, the audit statement should also include an opinion on the way in which financial statements have been prepared and presented. This should contribute to an improved control environment in the company.

Many countries have introduced measures to improve the independence of auditors and to tighten their accountability to shareholders. It is desirable for such an auditor oversight body to operate in the public interest, and have an appropriate membership, an adequate charter of responsibilities and powers, and adequate funding that is not under the control of the auditing profession, to carry out those responsibilities.

It is increasingly common for external auditors to be recommended by an independent audit committee, of the board or an equivalent body and to be appointed either by that committee body or by shareholders directly. Principles of Auditor Independence and the Role of Corporate Governance in Monitoring an Auditor's Independence states that, "standards of auditor independence should establish a framework of principles, supported by a combination of prohibitions, restrictions, other policies and procedures and disclosures, that addresses at least the following threats to independence: self-interest, self-review, advocacy, familiarity and intimidation.

The audit committee or an equivalent body is often specified as providing oversight of the internal audit activities and should also be charged with overseeing the overall relationship with the external auditor including the nature of non-audit services provided by the auditor to the company. Provision of non-audit services by the external auditor to a company can significantly impair their independence and might involve them auditing their own work. Total ban or severe limitation on the nature of non-audit work which can be undertaken by an auditor for their audit client, mandatory rotation of auditors (either partners or in some cases the audit partnership), a temporary ban on the employment of an ex-auditor by the audited company and prohibiting auditors or their dependents from having a financial stake or management role in the companies they audit.

An issue which has arisen in some jurisdictions concerns the pressing need to ensure the competence of the audit profession. In many cases there is "a registration process for individuals to confirm their qualifications. This needs, however, to be supported by ongoing training and monitoring of work experience to ensure an appropriate level of professional competence.

D. External auditors should be accountable to the shareholders and owe a duty to the company to exercise due professional care in the conduct of the audit.

The practice that external auditors are recommended by an independent audit committee of the board or an equivalent body and that external auditors are appointed either by that committee/body or by the shareholders' meeting directly can be regarded as good practice since it clarifies that the external auditor should be accountable to the shareholders. It also underlines that the external auditor owes a duty of due professional care to the company rather than any individual or group of corporate managers that they may interact with for the purpose of their work.

E. Channels for disseminating information should provide for equal, timely and cost-efficient access to relevant information by users.

Channels for the dissemination of information can be as important as the content of the information itself. While the disclosure of information is often provided for by legislation, filing and access to information can be cumbersome and costly. Filing of statutory reports has been greatly enhanced in some countries by electronic filing and data retrieval systems. Some countries are now moving to the next stage by integrating different sources of company information, including shareholder filings. The Internet and other information technologies also provide the opportunity for improving information dissemination.

F. The corporate governance framework should be complemented by an effective approach that addresses and promotes the provision of analysis or advice by analysts, brokers, rating agencies and others, that is relevant to decisions by investors, free from material

conflicts of interest that might compromise the integrity of their analysis or advice.

In addition to demanding independent and competent auditors, and to facilitate timely dissemination of information, a number of countries have taken steps to ensure the integrity of those professions and activities that serve as conduits of analysis and advice to the market. These intermediaries, if they are operating free from conflicts and with integrity, can play an important role in providing incentives for company boards to follow good corporate governance practices.

Concerns have arisen, however, in response to evidence that conflicts of interest often arise and may affect judgement. This could be the case when the provider of advice is also seeking to provide other services to the company in question, or where the provider has a direct material interest in the company or its competitors. The concern identifies a highly relevant dimension of the disclosure and transparency process that targets the professional standards of stock market research analysts, rating agencies, investment banks, etc.

2.4. DIRECTORS PERFORMANCE AND REMUNERATION

The Directors' Remuneration Debate

The last decade has seen considerable shareholder, media, and policy attention given to the issue of directors' remuneration. The debate has tended to focus on four areas : (i) the overall level of directors' remuneration and the role of share options (ii) the suitability of performance measures linking directors' remuneration with performance (iii) the role played by the remuneration committee in the setting of directors' remuneration (iv) the influence that shareholders are able to exercise on directors' remuneration.

The debate about directors' remuneration spans continents and is a topic that is as hotly debated in the USA as it is the UK. Indeed, the UK's use of share options as long-term incentive devices has been heavily influenced by US practice. Countries that are developing their corporate governance codes are aware of the ongoing issues relating to directors' remuneration and try to address these issues in their own codes. In the UK, the debate was driven in the early years by the remuneration packages of the directors of the newly privatized utilities. The perception that directors were receiving huge remuneration packages – and often, it seemed, with little reward to the shareholders in terms of company performance – further fuelled the interest in this area on both sides of Atlantic. The level of directors' remuneration continues to be a worrying trend and as Lee (2002) commented 'the evidence in the US is of many companies having given away 10 per cent, and in some cases as much as 30 percent, of their equity to executive directors and other staff in just the last five years or so. That is clearly not sustainable into the future; there wouldn't be any companies left in public hands if it were'.

It is interesting to not that a comparison of remuneration pay and incentives of directors in the USA and the UK gives a useful insight. Conyon and Murphy (2000) documented the differences in CEO pay and incentives in both countries for 1997. They found that chief executive officers in the USA earned 45 per cent higher cash compensation and 190 per cent higher total compensation. The implication is that, in the USA, the median CEO received 1.48 percent of any increase in shareholder wealth compared to 0.25 per cent in the UK. The difference being largely attributable to the extent of the share option schemes in the USA.

The directors' remuneration debate clearly highlights one important aspect of the principal-agent problem discussed at length. In this context, Conyon and Mallin (1997) highlight that shareholders are viewed as the 'principal' and managers as their 'agents' and that the economics literature, in particular, demonstrates that the compensation received by senior management should be linked to company performance for incentive reasons. Well-designed compensation contracts will help to ensure that the objectives of directors and shareholders are aligned, and so share options and other long-term incentives are a key mechanism by which shareholders try to ensure congruence between directors' and shareholders' objectives.

However, the debate is far from over and Bebchuk and Fried (2004) highlight that there are significant flaws in pay arrangements, which 'have hurt shareholders both by increasing pay levels and, even more important, by leading to practices that dilute and distort managers' incentives'.

Key elements of directors' remuneration

Directors' remuneration can encompass six elements :

- Base salary;
- bonus;
- stock options;
- restricted share plans (stock grants);
- pension;
- benefits (car, healthcare, etc.)

However, most discussions of directors' remuneration will tend to concentrate on the first four elements listed above and this text will also take that approach.

Base salary

Base salary is received by a director in accordance with the terms of his contract. This element is not related either to the performance of company or to the performance of the individual director. The amount will be set with due regard to the size of the company, the industry sector, the experience of the individual director, and the level of base salary in similar companies.

Bonus

An annual bonus may be paid, which is linked to the accounting performance of the firm.

Stock options

Stock options give directors the right to purchase shares (stock) at a specified exercise price over a specified time period. Directors may also participate in long-term incentive plans (LTIPs). UK share options generally have performance criteria attached, and much discussion is centered around these performance criteria, especially as to whether they are appropriate and demanding enough.

Restricted share plans (stock grants)

Shares may be awarded with limits on their transferability for a set time (usually a few years), and various performance conditions should be met.

Role of the remuneration committee

The Combined Code (2006) recommends that 'there should be a formal and transparent procedure for developing policy on executive remuneration and for fixing the remuneration packages of individual directors'. In practice, this normally results in the appointment of remuneration committee.

T Sykes (2002) points out that, although remuneration committees predominantly consist of a majority, or more usually entirely, of non-executive directors, these non-executive directors 'are effectively chosen by, or only with the full agreement of, senior management'.

The performance measures that the remuneration committee decides should be used are therefore central to aligning directors' performance and remuneration in the most appropriate way. Remuneration committees are offered some general guidance by the Combined Code (2006) recommendation that levels of remuneration should be sufficient to attract, retain and motivate directors of the quality required to run the company successfully, but a company should avoid paying more than is necessary for this purpose (principle).

Remuneration packages should have a balance between fixed and variable pay and between long and short-term incentives; performance-based remuneration arrangements should be demonstrably clearly aligned with business strategy and objectives; the remuneration committee should have regard to pay and conditions generally in the company, taking into account business size, complexity, and geographical location and should also consider market forces generally; share option schemes should link remuneration to performance and align the long-term interests of management with those of shareholders; performance targets should be disclosed in the Remuneration Report within the bounds of commercial confidentiality considerations.

Performance measures

Performance criteria will clearly be a key aspect of ensuring that directors' remuneration is perceived as fair and appropriate for the job and in keeping will the results achieved by the directors. Performance criteria may differentiate between three broadly conceived types of measures; (i) market-based measures (ii) accounts based measures, and (iii) individual based measures. Some potential performance criteria are :

- shareholder return;
- share price (and other market based measures);
- profit-based measures;
- return on capital employed;
- earnings per share;
- individual director performance (in contrast to corporate performance measures).

Another area that has attracted attention, is the area of 'golden goodbyes'. This is another dimension to the directors' remuneration debate because it is not only ongoing remuneration packages that have attracted adverse comment but also the often seemingly excessive amounts paid to directors who leave a company after failing to meet their targets. Large payoffs or 'rewards for failure' are seen as inappropriate because such failure may reduce the value of the business and threaten the jobs of employees. Often the departure of underperforming directors triggers a clause in their contract that leads to a large underserved pay-off, but now some companies are cutting the notice period from one year to, for example, six months where directors fail to meet performance targets over a period of time, so that a non-performing director whose contract is terminated receives six months' salary rather than one year's salary.

Remuneration of non-executive directors

The remuneration of non-executive directors is decided by the board, or where required by the articles of association, or the shareholders in general meeting. Non-executive directors should be paid a fee commensurate with the size of the company, and the amount of time that they are expected to devote to their role. Large UK companies would tend to pay in excess of £40,000 (often considerably more) to each non-executive director. The remuneration is generally paid in cash although some advocate remunerating non-executive directors with the company's shares to align their interests with those of the shareholders. However, it would not be a good idea to remunerate non-executive directors with share options (as opposed to shares) because this may give them a rather unhealthy focus on the short-term share price of the company.

2.5. THE RESPONSIBILITIES OF THE BOARD

The corporate governance framework should ensure the strategic guidance of the company, the effective monitoring of management by the board, and the board's accountability to the company and the shareholders.

Board structures and procedures vary both within and among OECD countries. Some countries have two-tier boards that separate the supervisory function and the management function into different bodies. Such systems typically have a "supervisory board" composed of non-executive board members and a "management board" composed entirely of executives. Other countries have "unitary" boards, which bring together executive and non-executive board members. In some countries there is also an additional statutory body for audit purposes. The Principles are intended to be sufficiently general to apply to whatever board structure is charged with the functions of governing the enterprise and monitoring management.

The board is chiefly responsible for monitoring managerial performance and achieving an adequate return for shareholders, while preventing conflicts of interest and balancing competing demands on the corporation. In order for boards to effectively fulfil their responsibilities they must be able to exercise objective and independent judgment. Another important board responsibility is to oversee systems designed to ensure that the corporation obeys applicable laws, including tax, competition, labour, environmental, equal opportunity, health and safety laws. In some countries, companies have found it useful to explicitly articulate the responsibilities that the board assumes and those for which management is accountable.

The board is not only accountable to the company and its shareholders but also has a duty to act in their best interests. In addition, boards are expected to take due regard of, and deal fairly with, other stakeholder interests including those of employees, creditors, customers, suppliers and local communities. Observance of environmental and social standards is relevant in this context.

A. Board members should act on a fully informed basis, in good faith, with due diligence and care, and in the best interest of the company and the shareholders.

B. Where board decisions may affect different shareholder groups differently, the board should treat all shareholders fairly.

In carrying out its duties, the board should not be viewed, or act, as an assembly of individual representatives for various constituencies. While specific board members may indeed be nominated or elected by certain shareholders (and sometimes contested by others) it is an important feature of the board's work that board members when they assume their responsibilities carry out their duties in an even-handed manner with respect to all shareholders. This principle is particularly important to establish in the presence of controlling shareholders that the factor may be able to select all board members.

C. The board should apply high ethical standards. It should take into account the interests of stakeholders.

D. The board should fulfil certain key functions, including.

1. Reviewing and guiding corporate strategy, major plans of action, risk policy, annual budgets and business plans; setting performance objectives; monitoring implementation and corporate performance; and overseeing major capital expenditures, acquisitions and divestitures.

   An area of increasing importance for boards and which is closely related to corporate strategy is risk policy. Such policy will involve specifying the types and degree of risk that a company is willing to accept in pursuit of its goals. It is thus a crucial guideline for management that must manage risks to meet the company's desired risk profile.

2. Monitoring the effectiveness of the company's governance practices and making changes as needed.

   Monitoring of governance by the board also includes continuous review of the internal structure of the company to ensure that there are clear lines of accountability for management throughout the organisation. In addition to requiring the monitoring and disclosure of corporate governance practices on a regular basis, a number of countries have moved to recommend or indeed mandate self-assessment by boards of their performance as well as performance reviews of individual board members and the CEO/Chairman.

3. Selecting, compensating, monitoring and, when necessary, replacing key executives and overseeing succession planning.

   In two tier board systems the supervisory board is also responsible for appointing the management board which will normally comprise most of the key executives.

4. Aligning key executive and board remuneration with the longer term interests of the company and its shareholders.

   In an increasing number of countries it is regarded as good practice for boards to develop and disclose a remuneration policy statement covering board members and key executives. Such policy statements specify the relationship between remuneration and performance, and include measurable standards that emphasise the longer run interests of the company over short term considerations.

   It is considered good practice in an increasing number of countries that remuneration policy and employment contracts for board members and key executives be handled by a special committee of the board comprising either wholly or a majority of independent directors. There are also calls for a remuneration committee that excludes executives that serve on each others' remuneration committees, which could lead to conflicts of interest.

5. Ensuring a formal and transparent board nomination and election process.

   These Principles promote an active role for shareholders in the nomination and election of board members. The board has an essential role to play in ensuring that this and other aspects of the nominations and election process are respected.

6. **Monitoring and managing potential conflicts of interest of management, board members and shareholders, including misuse of corporate assets and abuse in related party transactions.**

7. Ensuring the integrity of the corporation's accounting and financial reporting systems, including the independent audit, and that appropriate systems of control are in place, in particular, systems for risk management, financial and operational control, and compliance with the law and relevant standards.

   Ensuring the integrity of the essential reporting and monitoring systems will require the board to set and enforce clear lines of responsibility and accountability throughout the organisation. The board will also need to ensure that there is appropriate oversight by senior management. One way of doing this is through an internal audit system directly reporting to the board. In some jurisdictions it is considered good practice for the internal auditors to report to an independent audit committee of the board or an equivalent body which is also responsible for managing the relationship with the external auditor, thereby allowing a coordinated response by the board. It should also be regarded as good practice for this committee, or equivalent body, to review and report to the board the most critical accounting policies which are the basis for financial reports. However, the board should retain final responsibility for ensuring the integrity of the reporting systems. Some countries have provided for the chair of the board to report on the internal control process.

8. Overseeing the process of disclosure and communication.

The functions and responsibilities of the board and management with respect to disclosure and communication need to be clearly established by the board. In some companies there is now an investment relations officer who reports directly to the board.

E. The board should be able to exercise objective independent judgement on corporate affairs.

In order to exercise its duties of monitoring managerial performance, preventing conflicts of interest and balancing competing demands on the corporation, it is essential that the board is able to exercise objective judgement. In the first instance this will mean independence and objectivity with respect to management with important implications for the composition and structure of the board. Board independence in these circumstances usually requires that a sufficient number of board members will need to be independent of management.

The manner in which board objectivity might be underpinned also depends on the ownership structure of the company. A dominant shareholder has considerable powers to appoint the board and the management. However, in this case, the board still has a fiduciary responsibility to the company and to all shareholders including minority shareholders.

The variety of board structures, ownership patterns and practices in different countries will thus require different approaches to the issue of board objectivity. In many instances objectivity requires that a sufficient number of board members not be employed by the company or its affiliates and not be closely related to the company or its management through significant economic, family or other ties. This has led to both codes and the law in some jurisdictions to call for some board members to be independent of dominant shareholders, independence extending to not being their representative or having close business ties with them. In other cases, parties such as particular creditors can also exercise significant influence. Where there is a party in a special position to influence the company, there should be stringent tests to ensure the objective judgement of the board.

In defining independent members of the board, some national principles of corporate governance have specified detail. While establishing necessary conditions, such 'negative' criteria defining when an individual is not regarded as independent can usefully be complemented by 'positive' examples of qualities that will increase the probability of effective independence.

Independent board members can contribute significantly to the decision-making of the board. They can bring an objective view to the evaluation of the performance of the board and management. In addition, they can play an important role in areas where the interests of management, the company and its shareholders may diverge such as executive remuneration, succession planning, changes of corporate control, take-over defences, large acquisitions and the audit function. In order for them to play this key role, it is desirable that boards declare who they consider to be independent and the criterion for this judgement.

1. Boards should consider assigning a sufficient number of non-executive board members capable of exercising independent judgement to tasks where there is a potential for conflict of interest. Examples of such key responsibilities are ensuring the integrity of financial and non-financial reporting, the review of related party transactions, nomination of board members and key executives, and board remuneration.

   While the responsibility for financial reporting, remuneration and nomination are frequently those of the board as a whole, independent non-executive board members can provide additional assurance to market participants that their interests are defended. The board may also consider establishing specific committees to consider questions where there is a potential for conflict of interest. These committees may require a minimum number or be composed entirely of non-executive members. In some countries, shareholders have direct responsibility for nominating and electing non-executive directors to specialised functions.

2. When committees of the board are established, their mandate, composition and working procedures should be well defined and disclosed by the board.

   While the use of committees may improve the work of the board they may also raise questions about the collective responsibility of the board and of individual board members. In order to evaluate the merits of board committees it is therefore important that the market receives a full and clear picture of their purpose, duties and composition. Such information is particularly important in the increasing number of jurisdictions where boards are establishing independent audit committees with powers to oversee the relationship with the external auditor and to act in many cases independently. Other such committees include those dealing with nomination and compensation. The accountability of the rest of the board and the board as a whole should be clear. Disclosure should not extend to committees set up to deal with, for example, confidential commercial transactions.

3. Board members should be able to commit themselves effectively to their responsibilities.

   Service on too many boards can interfere with the performance of board members. Companies may wish to consider whether multiple board memberships by the same person are compatible with effective board performance and disclose the information to shareholders. Some countries have limited the number of board positions that can be held. Specific limitations may be less important than ensuring that members of the board enjoy legitimacy and confidence in the eyes of shareholders. Achieving legitimacy would also be facilitated by the publication of attendance records for individual board members (e.g. whether they have missed a significant number of meetings) and any other work undertaken on behalf of the board and the associated remuneration.

   In order to improve board practices and the performance of its members, an increasing number of jurisdictions are now encouraging companies to engage in board training and voluntary self-evaluation that meets the needs of the individual company. This might include that board members acquire appropriate skills upon appointment, and thereafter remain abreast of relevant new laws, regulations, and changing commercial risks through in-house training and external courses.

F. In order to fulfil their responsibilities, board members should have access to accurate, relevant and timely information.

Board members require relevant information on a timely basis in order to support their decision-making. Non-executive board members do not typically have the same access to information as key managers within the company. The contributions of non-executive board members to the company can be enhanced by providing access to certain key managers within the company such as, for example, the company secretary and the internal auditor, and recourse to independent external advice at the expense of the company. In order to fulfil their responsibilities, board members should ensure that they obtain accurate, relevant and timely information.

2.6.   THE RIGHTS OF SHAREHOLDERS AND KEY OWNERSHIP FUNCTIONS

A. Basic shareholder rights should include the right to: 1) secure methods of ownership registration; 2) convey or transfer shares; 3) obtain relevant and material information on the corporation on a timely and regular basis; 4) participate and vote in general shareholder meetings; 5) elect and remove members of the board; and 6) share in the profits of the corporation.

B. Shareholders should have the right to participate in, and to be sufficiently informed on, decisions concerning fundamental corporate changes such as: 1) amendments to the statutes, or articles of incorporation or similar governing documents of the company; 2) the authorisation of additional shares; and 3) extraordinary transactions, including the transfer of all or substantially all assets, that in effect result in the sale of the company.

The ability of companies to form partnerships and related companies and to transfer operational assets, cash flow rights and other rights and obligations to them is important for business flexibility and for delegating accountability in complex organisations. It also allows a company to divest itself of operational assets and to become only a holding company. However, without appropriate checks and balances such possibilities may also be abused.

C. Shareholders should have the opportunity to participate effectively and vote in general shareholder meetings and should be informed of the rules, including voting procedures, that govern general shareholder meetings:

1. Shareholders should be furnished with sufficient and timely information concerning the date, location and agenda of general meetings, as well as full and timely information regarding the issues to be decided at the meeting.

2. Shareholders should have the opportunity to ask questions to the board, including questions relating to the annual external audit, to place items on the agenda of general meetings, and to propose resolutions, subject to reasonable limitations.

In order to encourage shareholder participation in general meetings, some companies have improved the ability of shareholders to place items on the agenda by simplifying the process of filing amendments and resolutions. Improvements have also been made in order to make it easier for shareholders to submit questions in advance of the general meeting and to obtain replies from management and board members. Shareholders should also be able to ask questions relating to the external audit report. Companies are justified in assuring that abuses of such opportunities do not occur. It is reasonable, for example, to require that in order for shareholder resolutions to be placed on the agenda, they need to be supported by shareholders holding a specified market value or percentage of shares or voting rights.

This threshold should be determined taking into account the degree of ownership concentration, in order to ensure that minority shareholders are not effectively prevented from putting any items on the agenda. Shareholder resolutions that are approved and fall within the competence of the shareholders' meeting should be addressed by the board.

3. Effective shareholder participation in key corporate governance decisions, such as the nomination and election of board members, should be facilitated. Shareholders should be able to make their views known on the remuneration policy for board members and key executives. The equity component of compensation schemes for board members and employees should be subject to shareholder approval.

   To elect the members of the board is a basic shareholder right. For the election process to be effective, shareholders should be able to participate in the nomination of board members and vote on individual nominees or on different lists of them.

   The Principles call for the disclosure of remuneration policy by the board. In particular, it is important for shareholders to know the specific link between remuneration and company performance when they assess the capability of the board and the qualities they should seek in nominees for the board.

4. Shareholders should be able to vote in person or in absentia, and equal effect should be given to votes whether cast in person or in absentia.

D. Capital structures and arrangements that enable certain shareholders to obtain a degree of control disproportionate to their equity ownership should be disclosed.

   Some capital structures allow a shareholder to exercise a degree of control over the corporation disproportionate to the shareholders' equity ownership in the company. Pyramid structures, cross shareholdings and shares with limited or multiple voting rights can be used to diminish the capability of non-controlling shareholders to influence corporate policy.

E. Markets for corporate control should be allowed to function in an efficient and transparent manner.

   The rules and procedures governing the acquisition of corporate control in the capital markets, and extraordinary transactions such as mergers, and sales of substantial portions of corporate assets, should be clearly articulated and disclosed so that investors understand their rights and recourse. Transactions should occur at transparent prices and under fair conditions that protect the rights of all shareholders according to their class.

F. The exercise of ownership rights by all shareholders, including institutional investors, should be facilitated.

Institutional investors acting in a fiduciary capacity should disclose their overall corporate governance and voting policies with respect to their investments, including the procedures that they have in place for deciding on the use of their voting rights.

G. Shareholders, including institutional shareholders, should be allowed to consult with each other on issues concerning their basic shareholder rights as defined in the Principles, subject to exceptions to prevent abuse.

2.7. THE EQUITABLE TREATMENT OF SHAREHOLDERS

The corporate governance framework should ensure the equitable treatment of all shareholders, including minority and foreign shareholders. All shareholders should have the opportunity to obtain effective redress for violation of their rights.

Investors' confidence that the capital they provide will be protected from misuse or misappropriation by corporate managers, board members or controlling shareholders is an important factor in the capital markets. Corporate boards, managers and controlling shareholders may have the opportunity to engage in activities that may advance their own interests at the expense of non-controlling shareholders. In providing protection to investors, a distinction can usefully be made between ex-ante and ex-post shareholder rights. Ex-ante rights are, for example, pre-emptive rights and qualified majorities for certain decisions. Ex-post rights allow the seeking of redress once rights have been violated. In jurisdictions where the enforcement of the legal and regulatory framework is weak, some countries have found it desirable to strengthen the ex-ante rights of shareholders such as by low share ownership thresholds for placing items on the agenda of the shareholders meeting or by requiring a super majority of shareholders for certain important decisions. The Principles support equal treatment for foreign and domestic shareholders in corporate governance. They do not address government policies to regulate foreign direct investment.

One of the ways in which shareholders can enforce their rights is to be able to initiate legal and administrative proceedings against management and board members. Experience has shown that an important determinant of the degree to which shareholder rights are protected is whether effective methods exist to obtain redress for grievances at a reasonable cost and without excessive delay. The confidence of minority investors is enhanced when the legal system provides mechanisms for minority shareholders to bring lawsuits when they have reasonable grounds to believe that their rights have been violated. The provision of such enforcement mechanisms is a key responsibility of legislators and regulators.

A. All shareholders of the same series of a class should be treated equally.

1. Within any series of a class, all shares should carry the same rights. All investors should be able to obtain information about the rights attached to all series and classes of shares before they purchase. Any changes in voting rights should be subject to approval by those classes of shares which are negatively affected.

   The optimal capital structure of the firm is best decided by the management and the board, subject to the approval of the shareholders. Some companies issue preferred (or preference) shares which have a preference in respect of receipt of the profits of the firm but which normally have no voting rights. Companies may also issue participation certificates or shares without voting rights, which would presumably trade at different prices than shares with voting rights. All of these structures may be effective in distributing risk and reward in ways that are thought to be in the best interests of the company and to cost-efficient financing. The Principles do not take a position on the concept of "one share one vote". However, many institutional investors and shareholder associations support this concept.

   Investors can expect to be informed regarding their voting rights before they invest. Once they have invested, their rights should not be changed unless those holding voting shares have had the opportunity to participate in the decision. Proposals to change the voting rights of different series and classes of shares should be submitted for approval at general shareholders meetings by a specified majority of voting shares in the affected categories.

2. Minority shareholders should be protected from abusive actions by, or in the interest of, controlling shareholders acting either directly or indirectly, and should have effective means of redress.

   Many publicly traded companies have a large controlling shareholder. While the presence of a controlling shareholder can reduce the agency problem by closer monitoring of management, weaknesses in the legal and regulatory framework may lead to the abuse of other shareholders in the company. The potential for abuse is marked by, controlling shareholders to exercise a level of control which does not correspond to the level of risk that they assume as owners through exploiting legal devices to separate ownership from control, such as pyramid structures or multiple voting rights. Such abuse may be carried out in various ways, including the extraction of direct private benefits via high pay and bonuses for employed family members and associates, inappropriate related party transactions, systematic bias in business decisions and changes in the capital structure through special issuance of shares favouring the controlling shareholder.

In addition to disclosure, a key to protecting minority shareholders is a clearly articulated duty of loyalty by board members to the company and to all shareholders.

Other common provisions to protect minority shareholders, which have proven effective, include pre-emptive rights in relation to share issues, qualified majorities for certain shareholder decisions and the possibility to use cumulative voting in electing members of the board. Under certain circumstances, some jurisdictions require or permit controlling shareholders to buy-out the remaining shareholders at a share-price that is established through an independent appraisal. This is particularly important when controlling shareholders decide to de-list an enterprise. Other means of improving minority shareholder rights include derivative and class action law suits. With the common aim of improving market credibility, the choice and ultimate design of different provisions to protect minority shareholders necessarily depends on the overall regulatory framework and the national legal system.

3. Votes should be cast by custodians or nominees in a manner agreed upon with the beneficial owner of the shares.

   In some OECD countries it was customary for financial institutions which held shares in custody for investors to cast the votes of those shares. Custodians such as banks and brokerage firms holding securities as nominees for customers were sometimes required to vote in support of management unless specifically instructed by the shareholder to do otherwise.

4. Impediments to cross border voting should be eliminated.

   Foreign investors often hold their shares through chains of intermediaries. Shares are typically held in accounts with securities intermediaries, that in turn hold accounts with other intermediaries and central securities depositories in other jurisdictions, while the listed company resides in a third country. Such cross-border chains cause special challenges with respect to determining the entitlement of foreign investors to use their voting rights, and the process of communicating with such investors. In combination with business practices which provide only a very short notice period, shareholders are often left with only very limited time to react to a convening notice by the company and to make informed decisions concerning items for decision. This makes cross border voting difficult. The legal and regulatory framework should clarify who is entitled to control the voting rights in cross border situations and where necessary to simplify the depository chain. Moreover, notice periods should ensure that foreign investors in effect have similar opportunities to exercise their ownership functions as domestic investors. To further facilitate voting by foreign investors, laws, regulations and corporate practices should allow participation through means which make use of modern technology.

5. Processes and procedures for general shareholder meetings should allow for equitable treatment of all shareholders. Company procedures should not make it unduly difficult or expensive to cast votes.

The right to participate in general shareholder meetings is a fundamental shareholder right. Management and controlling investors have at times sought to discourage non-controlling or foreign investors from trying to influence the direction of the company. Some companies have charged fees for voting. Other impediments included prohibitions on proxy voting and the requirement of personal attendance at general shareholder meetings to vote. Still other procedures may make it practically impossible to exercise ownership rights. Proxy materials may be sent too close to the time of general shareholder meetings to allow investors adequate time for reflection and consultation. Many companies in OECD countries are seeking to develop better channels of communication and decision-making with shareholders. Efforts by companies to remove artificial barriers to participation in general meetings are encouraged and the corporate governance framework should facilitate the use of electronic voting in absentia.

B. Insider trading and abusive self-dealing should be prohibited.

Abusive self-dealing occurs when persons having close relationships to the company, including controlling shareholders, exploit those relationships to the detriment of the company and investors. As insider trading entails manipulation of the capital markets, it is prohibited by securities regulations, company law and/or criminal law in most OECD countries. However, not all jurisdictions prohibit such practices, and in some cases enforcement is 110t vigorous. These practices can be seen as constituting a breach of good corporate governance inasmuch as they violate the principle of equitable treatment of shareholders.

The Principles reaffirm that it is reasonable for investors to expect that the abuse of insider power be prohibited. In cases where such abuses are not specifically forbidden by legislation or where enforcement is not effective, it will be important for governments to take measures to remove any such gaps.

C. Members of the board and key executives should be required to disclose to the board whether they, directly, indirectly or on behalf of third parties, have a material interest in any transaction or matter directly affecting the corporation.

Members of the board and key executives have an obligation to inform the board where they have a business, family or other special relationship outside of the company that could affect their judgement with respect to a particular transaction or matter affecting the company. Such special relationships include situations where executives and board members have a relationship with the company via their association with a shareholder who is in a position to exercise control. Where a material interest has been declared, it is good practice for that person not to be involved in any decision involving the transaction or matter.

2.8. THE ROLE OF STAKEHOLDERS IN CORPORATE GOVERNANCE

The corporate governance framework should recognise the rights of stakeholders established by law or through mutual agreements and encourage active cooperation between corporations and stakeholders in creating wealth, jobs, and the sustainability of financially sound enterprises.

A key aspect of corporate governance is concerned with ensuring the flow of external capital to companies both in the form of equity and credit. Corporate governance is also concerned with finding ways to encourage the various stakeholders in the firm to undertake economically optimal levels of investment in firm-specific human and physical capital. The competitiveness and ultimate success of a corporation is the result of teamwork that embodies contributions from a range of different resource providers including investors, employees, creditors, and suppliers. Corporations should recognise that the contributions of stakeholders constitute a valuable resource for building competitive and profitable companies. It is, therefore, in the long-term interest of corporations to foster wealth-creating cooperation among stakeholders. The governance framework should recognise that the interests of the corporation are served by recognising the interests of stakeholders and their contribution to the long-term success of the corporation.

A. The rights of stakeholders that are established by law or through mutual agreements are to be respected.

In all OECD countries, the rights of stakeholders are established by law (e.g. labour, business, commercial and insolvency laws) or by contractual relations. Even in areas where stakeholder interests are not legislated, many firms make additional commitments to stakeholders, and concern over corporate reputation and corporate performance often requires the recognition of broader interests.

B. Where stakeholder interests are protected by law, stakeholders should have the opportunity to obtain effective redress for violation of their rights.

The legal framework and process should be transparent and not impede the ability of stakeholders to communicate and to obtain redress for the violation of rights.

C. **Performance-enhancing mechanisms for employee participation should be permitted to develop.**

D. **Where stakeholders participate in the corporate governance process, they should have access to relevant, sufficient and reliable information on a timely and regular basis.**

Where laws and practice of corporate governance systems provide for participation by stakeholders, it is important that stakeholders have access to information necessary to fulfil their responsibilities.

E. **Stakeholders, including individual employees and their representative bodies, should be able to freely communicate their concerns about illegal or unethical practices to the board and their rights should not be compromised for doing this.**

Unethical and illegal practices by corporate officers may not only violate the rights of stakeholders but also be to the detriment of the company and its shareholders in terms of reputation effects and an increasing risk of future financial liabilities. It is therefore to the advantage of the company and its shareholders to establish procedures and safe-harbours for complaints by employees, either personally or through their representative bodies, and others outside the company, concerning illegal and unethical behaviour. In many countries the board is being encouraged by laws and/or principles to protect these individuals and representative bodies and to give them confidential direct access to someone independent on the board, often a member of an audit or an ethics committee. Some companies have established an ombudsman to deal with complaints. Several regulators have also established confidential phone and e-mail facilities to receive allegations. While in certain countries representative employee bodies undertake the tasks of conveying concerns to the company, individual employees should not be precluded from, or be less protected, when acting alone. When there is an inadequate response to a complaint regarding contravention of the law, the OECD Guidelines for Multinational Enterprises encourage them to report their bona fide complaint to the competent public authorities. The company should refrain from discriminatory or disciplinary actions against such employees or bodies.

F. **The corporate governance framework should be complemented by an effective, efficient insolvency framework and by effective enforcement of creditor rights.**

Especially in emerging markets, creditors are a key stakeholder and the terms, volume and type of credit extended to firms will depend importantly on their rights and on their enforceability. Companies with a good corporate governance record are often able to borrow larger sums and on more favourable terms than those with poor records or which operate in non-transparent markets. The framework for corporate insolvency varies widely across countries. In some countries, when companies are nearing insolvency, the legislative framework imposes a duty on directors to act in the interests of creditors, who might therefore play a prominent role in the governance of the company. Other countries have mechanisms which encourage the debtor to reveal timely information about the company's difficulties so that a consensual solution can be found between the debtor and its creditors.

Creditor rights vary, ranging from secured bond holders to unsecured creditors. Insolvency procedures usually require efficient mechanisms for reconciling the interests of different classes of creditors. In many jurisdictions provision is made for special rights such as through "debtor in possession" financing which provides incentives/ protection for new funds made available to the enterprise in bankruptcy.

Barriers to the Performance of the Boards

In an effort to provide sustained value addition to shareholders' interests, boards need to concentrate on their own efforts. The word 'sustained value addition' means satisfaction that can be derived from the actual performance minus expectations :

Satisfaction = Performance Expectations

There are three barriers to performance of the board. The first barrier to the performance of the board is over-regulation of boards due to multiplicity of civil and criminal laws. As a result of these legal provisions, the directors do just the formality in compliance. This has taken away their attention to concentrate on their efforts for value addition. Secondly, the ideal non-legislative concept of self regulation is a Utopian expectation from the people as regards integrity, ethics and nurturing values. The third barrier is related to a balance which has to be struck/ reached between the three opposing forces, namely, the board (legislature), the regulator (judiciary) and the manager (executive). The regulator sits in judgment on the executive and the board even though it has the same information as the other two. This thought is based on the learning board model which we will discuss in the next chapter.

The board needs to ensure the directoral balance between board performance and board conformance. And, this is possible through four tasks of the board :

- Envisioning interesting formulation of policies
- Thinking strategically
- Supervising managing
- Accountability.

A board's performance is dependent upon the agreement between the CEO and the board and the ability to raise the right and correct questions. Likewise, conformance to the board is based on the agreement between the two bodies, related to implementation of the strategies and policies laid down by the board and observing the financial and legal provisions rules. This is based on the competence/ability to give the honest and correct answer.

An effectively designed and installed directoral board is helpful for the directors to get into the executive system and know how to question the executives as regards the planning and operational activities and oversee the operations.

In order to be effective in its performance, the board must understand the drivers and enable in last figures, which will be helpful in supporting sustainable performance objectives. Goal attainment is the prime driver and resource utilization is the main enabling force.

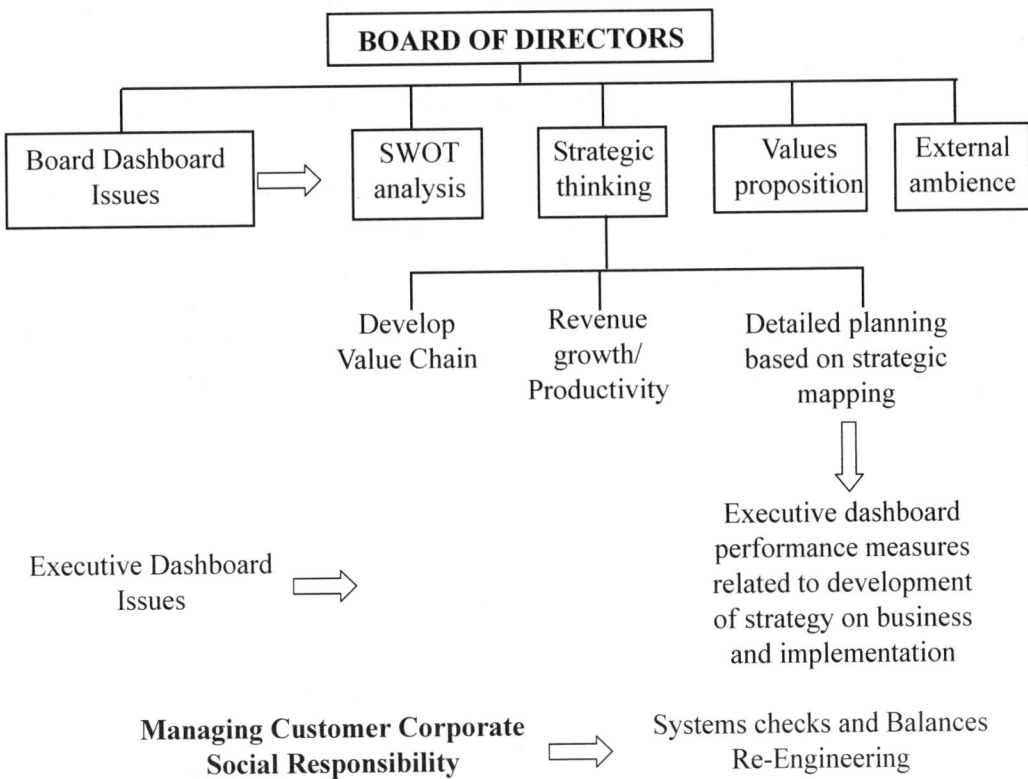

Strategic Thinking - A Model
(Adapted from Thin on Top by Bob Garratt, 2003)

Effectiveness and Total Board Performance

Effectiveness and total board performance depend upon the strategic thinking by the board. Based on the learning from the experience the board balances its course of action. The board's thinking process is based on the following sequence :

- Defining the strategy
- Translating the strategy
- Using the strategy
- Feedback

Defining the strategy   : Economic Value Addition (EVA)

                                     Policy defined (for integrated performance)

Translating strategy   : Laying the structure

Using the strategy   : Systemic procedures/methods/practices/technology

Feedback   : Through MIRs, built-in systems

Hard Data for Dashboard: In order to quantify the strategy and the expected results, certain data needs to be furnished to the dashboard of the directors. In this process, one needs to do a careful analysis of what are the strengths, weaknesses, opportunities and threats of the business. Next, one needs to understand what Bob Garratt terms as the key qualifiers of the business, viz., the core competencies and the operational strengths which are useful while analyzing the strengths and weaknesses as part of SWOT analysis of the company. The information that has to be furnished to the board can be grouped as follows:

That is :

- Essential for the Corporate Body
- Essential for the Board
- Essential for the Directors

While assessing the risks, directors will, in case of necessity, look at (a) the market potential, the trends and the uncertainties/the key qualifiers, (b) the internal potential (c) internal competence (d) market capability while setting the priorities and defining the strategies for the firm. The process involving (a to d) is called 'value diamond' by Garratt.

2.9. MEASUREMENT OF BOARD PERFORMANCE

The Changing Scenario

The need for measuring a board performance has recently become more important due to the mixed results of the functioning of the boards in the wake of the fast changing technology, the changing nature of the very concept of the corporation itself, besides the changing structure of the market. In the eighties, there were phenomenal changes in the macroeconomic environment of the developing economies like India and South East Asia. Most of these economies including India have committed their policies to initiate economic reforms as a measure for economic expansion. Economic liberalization has brought in high competition, improved technology and high foreign investments. This has paved way for rapid industrialization which will open windows of fresh opportunities for the industry to be competitive and global. Now to survive in the cut-throat competition, corporate wasn't their boards to perform at international standards.

In the backdrop of the above change, it is essential to take a look at the needs of the boards, to redefine their role and contributions so that they can perform better. Since, in the recent past pressures upon the corporations to give high performance have increased manifold, there is a corresponding pressure upon the boards to have a more prudent control over the organization in the interest of the organization. In fact, these factors highlight the need for revision of the pattern of ownership and the structure and, hence, the necessity for appraising the role and performance of the corporation.

In the Indian context, before defining a set of criteria for measurement of performance, there are many areas of corporate practices which have, hitherto, remained unexplored, e.g., definition of the main objectives of assessment, evolving set of criteria for assessment, developing criteria for assessment, developing methodology for evaluation of performance, etc., need to be addressed. The need for assessment of the performance of the boards arises because of the inherent limitations arising out of external and internal accountability of the board. We ought to take a look at the need for assessment from the angle of the multiplicity of the problems which the corporations come across. These problems are, to some extent, a reflection of the role and the contributions to be made by the board of directors. However, the need also exists because of multitude of recurring methods of :

- Role expansion
- Role distance
- Role erosion
- Role shrinkage

The boards are often subjected to many changes in their roles due to the recurring/ changing methods of role typology as above. These changing roles with recurring phases of shrinkage and expansion bring different kinds of pressures upon the corporates for performance. Hence, these are highly significant for performance and its effectiveness at the board level as we have discussed above. The need for evaluation of board performance should start from the quality and pattern of the role relationship between the management and the board. Besides, quality of the series of regular interactions and communications between the board and other stakeholders, and also the expectations that the staff have from the board, etc., are the essential factors which influence the need for evaluation.

In the Indian context, by and large, the theme about the board effectiveness has been analyzed under financial and regulatory criteria. It is relevant to note that recently some research studies have tried to link the subject of effectiveness to the fundamental aspects of the pattern of ownership and the companies' structures. And, as a corollary, these factors have been linked to the source and nature of the problems in performance.

A lot of efforts have been made to establish the ways and means to assess the board performance. But, there is no significant literature available that defines the performance measurement efforts that are being used by the board directors for improving the effectiveness of their own performance.

Performance Measurement

How to measure the performance of a director? Certainly, those who are responsible for managing cannot be given the responsibility to measure their own performance. It is just as a student cannot be given the responsibility to grade his own examination.

Further, if we want to measure the performance, we should keep in mind the purpose of the company as value creator on long-term basis. The purpose will, in effect, help delineation of the duty and rights of the directors as well as the shareholders, as also how they should be organized/structured and encouraged to perform and be evaluated in the best interests of the company. The meanings of the words long-term and value need to be classified. As regards the value, there is a wide spectrum of definitions related to the economic performance of a corporation.

Before we discuss the economic performance aspects, let us discuss the purpose of evaluation.

Purpose/Objectives of Evaluation: The purpose of evaluation is linked to the organizational objectives, and of the board, its vision and mission, board related systemic procedures, strategy, organizational communication processes and the practices with various stakeholders, scope of board independence, role of the Chairman, the contributions by the non-executive directors, besides, the nature and functioning of the sub-committees of the board, the methods of board reporting as well as the actual performance reporting to the stakeholders, etc. All these subsystems of the company's functioning have to be covered while examining the objectives of evaluation.

The emerging topics like the CSR (Corporate Social Responsibility), the ethical standards in decision making and the ongoing international agreements which have economic implications should also be covered while discussing the purpose of evaluation. And, the continuous enlargement of the purpose of assessment of the board puts the onus on the board of directors to define/redefine most suitable assessment measures.

There is an intellectual leadership role which requires that the board has to keep abreast with the national and international economic and regulatory environmental changes and keep the board members well updated as to the source, nature and their possible impact on the organization. These events may affect the purpose of the organization and so, the board has to be on guard as to when to make necessary changes in the purpose, if any. The board has to examine how the initiated changes in purpose vary from the original purpose of functioning of the board and how they will impact the overall goals of the corporation. Therefore, the subject, purpose of evaluation, is a big topic which has to be examined very carefully by the members of the board.

Evolving Criteria for Performance Evaluation: The criteria for assessing the board performance is derived from the purpose of the corporation and the board which is clearly defined in the organizational mission and vision statements enshrined in the corporation's strategic/corporate plan. The board has a stellar role in strategy formulation, and informs the management of the company, the framework for strategy formulation and review. This is a joint exercise of the board with the management/ executive. But, how effectively this is done is contingent upon :

- The clarity of the purpose
- Effective sharing and communication of the purpose
- The factors expected to adversely impact the purpose
- Clear understanding of the influencing factors
- Mutual understanding between the board and the executive.

The main point is not how the board evaluates itself. It is, whether the board accepts the fundamental need for performance evaluation, in the context of the emerging market conditions.

It follows from above that there are a whole gamut of issues to be taken into consideration before suggesting any criteria for measurement of a board's performance. The criteria must consider the explicit dimensions, namely, the purpose, structure, functions/roles, composition, practices, board strategies, etc. There are implicit dimensions which are intangible, viz., the value system and the cultural aspects which determine the effectiveness of all organizational change related plans and projects.

A board can develop certain criteria for measurement of performance, once it takes the above factors into consideration. The sources of criteria can be internal or external. The internal source is represented by the process available within the board, e.g., the internal accountability. The external sources may be too many factors depending upon the nature and pattern of the ownership of the corporation/ enterprise. The most common source is the multiplicity of the stakeholders.

Internal Accountability

Accountability as the Arbiter of Corporate Governance/Corporate Power

Corporate governance has to have checks and balances. Organizations can prove the legitimacy of their actions and credibility only through being actively accountable to a competent, independent and proactive agency, known as the board of directors, Companies are able to exercise vast power in our democratic setup. In his article, "The Corporation, How Much Power and What Scope", Carl Keysen has explained some ways to control or contain and limit the corporate power. He has cited three ways to overcome the situation: (a) limit the power through promotion of expansion of competitive market, (b) broaden the control of business influence/power through agencies not linked to the firm, and (c) institutionalize the firm's accountability/ responsibility for exercising influence/ power.

Keysen opines that efforts have so far been made through anti-trust policy, without any deliberate effort to limit the market influence and without damage to the business. The first method (a) is not acceptable since such a step needs commitment of faith in the proposal, with its resultant outcome. This is not widely accepted alternative either. A mix of alternatives (b) and (c) is highly acceptable.

Keysen has observed that he is not optimistic of the resultant prospects for corporate regulations. He is in favour of development of a mechanism which can change the internal functioning of the firm. So, there is a need for defining/ redefining the interests as well as the objectives to which the management of the corporation have to put all their energies.

Keysen has raised another pertinent question as to how one can be sure that a corporation is using its power in the best interests of the society.(2) And, how can we and the society measure performance of a corporate body? These two queries are significant and related to each other. Let us take a case study to make the point clear.

Measurement through Economic Performance

Use of Balance Sheet

Use of balance sheet and earning statements as per the GAAP (Generally Accepted Accounting Principles), besides the corporation's capacity to raise fund from external sources and the availability of working capital (cash) for the on-going needs of the organization are some of the known ways of measuring performance. In this context, Peter Drucker, a well known management expert, has observed that for the management, it is really difficult to understand/know the standards through which they value and appraise the performance. No doubt, one can take into account the share value on the stock market which is open to influence factors, including the market conditions, the way the economy moves, whims of security analysts, etc. But, do these factors truly reflect the company's real performance? This point is to be examined, and we will discuss it in the foregoing paragraphs, a little later. Meanwhile, let us first of all take a look at what Peter Drucker had to say of the perspectives on organizational evaluation.

Drucker had advised periodic auditing of business as a measure to know the perspective on evaluation of a corporation's performance. And, the audit should be done by expert external agencies.

As a concept, performance measurement needs to be flexible and open to change. And, the most significant aspect of measurement lies in setting the standards which should only be measured by those who are not the managers of the company, that is, it ought to be done by the directors and the shareholders who oversee the management are the best people who have at their heart the objective of long-term value creation of the corporation.

Good or Bad Corporation

One may raise a question to find out whether there is a good or bad organization. In this context, we can define a good corporation and how it can balance the economic interests with the social objectives.

Let us cite a case study to explain what is meant by a good organization, and maintaining a balance between the economic success and social responsibility.

By a good corporation, we mean the organization which has achieved excellence with regard to its main objectives/ purpose for which it exists. It also maintains integrity and the business is run on ethical values/ considerations in all spheres of operations.

The Measures of Corporate Governance The Economic Dimension

One general belief is that creation of value is the measure of corporate governance.

But, this is also difficult to do, since it is not possible to predict in real terms, the impact of present decisions on future value and the present value of the past decisions. However, the measures for this concept are many. In this context, let us take the case of GAAP (Generally Accepted Accounting Principles). These principles are generally accepted ones, not certified or accurate.

GAAP: It helps to give us the assets and liabilities of the company as reflected in the balance sheet. It also indicates the functioning of the assets and liabilities as outlined in the income statement. Accounting gives the contemporary performance in figure/numbers. It uses defined quantitative techniques. The accountants are able to write down the way the business progresses over a period of time.

Market Value: Fortune Magazine annually ranks companies operating round the globe, based on volume of sales, net earnings and the market value of the equity capitalization potential. Fortune classifies a company as the largest, based on the organization's worth of highest order. A company, covered under Fortune 100 or 500 is like owning a badge of respect and honour.

Market value of a company is based on statistical data. The value is given by the public in the open market. Judgment of the public is acceptable and is considered correct, it cannot be manipulated. There are many concepts attached to the notion of value, e.g., earnings per share, rate of return on capital, and the book value, which are based on accounting principles. But, the accounting principles are so flexible that they lack significance. However, market valuation is independent but it is not accurate in absolute sense. Market value only tells what a company is perceived to be worth.

To conclude, if a company's share which commands higher price in the market, has the power to raise additional capital through issue of sale of equity, today's strong market value is no guarantee for tomorrow's value. A good planning is an absolute necessity on all accounts to assure a good future value.

Earning Per Share (EPS): Earning per share is a simple factor, easily understood but is questionable for determining the financial health of a company. It is also subject to manipulation. EPS can go up through restructuring and weak balance sheet, due to play of accounting game/tactics with acquisition, switching convention and liquidating franchise, etc. But, none of these methods really help to improve the value of a company.

Economic Value Added (EVA): EVA is the estimate of the real economic profit after subtraction of the cost of capital. In 1993, Fortune Magazine had the cover story in which EVA was a hot subject. The story highlighted EVA as the real key to the creation of wealth.

EVA may be defined in terms of (ATOP-WACC): ATOP is post-tax operating profits. WACC is the weighted average cost of capital and TC (total capital). There is no formula to explain this in simple terms. Implications of EVA in fact change from company to company. Every company customizes its EVA as per its own needs and circumstances. One expert, who has done in-depth study on the subject has identified over one hundred and sixty significant adjustments that can be effected in balance sheet and the GAPP revenues/earning. Such adjustments are possible in such areas as goodwill amortization, reserves for bad debts, depreciation, costing related to inventories, etc. Ideally, organizations are advised to make some five to fifteen adjustments which can result in better accuracy and high performance. The main advantage of these is that the data are available readily and changes can be explained to those who are not finance experts. Besides, these changes result in better decisions which are also cost effective. EVA is used by companies such as Coca Cola, Sprint, Monsanto and others, although, it is a bit complicated method. As per Fortune Magazine, EVA is better than other methods such as return on equity, earning per share (EPS) and operating margin. EVA is one tool which is used to display what the investors want, i.e., the return on their capital invested. EVA thus helps the finance manager know how much the capital invested is earning.

The Intangible Assets

Human Capital: In CAAP, physical assets are still counted as the most important assets, although, investments in human capital have been going up for the past 35 years. The classical standards accounting norms/rules were designed when firm's most important assets were land, labour and capital. But, the time has since changed. Professor R.C. Crawford has recommended in his book in the era of Human Capital that the human capital be better included in the balance sheet of the company. His recommendation is also to include off-balance sheet intangible assets as well as the human capital assets in the balance sheet. Companies are increasingly aware of the need to account for the intangible capital.

As the concerns and focus in economies have changed from classical production of norms/standards of tangible materials and differentiation between production and consumption to the concept of the interlinked global/village economy, the focus is changed to providing knowledge service with higher mixture of consumer and the producer. In this, the prime thrust has changed from the physical capital assets to the human capital. As such, Professor Crawford's recommendation for inclusion of the intangible assets and the human capital in the balance sheet is meaningful and pragmatic.

At present, knowledge capital implies valuable assets like brands, patents, R&D, etc. Patents are valuable as a company's knowledge capital. Value of a patent is assessed by finding how many times the particular patent was meant for other applications. Likewise, the total knowledge capital cost can be estimated by taking the yearly normal revenues/ earnings and subtracting a number which is arrived through multiplication of the assets (which have been on the record) by the relevant post-tax returns that are expected. The residue is the revenue/earning through knowledge assets.

Value of a Company: Ability to Generate Cash

A board of directors' role is quite significant in energizing/promoting the capacity among the executive through proper direction, for generating cash on ongoing basis. This is the most important role of the board. This is an important criteria fur measuring the performance of the board.

A company is valued on the basis of the conviction of the people that the company's business will generate cash from its ongoing operations. Besides, a company's value just does not depend upon its assets but on how the market perceives the company's ability to manage the assets. This perception may entirely, be based on guess work and the rationality of the people to the company's past performance, its quality products and the competitive technology.

To sum up, an organization's worth depends upon the extent to which it can raise capital at a cost significantly lower than the resultant rise in earnings on account of fresh investment. A business organization with acquisition of capital at low cost is able to purchase goods/raw materials, equities and finance the operations at cheaper rates than its competitors. Such an organization has a cutting edge with its competitors. It is relevant to quote here that in 1980s US business was apprehensive of Japanese industry overtaking the rest of the world because of Japan's zero cost of capital.

A company can sustain on the basis of its revenues (in excess of the depreciation and dividend) from its operations. But, fast changing market conditions change the scenario. For instance, a debt which is attractive one year, turns out to be unattractive, the following year. Therefore, all organizations have the necessity to have access to the capital market for ensuring that they are making the best use of the capital on regular basis. One can say that financially successful enterprises are those which are able to generate higher returns on ongoing basis on the fresh investments in such a business which yields higher returns than the costs of obtaining the funds which is termed as externality costs.

The Non-economic Objectives of Corporations

According to Adrian Cadbury, this is a significant level of corporate responsibility which relates to the following points.

To what extent a corporate ought to give consideration to the societal priorities rather than its business? This is the most pertinent and inevitable question to be answered for appreciating the role of the corporate board.

In this context, David Engel has given some principles which can be the basis for developing principles of performance measurement of corporations. According to him, there are four specific facets where additional value maximizing purposes should be considered, as follows :

1. Obedience to the legal framework: Legal framework provides a sound basis for legitimate capitalism. It is essential for protecting and promoting healthy competition. Often, organizations prefer to violate the law on grounds of costs involved in following it. They weigh the benefits of circumventing the law, to the costs of obeying it. But, this is a foolhardy approach. Once caught on the wrong foot, the price is too high. Such organizations do so at the risk of subverting the legitimacy of the social and moral responsibility. A profitable company cannot afford to lose sight of this angle. We are all aware of the scandal of late 1950s-60s price-fixing in the electrical goods industry in the US. Executives of General Electric and Westinghouse were sentenced to jail. These companies had violated the anti-trust laws. GE was involved in law suits which cost the company $ 200 million and Westinghouse incurred $ 110 million. The electrical industry suffered in damages to the tune of $ 500 million.

2. Disclosure of information: Companies should disclose the impact of their operations on the society. The disclosure should be about more than what the basic minimum the law provides. Disclosure of the risks of their operations is also profitable for corporations. For instance, research results of tobacco companies that smoking is harmful to health have paid them dividends.

3. Corporates should involve in state policies on selective basis: They should involve only in those areas where it concerns them.

4. Observe the principle of 'Kew Gardens': A company should act only when it is in its own interest and not act against the interests of the society. Engel has drawn an analogy with one young lady, named Kitty Genoverse who was stabbed to death in the late 1960s in the presence of her neighbours who did not save her. The lady lived in Kew Gardens in the city of New York. She is considered as the symbol of deadly consequences for failing to act.

The above principles are meaningful and should provide a direction to corporate bodies when the question of the level of corporate responsibility is under consideration.

Let us examine the responsibilities of a corporation in terms of what is called as the externality cost of an organization, which is significant.

Externality Costs

The concept of 'corporate externality' can be explained as follows: operations of a company may have an adverse impact on the community living in its nearest vicinity /

the neighbouring locality. For illustration, let us take the case of Union Carbide factory located in Bhopal. In 1984, there was a gas leak in one of the sections of the factory in early hours in the day. The gas mixed with the wind, spread all over Bhopal. The worst victims happened to be those who lived near the factory. Thousands of people suffered vision deterioration and skin diseases. The impact of the gas tragedy will be felt for generations to come. The US parent company had to face prolonged litigation and crores of rupees of damages payable in compensation to those who suffered due to the negligent management of the corporation. Besides, the company was closed by the government orders, pending inquiry. The costs incurred by the corporation come under the category externality costs.

There are certain costs which a company incurs but does not account for in its Profit and Loss account. The costs have been incurred on account of confusion over how to value the intangible factors, success in having a government in power, professional auditors drafting audit accommodating rules, etc. But, all such costs come under corporate external costs. In some companies, efforts are made to reflect the real cost of operations, contrary to the rules under GAAP. In the US, a proposal was moved during Jimmy Carter's presidency to formulate a 'social accounting' system for corporations. The state was in favour of this system in view of the changing public expectations from the business. It was desired that corporations should communicate impact of their operations on the community, society, employees, physical environment, consumers and other affected bodies. Efforts were made to design a social responsibility accounting system as follows :

A corporate body had to furnish a social report each report should have a statement of objectives which permitted the assessment of :

  (a) Basis for data gathering and choice of the form of presentation.

  (b) The report should aim to fulfill the social responsibility in the spirit of high democratic values.

  (c) The report should relate to the objectives of a particular group.

  (d) The furnished information should be correct, based on the facts and audit.

The above is related to the social cost of operating a business and the social corporate citizenship responsibility of the business organization to the society. In addition, in Adrian Cadbury's words, this is the second level of a company's responsibility, i.e., the social responsibility as the cost of operations to the society.

To sum up, while measuring the board performance, all the above dimensions need to be examined critically and then assessed appropriately.

Employment/Compensation of Ceos

CEO's Contract: It is said that a CEO's annual compensation is like the tip of an iceberg. The proxy statement and the employment agreement will indicate the compensation which is theoretically supposed to be known to the public but it is not so. The details are given in the form of an attachment filled with company's OK but is not disclosed to the shareholders. It is a very time consuming and difficult process to find out (in the US system) when the contract is signed or altered. The contract is prepared by lawyers, head-hunters and compensation consultants. In fact, the same lawyer may have represented so many CEOs in negotiation and so he is picked up by the other companies and the company pays them the professional fee. As such, the contracts suffer from what is termed as "numbing sameness".

A remarkable contract of the best known CEO in the world, namely, Jack Welch of GE is worth discussing. It is brief to the point. It highlights the need to focus on the CEO's talent, time and attention to the business of the company. It does not guarantee any bonus, perks or payment or special protection. It includes the board's right to get rid of the CEO anytime. This is about the top most ranking CEO (who was employed in GE) in the world of industry.

Additional payments: Most of the contracts made between the corporation and the CEO provide for payment of 'excise tax'. This is done especially when the CEO owes extra tax burden to the exchequer.

Value addition is a significant point in compensation to the CEOs. Ideally, the contract of the CEOs should include incentives protection by linking performance with generation of value for the shareholders. Failure to incorporate this linkage sounds a wrong signal not only for the CEO but also across the board, staff, workers and the investors. This is a priority matter in the organizations.

It may be remembered that the contract of employment entered into between the company and the CEO is also an indicator of the priorities of the board. Shareholders, in reality, expect the CEOs to be a little greedy and aggressive. But, they (shareholders) expect the directors to ensure that the aggressiveness and greediness in CEOs results into value addition to the interests of the shareholders. The CEOs rightly expect the highest compensation. To meet such expectations, the directors can guarantee the CEOs that they will be compensated as per their wish but only when the stock prices double up. However, in case of a bad board, the shareholders have the duty to remind and reiterate their demand of accountability from the members of the board. This is the cycle which in fact should function in the best interests of all the agencies, viz., the CEOs, the board and the investors.

2.10. CORPORATE GOVERNANCE RATING

Corporate governance has come to be regarded as an important yardstick in investment decision making. A need has, therefore, begun to be felt to develop techniques to measure the quality of corporate governance in quantitative terms. The development of a corporate governance rating methodology, however, has to contend with two problems. First of all, there is no unanimity as to which particular elements of corporate governance should be subjected to evaluation. The second problem relates to the adoption of an appropriate rating methodology which could be free from subjectivity or bias. Notwithstanding these difficulties, attempts have been made by various agencies to evaluate or rate corporate governance practices.

Standard & Poor's Corporate Governance Scores

In its official publication, Standard and Poor has referred to corporate governance to mean 'the way in which a company manages itself in order to ensure fair and equitable returns to all shareholders and other financial stakeholders'. Corporate Governance has also been taken to include 'the rules and incentives by which shareholders control and influence a company's management so as to maximise profits and the value of the corporation'. High governance standards are an important factor in company's effort to attract and retain investors in global capital markets. Many companies highlight their good corporate governance standards to differentiate themselves from other capital seekers. Keeping in view the significance of good corporate governance, the Standard & Poor's Corporate Governance Services has developed Corporate Governance Scores and Corporate Governance Evaluations. This methodology analyses the interactions between a company's management, board of directors, shareholders and other stakeholders. The evaluation revolves around the following four components :

- Ownership Structure and Influence: It includes: (a) transparency of ownership structure, (b) concentration and influence of ownership

- Financial Stakeholders Relations: These include (a) regularity of access to and information to shareholders' meeting, (b) voting and shareholder meeting procedure, and (c) ownership rights

- Financial Transparency and Information Disclosure: This includes: (a) quality and content of public disclosure, (b) timing of and access to public disclosure, and standing of company's auditor

- Board and Management Structure and Process: This includes: (a) board structure and composition, (b) role and effectiveness of board, (c) role and independence of outside directors, (d) directors and executive compensation, evaluation and succession policies.

Corporate Governance Score Structure

It consists of two analyses as follows :

   (i) Company Governance: This focuses on the internal governance structure and processes at an individual company. Attention is focused on what a company does and not on what is the minimum required by local laws and regulations.

   (ii) Country Environment: This focuses on how external forces at a macro level can influence the quality of a company's corporate governance.

Both these micro and macro components are important to the practice of corporate governance. Specific scoring factors can be identified in order to analyse governance practices and facilitate objective and comparative analysis of corporate governance practices at individual companies.

Country analysis enables the individual company to make inter country comparison.

Meaning of Corporate Governance Score (CGS)

The Corporate Governance Score (CGS) of a company is an expression of Standard and Poor's current opinion about the extent to which a company adopts and conforms to codes and guidelines of good corporate governance practices that clearly serve the interests of its financial stakeholders. These scores are given on a scale from CGS-l0 (highest) to CGS-l (lowest). These score are awarded to the four individual components that contribute to overall CGS. These components can affect a company's ability both to honour contractual financial obligations to creditors and to maximize the value of a company's equity and distributions for its shareholders. Below is given a description of these components.

Stakeholder Relations

In a country, CGS analyses the extent to which a company adopts or exceeds codes and guidelines of good corporate governance practices. It covers the following :

   (a) Regularity of ease of access to, and information on shareholder meetings: Key analytical issues include shareholder meeting procedures, notices of meetings, documents sent to shareholders; information for shareholders, and attendance

   (b) Voting and shareholder meeting procedures: Shareholders representing at least majority of numbers voting rights should be able to call a special meeting and shareholders should have opportunities to ask questions of the board during the meeting and to place items on the agenda beforehand. A shareholder assembly should be able to control decisions through processes that ensure participation by all shareholders. Key analytical issues in this are: Charter provisions on calling meeting, arrangements for shareholders' participation in meetings, previous meeting minutes, shareholder information on voting procedures, any deposit agreement for overseas listing, proxy arrangements, charter provisions on voting thresholds.

(c) Ownership Rights - The criteria are as follows :

- There should be secure methods of ownership of shares and full transferability of shares

- A company's share structure should be clear and control rights attached to shares of the same class should be uniform and easily understood

- Shareholders' meetings should be able to exercise decisionrights in key areas, ensuring that minority shareholders are protected against dilution or other loss of value (e.g., through related party transactions on commercial terms)

- All shareholders should receive equal financial treatment including the receipt of any equitable share of profits.

Key analytical issues in this component are as follows :

- Charter provisions
- Arrangements with registrar
- Share structure class rights of common and preferred shares
- Charter provisions shareholder and board authorities
- Shareholders agreement
- Dividend history
- Examples of share repurchases and swaps

Financial Transparency & Information Disclosure

The standard of transparency and disclosure is an integral part of the corporate governance framework and one of the leading indicators of overall governance standards. It is a truism that companies that disclose more tend to be more open and investor friendly. Public companies all over the world are under pressure from institutional investors to disclose all material information.

Transparency involves the timely disclosure of adequate information concerning a company's operating and financial performance, and its corporate governance practices. It enables shareholders, creditors and directors to effectively monitor the actions of management and financial performance of the company. In certain countries, where accounting standards are limited, a commitment to transparency symbolises the adoption by the company of internationally recognised accounting standards.

From a board perspective, clear disclosure be made about directors, the basis of their remuneration and the extent to which they are independent of insiders:

Board Structure and Process

Board structure and process addresses the role of corporate board and its ability to provide independent oversight of management performance and hold management accountable to shareholders and other relevant stakeholders. Separation of authority at the board level is important, Boards with high accountability have strong base of independent outside directors who look after the interests of all shareholders. Conversely, the symptoms of poor accountability include presence of strong majority shareholders, dominance of few shareholders, heavy representation of company management on the board.

Management remuneration is another significant board governance factor. As regards the selection of management and board members and other voting matters, a cumulative voting structure can allow for board representation for minority shareholders. The board selection process is best when non-staggered to ensure the possibility of change.

The process by which outside directors are nominated and elected to the board, and the methods by which they are compensated for their board duties are important in assessment of board's accountability and practice.

(a) Board Structure and Composition - The board structure should ensure fair and objective representation of interests of all shareholders. The key issues are :

- Board size and composition
- Representation and constituencies

(b) Role and effectiveness of the board.

(c) Role and independence of non-employed directors.

(d) Board and executive compensation, evaluation and succession policies.

Procedure of Evaluation

The scoring is done by a scoring committee which includes the analytical team and other senior personnel from Standard and Poor's Corporate Governance Services. It may also include other individuals including local legal counsel and affiliate services staff. The Scoring Committee holds meetings with a company and prepares a detailed report covering main elements of the analysis. The CGS and individual scores for each of the four components are then assigned along with the logic underlying individual scores in the form of a scoring report. The report also includes descriptive information about the company's size, business segments, management, financial profile and ownership structure relating to abuse of shareholder rights on public record including pending items :

- Disclosure of board structure and composition
- Disclosure of company auditor
- Disclosure of major scale transactions in past three years (over 1096 of company's net assets)
- Identification of share register

II. Typical interviewees for the scoring process - The Standard and Poor's analysts then meet company's officers and individuals including :

- Chief Executive
- Finance Director
- Company Secretary / Corporate Counsel
- Board of Directors, particularly the Chairman and independent directors
- Key shareholders and creditors
- Company's auditor.

The report is given in the following format :

1. Executive summary rationale - The aggregate CGS with a rationale and summaries of key features of component scores are provided to highlight the main strengths and weaknesses of each.

2. Company Description - It contains basic operating, financial, management and ownership information.

3. Methodology - Showing scores and analysis for each of the following :

   - Transparency of ownership structure
   - Financial stakeholder relations
   - Financial transparency and information disclosure. Board Structure and Process

Information required prior to a corporate governance scoring meeting

The analyst from Standard and Poor's visits the company to inspect the company documents including the following :

- Company's annual and intra-year reports
- Company Charter/By-laws

- Filings with Government Regulatory Agencies
- Records of past three shareholders meetings - general and extraordinary.
- Minutes of board meetings of past three years
- Disclosure of new share issuance including options by the company or its subsidiary
- Identification of key shareholders (over 1096) and creditors
- Records of any penalties, fines and other violations.

3

THE BOARD CHAIRMAN AND THE CEO

Chairman constitutes the links between the board and the management. They represent the company before the shareholders, Government and public institutions. If the chairman is the Chief Executive, he has the duty to ensure effective participation of directors in board meetings as well as good performance by the management team. This chapter has discussed the respective roles of chairman and CEO in corporate governance, and the desirability or otherwise of separating these two positions.

CHAIRMAN THE CONCEPT

In the company law, there is no such thing as a chairman but only directors who from among themselves elect a chairman. The chairman is not, separate from the other directors who are equally responsible for conducting the affairs of the company. In actual practice, he is an important person, whether he presides over meetings of the board or the general meetings. He settles the agenda for the board meeting, determines the manner of presentation of information, elicits the views of directors and helps the board take timely and informed decisions. However, it is at the general meetings that he gets exposed to a wider world. The general meeting is a forum where he is confronted with difficult and sensitive situations out of which he has to steer clear.

Types of Chairman

 i. Chairman of Board Generally, the Articles provide that the directors may, from time-to-time, elect from among themselves, a chairman for presiding over the board meetings. Some of the companies name the chairman of the board in their articles.

ii. Chairman of Board Committees A Board Committee may elect a chairman at its meeting. As a matter of convention, the chairman of the board may also preside at board committee meetings.

iii. Chairman of general meetings Under the articles, the chairman of the board may preside at the general meetings. If the articles do not contain any provision, section 175(1) of the Companies Act would be attracted and the members personally present at the general meeting are required to elect one of them to be the Chairman.

The Chairman may be appointed either on a full time or a part-time basis. However, it is better to have a full time chairman of the Board. The appointment of a part-time chairman of the Board may cause this position to, degenerate into a sinecure offering ample opportunities for patronage. Such an appointment will lack the advantage of supplementing the experience and expertize of a CEO which can come only through a full-time chairman.

Qualities of Chairman

1. Firstly, the chairman being the man at the top must be a conceptualiser so that he can look far enough into the future and keep his organisation on an even keel. If he is a conceptualiser, he will make a better organisation, improve its capacity, bring out of its team, what it may not have thought it possessed.

2. Another quality and which should perhaps rank the foremost is character. It is the integrity of the man, his intellectual honesty and objectivity, and courage that makes the organisation trust him. His decisions may seemingly be questionable but they must be well thought out, honest and objective. These qualities provide him the strength to do what he thinks right, resisting pressure from inside or outside.

3. A quality of prime importance for a chairman is sound health, physical and mental, to enable him to take the strains at the top which can be telling.

4. He should also acquire a certain degree of aloofness, i.e., keeping his own counsel which, if conveyed even to his close colleagues may create anxities. Like a shock absorber, he must absorb the impacts from without and not transmit them to the organisation. Only when he feels that things have come to a pass where he should share his anxieties, he should formally present them to his colleagues so that they pull together and come out of the crisis.

5. The quality of listening to criticism is a desirable virtue but pains takingly acquired.

It means shunning the desire to know how he is doing. Let the events show him the mirror since the desire to know leads to a fatal temptation of knowing only the nice things. There have been seen chairmen who feel happy to listen to pleasant feedback and close the ears if the feedback is unpalatable.

6. He must be his own critic and judge. This objective can be well achieved if he introduces a consultation machinery and proper planning and appraisal systems. It is no longer necessary for the king to go around incognate in the streets of his kingdom at night. Management information systems coupled with personal nuances will help him provided he learns to listen with the third ear where 'you often pick up not just what is being said but what is not being said'.

7. An ideal chairman should not be too prudent, rather he should be an innovative risk-taker, lest his board becomes a mere caretaker and administrator of the status quo.

To sum up, the most important qualities required in a chairman are those collectively referred to as the qualities of leadership: vision, courage, good judgment, diplomacy and the ability to communicate with his board and management.

Functions of Chairman of the Board

The functions of the Chairman can be described under the following headings :

(i) **Management Functions:** The primary task of the chairman is to ensure that the board properly discharges its responsibilities. He presides over the meetings of the board and allocates the time of the meeting between different board members. The agenda items are put for deliberations in such a manner that board decisions are made in the time available. He also maintains close relations with CEO.

(ii) **Approval Functions:** He secures the approval of the board to various matters such as budget, strategies, important appointments and policies relating to running of the organisation.

(iii) **Leadership functions:** The chairman provides leadership to the board and constitutes the members into a cohesive collectivity. He is instrumental in the development of sound board system and procedures. He also ensures that the board properly discharges its responsibilities. He motivates individual directors, sets the agenda, and organises the work to be done by committees.

(iv) **Consultative functions:** He is instrumental in devising appropriate techniques for handling and avoidance of conflict of interests of board members or the management with the interests of the company. Freed of executive responsibilities, he also acts as a confidential consultant to the CEO.

(v) **Spokesperson:** Chairman represents the company and the Board in their relation with external agencies such as the Government and other corporate or non-corporate bodies. He also interacts with the shareholders for redressing their problems.

(vi) **Initiative functions:** The chairman takes initiative with regard to the following matters :

    (a) formation of corporate strategies,

    (b) providing relevant information to outsiders,

    (c) developing external relations.

(vii) **Special invitees** Considering whether any additional people need to be invited for all or a part of the discussions as non-voting members is one of the chairman's functions.

(viii) Handling of matters by securing consent of board members through phone or circulation of documents for communication and approval, particularly when these do not require time and attention of full board meeting.

(ix) Planning and Running of Meetings Chairman's function in relation to board meetings is as follows :

- Determination of purpose of the meeting and what really needs to be achieved.
- Creation of agenda besides the sequence and time allocated to each of the item.
- Deciding the location - not necessarily the boardroom.

During meetings, the Chairman establishes a business like atmosphere in which issues can be faced, explored in depth, alternative viewpoints put forth, and resolutions reached.

(x) Drafting of minutes The chairman ensures the recording of minutes comprising a summary of decisions. He should review each member's contribution. Under law, he may also be required to sign the minutes.

Role of Chairman

Effective corporate governance is founded on the bedrock of strong and independent board. The reins of the board are in the hands of the chairman. He stands at the pinnacle of the organisational pyramid. No wonder, the Chairman's role is fundamental within the system of corporate governance. Chairing a board is a difficult and demanding task. This is particularly true of a unitary board consisting of both executive and non-executive directors. He has to draw the best out of the inside knowledge of executive directors and the external perspective of non-executive directors. He is primarily responsible for the working of the board, and the maintenance of balance between the board and management. It is his job to ensure that the board apply themselves solely to those issues which the board alone can address. With respect to boards meetings, the Chairman ensures that all relevant issues are incorporate in the agenda, and that all types of directors are enabled to play their proper role in the making of decisions by the Board. To accomplish this objective, the chairman provides a climate wherein each viewpoint can be established and yet board members remain united in their purpose. He should also be prepared to define the problem and ensure that every director makes his contribution recognising fully well that some of the members may tend to dominate whereas the more reticent of them may have to be implored to make their contribution. Differences of personalities and political abilities among board members must be treated realistically. The chairman must manage conflicts, simulate new insights and avoid rushing through the agenda.

His role consists in evolving proper systems and procedures with a view to attract the right talent. All board members must be given a sense of participation besides an assurance that there is no riding roughshod. It will ensure free and frank discussions besides giving the board members a sense of direction.

Relationship between Chairman of Board and the CEO

Long-term development of an organisation can be enabled by designing procedures for checks and balances between the board and management. When two separate individuals occupy the positions of Chairman of the Board and, CEO, the following is the shape of their relationship :

1. Complementing the CEO in areas in which he has no time or in which he lacks expertise such as formation of international contracts, or forging relationship with the Government.

2. Directing the CEO and his team of senior executives to formulate the overall corporate objectives and strategy.

3. Consulting and coaching the CEO whenever he seeks the advice of Board Chairman in arriving at a decision.

4. Reviewing and Controlling The chairman supervises the CEO's decisions regarding allocation of resources, review of financial results of the company and overall performance of the CEO.

5. Restricting the power of the CEO by laying down appropriate procedures and rules.

The actual contours of their relationship are determined on the basis of organizational structure and the respective personalities of the people concerned. The leadership of the chairman has significant impact on the activities of individual directors and the board performance as a whole.

The actual role of Chairman of Board, who is not a CEO, remains limited in scope and does not include the managing of board activities. Chairmen tend to view their role with respect to CEO as one of consultant and advisor. The formulation of strategy of the company remains the prime responsibility of the CEO irrespective of whether he is also the Chairman of the Board or its president. In comparison, the role of the Board and its Chairman is more in relation to review and evaluation of the proposed strategy before its implementation. The chairmen who are not CEOs tend to perform "passive functions" and give a lower priority to initiative functions.

Infosys Technologies have carved separate positions of Chairman and Chief Mentor; a Chief Executive Officer; President and Managing Director; Chief Operating Officer, and Deputy Managing Director. The demarcation of responsibility and authority between the following positions is given hereunder :

- Chairman and Chief Mentor: He is responsible for monitoring Infosys' Core Management team in transforming the company into a world class, next generation organisation. He also interacts with global thought leaders to enhance the leadership position of the company. As chairman of the board, he is responsible for all board matters.

- The CEO, President and Managing Director: He is responsible for corporate strategy, brand equity, planning, external contacts, new initiatives and other management matters. He is also responsible for achieving the annual business plan.

- The COO and Deputy Managing Director: He is responsible for all customer service operations.

The Chairman, CEO, COO and other executive directors and the senior management make periodic presentations to the board on their responsibilities, performance and targets.

Separation of positions of Board Chairman and the CEO

The Corporate governance structure once epitomized by forceful business leadership has now become a relic of the past. Dey Committee Report (Canada), headed by Peter Dey, Chairman of Morgan Stanley, Canada had strongly advocated the splitting of CEO and chairman positions. The first principle of this report is; 'don't ask the foxes to patrol the hen house'. A rising number of codes and investor guidelines are concluding that in most circumstances, a company is more stable and less risky over the long-term, if it is inoculated against domination by a single all-powerful leader. The proponents of agency theory have also strongly advocated the separation of the position of Chairman and CEO so as to provide for a system of checks and balances.

International Practices

In Germany and Netherlands, the large companies are legally required to have supervisory boards chaired by non-executives. In Japan, the coronations have separate executives holding the jobs of chairman, and shacho (i.e., President or CEO) though the latter usually acts as the defacto board chairman. The doctrine of splitting the jobs which was first enshrined in the Cadbury Code has gained rapid momentum in its application. Corporate governance codes in Australia, Britain, Canada, Japan, Malaysia and South Africa all favour the splitting of roles. Others such as France's AFG-ASFFI, Spain's Olivencia and Belgium's Cardon reports have recommended balancing by non-executives in cases where the posts of chairman and CEO are combined. The CII code has suggested that where the posts of Chairman and CEO are combined, half the board should consist of non-executives as compared to 30 per cent, if there is non-executive chairman. The USA's Council of Institutional Investors Code have urged a lead non-executive director where the posts are combined. The Business Roundtable requires a substantial majority of independent directors as the balancing factor in case of a combination of positions.

Despite clear cut global unanimity on the need for checks on CEO's power, the trend towards CEO-Chairman duality is not universal. Some highly successful companies in North America, Europe and Australia have been led decisively by one person over a considerable period.

Objectives of separation of positions of Chairman and CEO

The separation of the positions of Chairman and CEO serves following objectives :

(i) It reaffirms the established role and legal authority of the board as a crucial interface between company's owners, its CEO and management.

(ii) It ensures the board's direct accountability to the shareholders for the performance of management.

(iii) It recognises and restores the key role of the chairman as the architect and leader of a strong and independent board that is capable of fulfilling its fiduciary duties and responsibilities.

(iv) It ensures the appointment by the board of a CEO who is competent to lead and manage the company effectively. Such a scheme of things subordinates the CEO and makes him liable to the board which also has the right to remove the CEO.

Therefore, if board really represents shareholders, where is the need to have CEO on it. Does it not lead to conflict of interest? The CEO is a professional manager and cannot represent shareholders and impartially sit in judgment of himself. Where CEO is also the chairman, there is too great a temptation to tilt things towards protecting CEO's career interest.

Not much empirical work has been done on this issue. A study by P.L. Rechner and D.R. Dalton. has found that companies with separate CEO and Chairman consistently outperform those companies which have combined these roles. The authors have advocated separation of positions of CEO and chairman on three grounds, viz. : (i) more objective evaluation of CEO, (ii) creation of greater accountability and (iii) better performance. When these positions are combined, it should not be concluded that CEO will inevitably manipulate board but it does give him that opportunity.

Clause 49 of the Listing Agreement, however, has not made any mandatory stipulation as to separation of the positions of Chairman and CEO. The listed companies may opt to have either an executive or non-executive chairman. In case of an executive chairman, the Board shall comprise of more than 50 per cent independent directors. The proportion of independent directors may be lower at one-third of the total board strength in case there is a non-executive chairman of the Board.

Grounds for Dichotomy between Board and Management Teams

There is a fundamental distinction between board and management. Management is hierarchical and authoritarian with responsibility defined in terms of an individual. The Board's responsibility is to direct, control and approve the course to be taken as well as to ensure that it is followed. As head of the management team, the Chief Executive is responsible to the board for management of business. However, it is the board alone which has the total responsibility towards shareholders for the activities of the enterprise. Board appraises the management team and in a sense controls the direction of a company's operations. But it should not mean that there is an adversary relationship between the board and management. The obligation for the creation and maintenance of climate of success is the joint responsibility of the board and management team. In actual practice, a lot of initiative to the board is provided by the management through the Chief Executive.

The duties of board are more in the nature of broad planning and supervising than executing details of operations. After setting the criteria for measuring effectiveness of management, the board considers it more appropriate to refrain from meddling with the management of the company. The management's duty is to keep the board fully informed of company's operations. Undoubtedly, a smooth working relationship between board and top management can result in continuous success and prosperity.

According to Hugh Parker, where a company has both a Managing Director (CEO) and a Chairman, the most effective division of responsibilities can be achieved if both these positions are segregated and assigned to two separate individuals. Combining of these two positions into a single individual will render it difficult to expect dispassionate evaluation of management performance from the chairman who is also the head of management team. Such a separation is also essential to ensure that trusteeship position of the board is kept separate from operating management. The Board is a bridge between shareholders and the Chief Executive. The CEO needs a buffer, a cushion a sort of shock absorber in the form of chairman. When both the positions are clearly separated, the CEO can concentrate on internal management, leaving enterprise environment interface largely to the chairman.

There is another reason for the maintenance of difference between the positions of CEO and chairman. Board can be likened to the legislative wing of the Government whereas 'Management' constitutes the executive wing of the Government, in the case of a company. The former lays down the policies, determines priorities and suggests strategies. The function of the management is to implement these. Absence of distinction between these two wings may denigrate the Board to a mere ratification agency, kowtowing to the whims of management. In such a situation, the company is likely to lose its sense of direction and motivation. The maintenance of this distinction would clarify the respective roles of Chairman and the Chief Executive in a better way. When such a separation is made, the Chairman would act as primus inter pares (first among equals) whereas the CEO would be responsible to the board as a whole.

The situation becomes particularly worrisome if the company has a part-time Chairman along with a full-time Managing Director. Very often, part-time chairmen are content to restrict their role to conducting board meetings but some of them may begin to project themselves as an authority parallel or superior to that of CEO. Such a situation is ripe for a conflict. There may come about rival centres of power at board meetings. To the extent possible, such a situation should not be allowed to take place or otherwise should be resolved through mutual dialogue. But this possibility should not detract from the desirability of keeping both the positions separate.

The CEO and his Role

The Chief Executive Officer or the 'Managing Director', as he is popularly called in the context of Indian companies, is responsible for operational management and profitability of the company. A dynamic CEO galvanises the entire organisation by providing inspiration, reorganisation, restructuring, induction of fresh talent and elimination of deadwood.

His right to manage business within the limits stipulated by the Board should be unrestricted. The very nature of his responsibility makes it essential that his position be kept separate from that of the board chairman. Before describing the logic behind this thinking, it is significant to firstly examine his role.

The areas earmarked for the CEO are quite clear. Matters such as direction of company's growth, adoption of new technology, share of market to be captured, financial viability of an investment proposal, procurement of finance, recruitment and placement of key personnel, etc., are decided in consultation with the CEO. A CEO with drive, imagination and great personal attributes very often dominates the scene and may carry the company to great heights. Conversely, if he has dictatorial tendencies and an intolerant attitude towards viewpoints of others, he may be considered to have short-sightedness which will simply enable him to end up scheming good viewpoints and thereby sealing his own as well as his company's future.

He is truly the fountainhead of plans and ideas that flow in the organisation. He is the head of the management team and may be regarded as the hub of the company. His performance should be evaluated annually by the board as a whole, and the result be communicated to the CEO by the Chairman (preferably non-executive chairman). The evaluation should be based on objective criteria such as performance of business, accomplishment of long-term strategic goals, development of management and the like.

The very nature of his responsibility makes it essential that his position be kept separate from that of board chairman. Since he has the responsibility to execute corporate strategy, charging him with the duty of monitoring the management on behalf of shareholders is likely to create an untenable situation. Such a bar is not designed to emasculate the CEO but to rationalise his powers. The responsibility for his appointment rests with the board. The board selects that person as the CEO who is competent and capable of managing the company effectively, Yet he is actually a subordinate, accountable to the board for his actions and removable by the board. Thus, for promoting good governance, the position of CEO must be kept separate from that of the chairman.

Should the CEO be given unlimited powers? No doubt an efficient CEO contributes to a company's success to a significant extent, yet it may not be prudent to entrust unbridled powers to him. Generally, the CEO is expected to function optimally and take a balanced view of the interests of all the stakeholders. Even then he will need the guidance, counsel, praise and right dose of criticism of the board. The CEO or the board cannot be the sole arbiter of the quality of corporate governance. It is determined through the combined contribution of both. Even the most competent CEOs need to be properly guided so that they do not become larger than their real life. Without such reining, even the highly gifted may become whimsical and end up taking an unbalanced decision. The Robert Maxwell debacle in U.K. has rightly brought home the lesson that an unchecked CEO is a legitimate matter for risk management.

Responsibilities of CEO

1. To get the systems and procedures reviewed periodically by the professionals, especially when cracks appear therein.
2. To seek the opinion of the Chairman in case of doubt and bring the unresolved differences of opinion at the board for debate and resolution.
3. To establish harmonious relations with the non-executive directors, and to provide information needed by them for taking decisions.
4. Not to set up the executive directors as a group against the chairman or against the non-executive directors.
5. To spot talent amongst the executives and encourage their promotion to the board.
6. To submit his performance for evaluation by the board.
7. To discharge his liabilities by keeping within the limits prescribed by the board.

CEO Dominance and Voluntary Corporate Disclosures

In a study about the link between CEO dominance, NED quality and voluntary disclosure of Hongkong companies, it has been suggested that combining of position of CEO and chairman of the board in one person (CEO dominance) creates a strong individual power base which could erode the board's ability to exercise effective control. This phenomenon can constrain board independence and reduce its ability to execute its oversight and governance roles. The study has suggested that the negative relationship between CEO dominance and voluntary disclosures is mitigated by non-executive director quality.

For Example : Executive Chairman of ITC Ltd.

The ITC Ltd. has an Executive Chairman who operates as the Chief Executive of the company as a whole. He functions as the Chairman of the Board and the Corporate Management Committee. His primary role is to provide leadership to the Board and Corporate Management Committee (CMC) for realising company goals in accordance with the charter approved by the Board. He is responsible for the working of the Board, for its balance of membership (subject to Board and shareholders' approvals), for ensuring that all relevant issues are on the agenda, for ensuring that all directors are enabled and encouraged to play a full part in the activities of the Board. He is to keep the Board informed on all matters of importance. He presides over the general meetings of shareholders.

CEO's Succession Planning

The succession of CEO demands careful planning and management. For this purpose, every board should have a separate committee that plans for the succession process. In many cases, this task is handled by the compensation or nomination committee. Either way, there should be regular executive sessions for discussing the issue of succession. When a new CEO is being appointed, key members of the selection committee should hold individual interviews with candidates and then the whole board can meet to make a final decision.

Once the CEO has been appointed, the whole board is responsible for helping him to assimilate. The new CEO needs to be informed of what the board expects in terms of performance and also in terms of communication.

The 1998 NACD Blue Ribbon Commission on CEO Succession has suggested that when selecting a new CEO, the companies should develop a 'profile of success' establishing the specifications and requirements of the CEO position to be filled. The board needs to make a strategic assessment of company and determine what competencies and skill sets fit the position. These must be time specific. The necessary qualifications of the CEO may vary according to the particular challenges which the company would face at a particular time.

A Conference Board report has suggested the following three basic considerations in determining a company's current requirements and the mix of personal qualities a prospective CEO will need to carry out the board's mission:

(i) The degree of business dynamism the company wants to project.

(ii) The particular issues which the new CEO will confront.

(iii) The personal qualities sought by the board.

The board should also decide the manner in which the success of a CEO will be determined. These goals should be based on financial and non-financial targets. Such evaluation should be done on an annual basis. Moreover, these goals should be regularly reviewed and revised. It will keep the succession plan routed in the context of company's strategic and business plans. Other issues to be considered should include CEO's leadership abilities, and relationship with board members and senior management.

If the CEO is not performing well enough, then the board will have to consider replacing him.

Simultaneously with CEO evaluation, annual evaluations of company's top prospects must also be conducted for considering the replacement of current CEO. This will build a level of crisis management into the succession plan so that an emergency step in will be available if the current incumbent suddenly leaves.

Board Committees

WHILE THE CORPORATE board remains the ultimate decision making authority within the company, it is customary for boards to constitute various committees comprising of their members that can undertake detailed scrutiny and evaluation of delegated matters. This paves the way not only for better management of the full board's time and work, but also, perhaps even more importantly, for greater in-depth scrutiny and attention to the key elements of successful implementation of board policies. It enables the board to have the wherewithal to discharge its onerous oversight responsibilities in governing the corporation for the benefit of all its shareholders. As the Bosch Committee in Australia noted in 1990, "The effectiveness of the board, and particularly of the non-executive directors, is likely to be enhanced by the establishment of appropriate board committees. They can distribute the board's workload and enable more detailed consideration to be given to important matters and, where sensitive issues (such as the appointment of auditors) have to be considered, an appropriately constituted committee may give independent consideration which will be valuable." In the US, board reliance on committees and outside experts increasingly became the norm after the seminal judgement in the case of Smith vs Van Gorkom. Here, the Delaware Supreme Court handed down a landmark ruling that the company's board had not adequately discharged its fiduciary duty of care, in a matter that involved the company in a leveraged buyout by an external corporate takeover specialist, at a price calculated not on any robust valuation principles but on what would allow for easily managing the required debt-servicing needs. In the event, the board approved the proposal after a twenty-minute presentation by the CEO at a meeting that lasted two hours, with no advance notice of the agenda or any supporting documentation. The decision that spelt a significant departure from precedent litigation on the subject reportedly sent "shock waves through corporations across the country" and led to intense lobbying and other reactive measures to restrict wider application of this decision. In order to demonstrate that appropriate and thorough consideration was given to important matters, boards began delegating responsibilities to specially constituted committees whose members had the time, domain expertise and experience to deliberate matters in greater depth than would be possible at a full board meeting. Where necessary external specialist advice was sought and the course of action to be adopted was recommended to the board based on this detailed consideration.

In India, the Working Group on the Companies Act 1956 had recommended audit committees in 1997 but stopped short of suggesting legislation, on the basis that compliance in that case would be more in letter than in spirit. The influential industry chamber, Confederation of Indian Industry, in its Desirable Corporate Governance A Code, published in 1998 recommended to its members that by 1998-99, listed companies with a sales revenue of one billion rupees or a paid up capital of two hundred million rupees, "should set up an audit committee within two years, with at least three members, all drawn from the company's non-executive directors, who should have adequate knowledge of finance, accounts and basic elements of company law." However, it took much longer and a regulatory mandate through the listing agreements with stock exchanges for such committees to be actually formed in most companies.

While it is up to the boards to constitute as many committees as they deem necessary and appropriate, three such committees are quite important and are even mandated in some jurisdictions. These are the audit committee, the compensation or remuneration committee, and the governance and/or nominations committee. We will first discuss some general issues common to all board committees and follow this up with a discussion of matters specific to each of these three committees.

Board Primacy Over Committees

Committees are creatures of the board, and as such they are hierarchically subordinate to the main board. Committees derive their authority from the powers delegated to them by the board. The board, on reviewing the recommendations of the committee may decide in its wisdom, to accept the recommendations with or without modifications, and even reject them, though this would be under very unusual circumstances indeed. After all, the main purpose of the committee structure is to avail of the time and expertise of the committee members on the delegated matters, and it would hardly be sensible not to abide by their recommendation. On the other hand, where the committee is vested with 'approval' powers, and accords its 'approval' in a matter within the terms of delegation, it will not be open to the board to rescind that approval with retrospective effect. In other words the transactions or matters so approved by the committees will continue to be binding upon the company. In such cases, even if approved matters are placed before the board, it can only be for information and noting. Of course, this will not preempt the board from revisiting the decision, and modifying or rescinding it prospectively, if in its wisdom such correctives are called for.

There are some important caveats to this general principle. Country mandates by legislation or capital market regulations and stock exchanges in the case of listed corporations, take precedence over company decisions, and hence their requirements will continue to apply, and boards or company charter documents may not override them, but certainly can improve upon them. A typical example is the Indian provision relating to audit committees that mandates, "the recommendations of the Audit Committee on any matter relating to financial management including the audit report, shall be binding on the Board," and that, "if the Board does not accept the recommendations of the Audit Committee, it shall record the reasons therefore and communicate such reasons to the shareholders," presumably in their Annual Report to the shareholders. Given the fact that the audit committee is a creature of the full board and derives its authority from it, it may be difficult for the board to accept a situation where the delegating body becomes a captive of the delegated body, but if such is the law, then there is little that the board can do but to accept the position. Of course, it does have the right to disagree and explain its reasoning to the shareholders.

Directorial Liability for Committee Decisions

The related question then is whether the other members of the board are absolved of any responsibility in matters concerning the financial management of the company (or even the other decisions of committees). This clearly runs counter to the concept of collective responsibility of the board towards the affairs of the company. It is a ground reality that boards usually rely upon board committees, and in turn, committee members often rely on the advice of specialists and experts including those from within the executive ranks. The key of course is to ensure that this reliance is well placed and justified, and also subject to any applicable legal or regulatory limitations. This need to be assured of the appropriateness of the reliance placed on committee recommendations and expert advice flows directly from the fiduciary responsibility that directors owe to the corporation and its shareholders. A key element in the discharge of this responsibility is the duty of care. Care is evidenced "by the directors committing time, by regularly attending meetings, by being adequately informed as to board decisions, and when circumstances suggest, by making appropriate inquiry into the operations of the business or into facts that will inform their decisions." Unless they have knowledge that makes such reliance unwarranted, directors are entitled to rely on information provided by officers, board committees, counsel and public accountants, and so on, so long as they have assured themselves that the information or advice proffered is within the fields of their expertise and in case of delegated matters (such as to a board committee), they are delegable according to the legal, regulatory and in-house charter frameworks.

Each director on the board and board committees has a statutory duty of care arising from the fiduciary relationship between them on the one hand, and the corporation and its shareholders on the other. Fiduciary responsibility can be delegated (and relied upon as a defence) only if the delegation is clear and the delegates' actions and inactions are regularly and carefully monitored. That is why committee charters need to be in writing, minutes of meetings must be made available to all the directors, and committee chairs' feedback briefings to the full board must be a regular agenda item, fully recorded in board minutes. While these may at first sight appear to be routine, even clerical chores, they are the only tangible support that would help directors to demonstrate their application, and due diligence and care in monitoring the delegates' work and recommendations, and their own final approvals of such recommendations. As Braiotta points out, "since the directors serve the corporation in a fiduciary capacity, their statutory duty of care cannot be delegated because of the personal nature of the director's relationship with the corporation. Hence, although the audit committee [for example] can make recommendations to the entire board, the final decisions are made by the board because it has overall responsibility for the committee's actions. In short, the standing committees of the board cannot eliminate each director's duties and obligations because of the fiduciary principle."

Landfeld and his co-authors make another insightful point relating to intra-board and inter-committee communications. They highlight the importance of ensuring that relevant information available to a member of the board, even if he or she is not a member of a committee deliberating upon a matter, or information available to one committee which may be relevant to another committee with overlapping charter responsibilities, should not be allowed to slip between the cracks, and the concerned directors or committee should share that information with the committee dealing with the matter. Further, especially in matters of overlapping responsibilities, the deliberating committee should seek from the other committee, such information as may be relevant for their consideration. This also emphasizes the need for board and committee members to carefully peruse the meeting agenda and minutes of board committees even where they are not members of those committees, because of the need and obligation to share known information that may prove material to the committee's deliberations.

At least in the US context, this emphasis on processes relating to "being adequately informed" has attracted derisive criticism of the business judgment rule that governs directors' liability for negligence and breach of duty. It is considered ironic that directors are not penalised for "wrong" decisions.

3.1. AUDIT MECHANISM

Most of factors that emphasize the need for better financial oversight and audit practices include; fiduciary failure of board of directors, high risk accounting practices; conflicts of interest in financial matters; failure of internal audit and internal control systems;

excessive compensation to board members and lack of independence of board members; and audit committees and auditors.

The Accounting process and Accounting Profession

The accounting profession has been endeavoring to develop techniques to meet the demand of financial disclosures. A good audit report is derived from financial statements that are based on sound accounting principles enunciated by professional bodies.

The level of transparency in financial transactions, the norms for disclosure and the way accounting transactions are recorded, are central to the issue of ensuring accountability in corporate governance.

Preparation of financial statements in consonance with the accounting standards facilitate interpretation, reliability and comparability of information across enterprises enabling investors to identify trustworthy firms.

Internal Audit

In the modern times, the task of internal audit has been extended to a review of whether the resource utilization of the enterprise is efficient and economical.

"Internal auditing is an independent, objective assurance and consulting activity designed to add value and improve an organisation's operations. It helps an organization accomplish its objectives by bringing a systematic, disciplined approach to evaluate and improve the effectiveness of risk management control and governance".

Reasons for internal Audit

     i. Assurance to the board of directors and management that control functions are working effectively that risks have been adequately managed.

    ii. Ensuring growth of the enterprise through extensions in new markets, mergers and acquisitions, internal growth, product diversification etc.

   iii. Performance of duty of accountability to stakeholders from whom capital resources have been raised.

   iv. Use of information technology in day-to-day operations necessitating a confirmation of data integrity, transfer of data and associated security systems.

    v. Toughened regulatory regime and disclosure requirements.

   vi. Independent and objective evaluation of activities connected with the management of risks, enhancement of transparency and improvement of corporate governance.

Internal audit function should be so structured as to achieve organizational independence; besides ensuring unrestricted access to top management, audit committee and the board. The audit committee should review the internal audit function's character and assess the budget and resources allocated. The decision to hire or terminate the chief of internal audit should require endorsement by the chairman of audit committee. The audit committee should regularly provide the chief of internal audit, the opportunity to confer privately with the committee. The practice of internal audit must be governed through a system of self-regulation based on widely accepted standards, ethical principles and other best practices.

Scope of Internal Audit

Internal Audit has to function in tandem with the board and management for controlling misfeasance, malfeasance and other uncalled for or unethical practices. The management has to continually demonstrate its commitment to high ethical standards and the internal audit has to concretize this commitment.

Key Consideration in internal Audit Function for Enhancing Corporate Governance and Control.

Internal audit has been considered a part of internal control system. In India the need for installing internal audit has not been statutorily prescribed. But the role of internal auditors is to assist the board to oversee the senior management activities to secure assurance about the organization's system of internal control.

The following need to be given due consideration :

1. Improvement of the quality of internal audit.
2. Independence an objectivity of internal auditor.

    It is the duty of a board and the audit committee to decide on the independence and objectivity of internal auditor.

3. Greater accountability.

    This is the main problem area to ensure transparency of the board of directors.

4. Internal audit must coordinate with external audit to allow a proper coverage of financial compliance and controls.

5. Business process risk assessment.

    Internal auditor develops such a business process model to minimize the risk management.

6. Audit team must possess requisite skills.

7. Reporting system.

The internal audit pre-supposes proper reporting to the shareholders on the investing public. Thus the report on internal control should draw attention to different sources.

    (a) Independent evaluation performed by internal auditor.

    (b) Review of internal controls by external audit. Management opinion must be carried out to ensure controls and checks.

    (c) Lastly results of review and checks must be the responsible of the board.

The External Audit Function

The statutory audit or external audit is regarded as the cornerstone of the Corporate Governance. It is the only medium through which the success of the board of directors is accessed, because audit provides an objective check on financial statement.

The glaring example of ENRON IMBROGLIO has brought into sharp focus the great dangers faced by the investing community on account of the existence of cosy relationship between the company and the auditors. The failure of company's auditors to honestly report on the financial transaction to the shareholders, Arthur Anderson, the well-known audit company incurred heavy criticisms of the public and the community.

It is but natural to evince a serius interest in the need to examine the relationship of companies with the statutory auditors, the weak spots in such relationship and ways to put them on an even keel. Financial reports must provide financial information of the highest quality. The quality of such financial information are :

    i. Companies' accounting department must maintain true and fair financial records.

    ii. Internal Audit must be done on financial records along with review of the directors (or Supervisory Boards and the Audit Committee).

    iii. Proper approval procedure by the relevant bodies within the company according to the standardized quality assurance system's conformance to the norms of capital market regulators and stock exchanges.

    iv. It is the obligation of the company to seek opinion from the financial analyst rating agencies as part of ethical obligations of the company.

Auditor-Company Relationship

The auditor's role is basically to report whether the financial statements, give a true and fair view of the existing state of affairs. Audit mechanism is designed to provide a reasonable assurance that the financial statements are free of material mis-statement. The auditor's failure to abide with this norm causes the so caused expectation gap. This gap occurs because of the difference between what auditors do achieve and what they caught to achieve in order to narrow the expectation gap. There is a need to clarify the respective responsibilities of directors and auditors for preparing and reporting of the financial statements.

Financial statements must be drawn up with utmost impartiality. This will reduce the susceptibility of the auditors to pressures from management. The law has recognized the need to maintain the independence of the directors by means of appointment by the shareholders rather than the board. But in actual practice shareholders simple endorse the names of the auditor suggested by the management. The causality occurs because of the lack of proper disclosure of the nexus between the management and the company. East Asian Financial Crisis occurred due to such lack of proper disclosure.

Different financial frauds and scams occurred due to the following :

    I. Concealment of the actual size of the death related to transaction.

    II. Non disclosure of short term foreign currency borrowings and consequent foreign exchange risk exposure.

    III. Absence of proper information in the financial statements especially of assets and liabilities of the company.

    IV. Absence of information on loan guarantee, particularly foreign currency loan.

In view of the above apprehensive steps towards financial stalemates have been taken.

Audit committee requirements under this act have empowered to enquire into the financial transaction made by the management.

Secondly each members of the committee is independent like the independent directors.

Thirdly the Audit Committee has ultimate responsibility of approving all audit and unaudit services. Under GAAP the Audit Committee has established procedures to receive an address complaints regarding accounting, internal control and audit issues.

Audit Committee

Audit Committee has a critical role to play in ensuring the integrity of financial management of the company. This Committee adds assurance to the shareholders that the auditors, who act on their behalf, are in a position to safeguard their interests.

The Board shall set up, a qualified and independent Audit Committee and give its terms of reference. While the requirement of clause 49 applies to all listed companies, provisions of section 292A of the Companies Act, 1956 apply to every public limited company having a paid-up share capital of rupees five crore or more.

Constitution of the Audit Committee

The Audit Committee contemplated under Clause 49 of the Listing Agreement shall have at least three directors as members. Two-third of the members of the Audit Committee must be Independent Directors. The Chairman of the Audit Committee must be an Independent Director. The Company Secretary must act as the Secretary to the Committee. All members of Audit Committee shall be financially literate and at least one member should have accounting or related financial management expertise. The clause further defines the term "financially literate" to mean the ability to read and understand basic financial statements i.e., balance sheet, profit and loss account, and statement of cash flows. A member will be considered to have accounting or related financial management expertise if he or she possesses experience in finance or accounting, or requisite professional certification in accounting, or any other comparable experience or background which results in the individual's financial sophistication, including being or having been a chief executive officer, chief financial officer, or other senior officer with financial oversight responsibilities.

In case a vacancy arises in the position of a committee member, the Board should file such intimation with the stock exchange. If the vacancy is filled up at the meeting of the committee held immediately alter the vacancy arose, it should be construed as compliance with the conditions of Corporate Governance.

The Audit Committee may invite such of the executives as it considers appropriate (and particularly the Head of the finance function) to be present at the meetings of the Committee, but, on occasions it may also meet without the presence of any executives of the company. The finance director, Head of Internal Audit and a representative of the Statutory Auditor may be present as invitees for the meetings of the Audit Committee.

As a good governance practice it may be considered that since the functions of the Audit Committee relate primarily to the finance department, it is desirable to designate a senior officer from the finance department, who should function with respect to the Audit Committee in close coordination with the company secretary. Such officer and an officer from the secretarial department of the company could be made jointly responsible for the operation of and support services to the Audit Committee. It is worthwhile to note that since the Audit Committee is a committee of the Board, the provisions of the Companies Act, 1956 and the Articles of Association of the company regarding committees will be applicable to Audit Committee with regard to items such as Notice, Quorum, Minutes, etc. However, if the company has set up an audit committee pursuant to provision of the Companies Act, the said committee shall have such additional functions/features as is contained in this clause.

It is good governance practice· to disclose in the Annual Report, the date of constitution of the Audit Committee along with description of terms of reference. In case, the company has reviewed the composition and working of Audit Committee to conform to other Corporate Governance Code(s) or to conform to changes made in the Companies Act, 1956, the details thereof may also be indicated.

In case the Board does not accept the recommendations of the Audit Committee, it shall record the reasons therefor. These reasons should be communicated to the shareholders.

It is proposed that such a communication be made through the Corporate Governance Report.

Authority of Audit Committee

The Audit Committee, constituted in accordance with section 292A of the Companies Act, shall have authority to investigate into any matter in relation to the items specified in the said section 292A or referred to it by the Board. For accomplishing these purposes, the Committee shall have full access to information contained in the records of the company and can seek external professional advice, if necessary.

Default

If a default is made in complying with the provisions of section 292A of the Companies Act, 1956, the company and every officer who is in default, shall be punishable with imprisonment for a term which may extend to one year, or with fine which may extend to fifty thousand rupees or with both.

Powers of the Audit Committee

As per Clause 49-II(C), the powers of the Audit Committee shall include the following :

    (a) To investigate any activity within its terms of reference

    (b) To seek information from any employee

    (c) To obtain outside legal or other professional advice

    (d) To secure attendance of outsiders with relevant expertise, if it considers necessary.

The powers of the Audit Committee specified above are illustrative and apart from the above, the Board may delegate such other powers, as it may deem fit and proper.

Role of Audit Committee

As per Clause 49-II (D), role of Audit Committee shall include the following :

(a) Oversight of the company's financial reporting process and the disclosure of its financial information to ensure that the financial statement is correct, sufficient and credible.

(b) Recommending to the Board the appointment, reappointment and, if required, the replacement or removal of the statutory auditor and the fixation of audit fees.

(c) Approval of payment to statutory auditors for any other services rendered by the statutory auditors.

(d) Reviewing with management the annual financial statements before submission to the Board for approval with particular reference to :

   • Matters required to be included in the Director's Responsibility Statement to be included in the Board's report in terms of clause (2AA) of section 217 of the Companies Act, 1956.

   • Changes, if any, in accounting policies and practices and reasons for the same

   • Major accounting entries involving estimates based on exercise of judgment by management. Significant adjustments made in the financial statements arising out of audit findings Compliance with listing and other legal requirements relating to financial statements

   • Disclosure of any related party transactions

   • Qualification in the draft audit report.

(e) Reviewing, with the management, the quarterly financial statements before submission to the board for approval.

(f) Reviewing with the management, performance of statutory and internal auditors, adequacy of internal control systems.

(g) Reviewing the adequacy of internal audit function, if any, including the structure of the internal audit department, staffing and seniority of the official heading the department, reporting structure coverage and frequency of internal audit.

(h) Discussion with internal auditors any significant findings and follow up thereon.

(i) Reviewing the findings of any internal investigations by the internal auditors into matters where there is suspected fraud or irregularity or a failure of internal control systems of a material nature and reporting the matter to the Board.

(j) Discussion with statutory auditors before the audit commences about the nature and scope of audit as well as post-audit discussion to ascertain any area of concern.

(k) To look into the reasons for substantial defaults in payment to, depositors, debenture holders, shareholders (in case of non-payment of declared dividends) and creditors.

(l) To review the functioning of the Whistle Blower mechanism, in case the same is existing.

(m) Carrying out any other function as is mentioned in the terms of reference of the Audit Committee.

Auditor's disclosure of qualifications and consequent action: In this context the Committee observed that the Companies Act makes it more difficult to replace an auditor than to reappoint him. While this is as it should be, corporate governance would benefit from disclosing the reasons for replacement. If the management were to be more accountable to the shareholders and the Audit Committee, in the matter of replacing auditors, this is likely to make auditors more fearless.

Management's certification in the event of auditor's replacement: The Naresh Chandra Committee noted the provisions of section 225 of the Companies Act and suggested its amendment so as to require a special resolution of shareholders, in case an auditor, while being eligible for re-appointment, is sought to be replaced. The Committee further recommended that the explanatory statement accompanying such a special resolution must disclose the management's reasons for such a replacement, on which the outgoing auditor shall have the right to comment. Pointing to the role of Audit Committee in this regard, the Committee suggested that the Audit Committee would have to verify that this explanatory statement is 'true and fair'.

Auditor's annual certification of independence

The Committee observed that it would be good practice for the audit firm to annually file a certificate of independence to the Audit Committee and/or the Board of Directors of the Client Company. This will help in ensuring that the auditors have retained their independence throughout the period of engagement.

In the above context, Naresh Chandra Committee recommended that a certificate of independence must be submitted to the Audit Committee or to the Board of Directors of the client company, by the audit firm before agreeing to be appointed the auditor, certifying that the firm, together with its consulting and specialized services affiliates, subsidiaries and associated companies :

1. are independent and have arm's length relationship with the client company;

2. have not engaged in any non-audit services listed and prohibited as mentioned in recommendation 2.2 above,

3. are not disqualified from audit assignments by virtue of breaching any of

the limits, restrictions and prohibitions as mentioned in recommendation above.

In the event of any inadvertent violation relating to recommendations the audit firm will immediately bring these to the notice of the Audit Committee or the Board of Directors of the client company, which is expected to take prompt action to address the cause so as to restore independence at the earliest, and minimize any potential risk that might have been caused.

The Committee felt that the Audit Committee should be allowed to be true to their name by ensuring that they have a larger role with regard to audit. In fact this should be the starting point in empowering Audit Committees.

CEO and CFO certification of annual audited accounts

Taking cue from above practice, Naresh Chandra Committee has recommended that for all listed companies as well as public limited companies whose paid-up capital and free reserves exceed Rs. 10 crore, or turnover exceeds Rs. 50 crore, there should be a certification by the CEO (either the Chairman or the Managing Director) and the CFO (whole time Finance Director or otherwise) which should state that, to the best of their knowledge and belief:

- they, the signing officers, have reviewed the balance sheet and profit and loss account and all its schedules and notes on accounts, as well as the cash flow statements and the Directors' Report. These statements do not contain any material untrue statement or omit any material fact nor do they contain statements that might be misleading.

3.2. AN OVERVIEW OF IMPORTANT CORPORATE GOVERNANCE COMMITTEES

Cadbury Committee, 1992

The Cadbury Committee investigated the accountability of the Board of Directors to the shareholder and to society. It has 19 recommendations in the nature of guidelines for the board of directors, executive directors, non-executive directors, and such other officials.

Important recommendations include enhanced information to the shareholder and the setting up of the audit committee with independent members. Its model is one of self-regulation. The most controversial and revolutionary requirement was the 'Code of Best Practice' proposed by the committee. It required that :

The directors should report on the effectiveness of a company's system of internal control. The services of the directors should not extend beyond three years without the approval of the shareholders.

Every listed company should create an audit committee with a minimum of three non-executive directors.

It was the extension of control beyond the financial matters that caused the controversy.

The Paul Ruthman Committee

The committee mitigated the controversial element of reporting beyond the financial matters by limiting the reporting to internal financial control. However, it continued the thread of the progressive element by extending the directors' responsibilities to 'all relevant control objectives including business risk assessment and minimizing the risk of fraud.'

The Greenbury Committee, 1995

The committee was set up to provide an answer to the general concern about accountability and the directors' remuneration.

The committee believed in strengthening accountability by proper allocation of responsibilities for determining directors' remuneration, accurate reporting to shareholders, and greater transparency in the process. The committee did not endorse statutory control for achieving the best practice.

The Code of Best Practice devised by the committee was divided into four sections: Remuneration Committee; Disclosures; Remuneration Policy; Service contracts; Compensation.

All public limited companies were expected to produce annual compliance statements and encourage maximum implementation of the Code of Best Practice.

The Hampel Committee, 1995

The committee further developed the Cadbury Committee Report.

It recommended that :

The auditors should report on internal control privately to the directors.

The directors should maintain and review all reports, not just financial controls. Companies that do not already have an internal audit function should, from time to time, review their need for one.

The committee introduced a combined code that consolidated earlier reports, specifically Cadbury and Greenbury reports.

The Combined Code, 1998

The Combined code was basically derived from the Hampel report, the Cadbury report, and the Greenbury report. The recommendations aimed at :

Maintaining a sound system of internal control to safeguard shareholders' investment and company's assets.

To review at least annually the internal controls covering financial, operational, and compliance and risk management and report to shareholders that they have done so. Importance of risk management was highlighted as an important ingredient in the success of the corporate.

It was mandatory for all listed companies to adhere to the combined code.

Turnbull Committee, 1999

The committee was set up by Institute of Chartered Accountants in England and Wales to assist in implementing the requirements of the Combined Code relating to internal controls.

It recommended that internal audit should be carried out annually, specially in companies that did not have an internal audit function.

To manage risks the board of directors were advised to confirm the existence of procedures of evaluating and managing key risks.

Sarbanes Oxley Act, 2002

The Sarbanes Oxley Act or SOX Act, 2002 is one of the most comprehensive Acts to control fraud and achieve quality governance. For the first time an Act tried to monitor minute details of organizational governance. It was aimed at protecting the investors and other stakeholders from corporate failures. Therefore, it provided detailed recommendations on various aspects :

1. Establishment of Public Company Accounting Board (PCAB)
2. Audit Committee
3. Conflict of Interest
4. Audit Partner Rotation
5. Improper influence on conduct of audits
6. Prohibition of non-audit services

7. CEOs and CFOs required to affirm financials

8. Loans to Directors

9. Attorneys

10. Securities Analysts

11. Penalties: Studies should be conducted by the Securities and Exchange Commission (SEC) or the Government Accounting Office in :

   (a) Auditor's rotation

   (b) Off-balance sheet transactions

   (c) Consolidation of accounting firms and its impact on the accounting industry

   (d) Role of credit rating agencies

   (e) Study of violators and violation during the years 1998-2001

   (f) SEC enforcement actions over the past five years

   (g) Role of investment banks and financial advisers

   (h) 'Principle based' accounting

Indian Committees

The Companies Act of 1956 was rooted in an environment of License and Permit Raj. The deficiencies were :

1. Though non-executive directors can play a significant role in providing independent and objective opinion, the Act does not assign any formal role between executive and non-executive directors. The effective control was in the hands of the executives, whole time directors, and the MD.

2. Non-executive directors have only ornamental value as no commitment from them is expected, since the Act allows them to be members of as many as 20 companies at the same time. Financial reporting was not transparent.

3. No formal qualification for a director of a company was laid down.

4. Formal provision of auditors to be appointed by shareholders was available, but auditors worked in collusion with the management.

5. Hardly any service was provided to investors.

CII'S Recommendation, 1998

This body recommended

1. A single board can maximize long-term shareholder value. It should meet at least six times a year, preferably at an interval of two months.

2. A listed company with a turnover of Rs 100 crores and above should have professionally competent and recognized independent non-executive directors who should constitute: at least 30 per cent of the board if the chairman of the company is a non-executive director, or at least 50 per cent of the board if the chairman and MD is the same person.

3. A person should not hold directorship in more than ten listed companies which excluded directorship in subsidiary companies where over 50 per cent stake was held by the group company, and directorship in associate companies where more than 25 per cent but less than 50 per cent equity stake was held by the group company.

4. Non-executive directors should :

    (a) Become active participants in boards, not just passive advisors.

    (b) Have clearly defined responsibilities within the board, such as the audit committee.

    (c) Know how to read a balance sheet, profit and loss account, cash flow statements, and financial ratios and have some knowledge of various company laws. This excludes those invited as experts in other fields such as science and technology.

5. To secure better efforts from non-executive directors, companies should :

    (a) Pay a commission over and above the sitting fees for the use of the professional inputs. Commissions are rewards on current profits.

    (b) Consider offering stock options, so as to relate rewards to performance. Stock options are rewards contingent upon future appreciation of the corporate value.

6. While reappointing members of the board companies should give the attendance record of the concerned directors. If the director has not been present, i.e., absent with or without leave, for 50 per cent or more meetings, then this should be explicitly stated in the resolution that is put to vote. One should not reappoint such directors.

7. Key information that needs to be placed before the board must contain :

    a. Annual operating plans and budgets as well as updated long-term plans, capital budgets, manpower, and overhead budgets.

    b. Internal audit reports including cases of thefts and dishonesty of material nature, show cause, demand, and prosecution notices from revenue officers.

c. Fatal and serious accidents, dangerous occurrences, any effluent and pollution problems.

d. Default in payment of interest or non-payment of the principal on any public deposit and/or any secured creditor or financial institution.

e. Default such as non-payment of inter-corporate deposits by or to the company or materially substantial non-payments for goods sold by the company.

f. Any issue that relates to public or product liability claim of a substantial nature.

g. Details of any joint venture or collaboration agreement.

h. Transactions that involve substantial payment towards goodwill, brand equity, or intellectual property.

i. Recruitment and remuneration of senior officers just below the board level, including appointment or removal of the CFO and the Company Secretary.

j. Labour problems and their proposed solutions.

k. Quarterly details of foreign exchange exposure and the steps taken by management to limit the risks of adverse exchange rate movement.

8. For all companies with paid up capital of Rs 20 crores or more, the quality and quantity of disclosure that accompanies a GDR issue should be the norm of any domestic issue.

   Listed companies with either a turnover of Rs 100 crore or a paid up capital of Rs 20 crore, whichever is less, should appoint an Audit Committee within two years.

9. Under 'Additional Shareholder's Information' listed companies give data on the following :

   (a) High and low monthly averages of share prices in a major stock exchange where the company is listed for the reporting year

   (b) Greater detail on business segments up to 10 per cent of turnover, giving a share in sales revenue, review of operations analysis of markets, and future prospects

10. Companies that default on fixed deposits should not be permitted to accept

further deposits and make inter-corporate loans or investments or declare dividends until the default is paid.

11. Major Indian Stock Exchanges should insist upon a compliance certificate signed by the CEO and the CFO which should clearly state :

   (a) The company will continue in business in the course of the following year.

   (b) The accounting policies and principles conform to the standard practice.

   (c) The management is responsible for the preparation, integrity, and fair presentation of financial statements, and other information contained in the annual report.

   (d) The board has overseen the company's system of internal accounting and administrative controls either directly or through its audit committee.

SEBI's Initiatives

SEBI appointed a committee on 7 May, 1999 under the chairmanship of Kumara Mangalam Birla with a view to promoting and raising the standard of Corporate Governance.

Mandatory Recommendations

Applicability of all listed companies with paid-up share capital of Rs 3 crore and above had to comply.

1. **Board of directors:** Optimum combination of executive and non-executive directors was suggested. The number of independent directors should be at least one-third in case the company has a non-executive chairman and at least half of the board in case the company has an executive chairman.

2. **Independent directors:** Those who, apart from receiving director's remuneration do not have any material pecuniary relationship or transaction with the company, its promoters its management, or its subsidiaries, which in the judgement of the board, may affect the independent judgement of the directors.

3. **Audit committee:** Qualified and independent audit committee to enhance credibility and encourage transparency had to be created.

   The audit committee should have a minimum of three members, all being non-executive directors with a majority being independent and at least one director having financial and accounting knowledge.

   The audit committee should invite such executives as it considers appropriate, besides which the required head of internal audit and external audit representative should be present as invitees for meetings of the committee.

The audit committee should meet at least thrice a year with a gap of not more than six months.

The quorum should be two members or one-third, whichever is higher with a minimum of two independent directors.

As in the earlier recommendation, this committee also required confirmation that :

(a) The company will continue business in the course of the following year.

(b) The accounting policies and principles conform to standard practices.

(c) The management is responsible for the preparation, integrity, and fair presentation of financial statements and other information contained in the annual report.

(d) The chairman should be an independent director and must be present at the AGM to answer shareholder questions.

4. **Remuneration committee:** The board should decide the remuneration of the non-executive directors.

    Full disclosure of the remuneration package of all the directors including salary benefits, bonuses, stock options, pension fixed component, performance linked incentives, along with the performance criteria, service contracts, notice period, severance fees, etc., should be made available in the section on Corporate Governance of the annual report.

5. **Board procedures:** The board meetings should be held at least four times a year with a maximum time gap of four months. Minimum information on annual operating plans and capital budgets, quarterly results, minutes of meeting of audit committee, and other committees, information on recruitment and remuneration of senior officers, significant labour problems, material default in financial obligations, statutory compliance, etc., should be placed before the board for its deliberation.

    A director should not be member in more than ten committees and act as chairman of more than five committees across all companies in which he is a director.

6. **Management:** Management discussions and analysts report about industry structure, opportunities and threats, segment-wise or product-wise performance, risks, internal control systems, etc., are to form part of the director's report or an addition to it.

    The management must make disclosure to the board relating to all material, financial, and commercial transactions where they have personal interest that may have potential conflict with the interest of the company.

7. **Shareholders:** In case of a new director or re-appointment of existing director, information on resume, qualification, companies where he holds directorship and committee membership for shareholders' perusal should be available. Sharing information of quarterly results on website for the benefit of shareholders.

A board committee under the chairmanship of a non-executive director should be created to look into redressing shareholder grievances.

For share transfer authority should be delegated to officer, committee, registrar, and share transfer agents to attend to issues at least once in a fortnight. Manner of Implementation: A separate section on Corporate Governance in annual reports is to be introduced.

Non-mandatory recommendations

1. **Chairman of the board:** The chairman's role, in principle, should be different from that of the chief executive. A non-executive chairman should be entitled to maintain an office at the company's expense.

2. **Remuneration committee:** Credible and transparent policy should be put in place. Remuneration should be good enough to attract, retain, and motivate. The committee should comprise of at least three directors, all of whom should be non-executive directors, the chairman being an independent director. The chairman should attend the AGM to answer shareholder queries.

3. **Shareholder's right:** Half-yearly declaration of financial performance including summary of the significant events of the six months should be sent to the shareholders.

4. **Postal ballot:** The following important activities should be decided by postal ballot to ensure shareholder participation :

    (a) Matters relating to alteration in the memorandum of association. e.g., change in name, address of registered office, etc.

    (b) Sale of whole, or substantially the whole, of the undertaking.

    (c) Sale of investments in the companies where the shareholding or the voting rights of the company exceeds 25 per cent.

    (d) Making a further issue of shares through preferential allotment or private placement basis.

    (e) Corporate restructuring.

    (f) Entering a new business not germane to existing business.

    (g) Variation of rights attached to class of securities.

    (h) Matters relating to change in management.

In 2000 SEBI adopted the recommendation that the stock exchanges should modify the listing requirements by incorporating in them a new clause Clause 49 so that disclosure is made in the following areas: board of directors; audit committee; remuneration of directors; board procedure; management; shareholders; report on corporate governance; compliance certificate from auditors.

SEBI's Code of Corporate Governance requires that the following information by a company should be made available to the board of directors periodically :

(a) Annual operating plans and budgets and any updates

(b) Capital budgets and any updates

(c) Quarterly results for the company and its operating division or business segment

(d) Minutes of audit committee meetings

(e) Information on recruitment and remuneration of senior officers just below the board level

(f) Material communications from government bodies

(g) Fatal or serious accidents, dangerous occurrences, or any material effluent pollution problems

(h) Details of any joint venture or collaboration agreement

(i) Labour relations

(j) Material transactions that are not in the ordinary course of business

(k) Disclosures by the management on material transactions, if any, with potential for conflict of interest

(l) Quarterly details of foreign exchange exposures and risk management strategies

(m) Compliance with all regulatory and statutory requirements.

Naresh Chandra Committee Report, 2002

Department of Company Affairs appointed the Naresh Chandra (NC) Committee.

Recommendations mainly concerned: (a) The auditor company relationship; (b) disqualifications for audit assignments; (c) list of prohibited non-audit services; (d) independence standards for consulting; (e) compulsory audit partner rotation; (f) auditor's disclosure of contingent liabilities; (g) auditor's disclosure of qualifications and consequent action; (h) managements' certification in the event of auditor's replacement; (i) auditor's annual certification of independence; (j) appointment of auditors; (k) certification of annual audited accounts by the CEO and CFO; (l) auditing the auditors; (m) setting up of the independent quality review board; (n) proposed disciplinary mechanism for auditors; (o) independent directors; (p) audit committee charter; (q) exempting non-executive directors from certain liabilities; (r) training of independent directors; (s) establishment of corporate serious fraud office and; (t) SEBI and subordinate legislation, etc.

The difference between the two committees: The NC Committee made no distinction between a board with an executive chairman or with the non-executive chairman. It has recommended that all boards should have at least half of its members as independent directors.

About audit committee, the KM Birla Committee suggested it should have non-executive directors as its members with at least two independent directors. The NC committee recommended all should be independent directors.

The NC Committee was strict about the relationship between auditors and their clients.

The committee has recommended that along with its subsidiary, associates or affiliated entities, an audit firm should not derive more than 25 per cent of its business from a single corporate client. This move could affect small audit firms, because new firms which are usually small, may not be able to get business from a number of companies as they would have to establish their credentials in the market. Therefore, if it is regulated that not more than 25 per cent of business can be earned from one business, these small firms would never be able to establish themselves and would be taken over by bigger firms in the market who would continue to hold monopoly sway. This monopoly may become the breeding ground for corruption and ultimately impact the socio-economic structure of a country.

To ensure transparency, the proposal for compulsory rotation of audit firms was suggested. It stressed the partners and at least 50 per cent of the audit team working on the accounts of a company need to be rotated by the audit firm once every five years.

The NC Committee drew up a list of prohibited non-audit services.

It said nominees of institutions (FIs) cannot be counted as independent directors. The committee aimed at tightening the noose around the auditors by asking them to make an array of disclosures.

By calling upon the CEO and the CFO of listed companies to certify their company's annual accounts it brought in accountability.

It suggested setting up of quality review boards by the Institute of Chartered Accountants of India, Institute of Company Secretaries of India and the Institute of Cost and Works Accountants of India, instead of a public oversight board, as in the US.

To attract quality independent directors the committee recommended that these directors should be exempt from criminal and civil liabilities under the Companies Act, the Negotiable Instruments Act, the Provident Fund Act and the Employees State Insurance Act, the Factories Act, The Industrial Disputes Act. and the Electricity Supply Act.

The Narayana Murthy Committee Report, 2003

The terms of review were :

1. To review the performance of Corporate Governance.

2. To determine the role of companies in responding to rumour and other price sensitive information circulating in the market in order to enhance the transparency and integrity of the market.

The committee agreed with the NC Committee on :

- Disclosure of contingent liabilities
- Certification by CEOs and CFOs
- Definition of independent directors
- Independence of audit committees

Mandatory recommendation :

1. Audit committee: Should review:

    (a) Financial statements and draft audit reports including quarterly and half yearly information.

    (b) Management discussion and analysis of financial condition and the result of operations.

    (c) Report relating to compliance with laws and risk management.

    (d) Management letters of internal control weaknesses issued by statutory internal auditors.

    (e) Records of related party transactions.

2. Related party transactions: A statement of all transactions with related parties including their bases should be placed before the audit committee.

3. Proceeds from initial public offerings: Companies raising money through IPO should disclose to the audit committee the uses and application under major heads on a quarterly basis. Each year the company shall prepare a statement of funds utilized for purposes other than those stated in offer document/ prospectus.

4. Risk management: Procedures to inform the board members of the risk assessment and minimization procedures.

   Management should place a report before the entire board every quarter documenting the business risks faced by the company, measures to address and minimise such risks, and any limitation to the risk taking capacity of the company. The board should formally approve this document.

5. Code of conduct: The code should be posted on the company's website and all board members and senior management personnel shall affirm compliance with the code on an annual basis. The annual report of the company shall contain a declaration to this effect signed by the CEO and COO.

6. Nominee directors: Recommended removing the concept of nominee directors. Shareholders should appoint any director. An institutional director, if appointed, would have the same responsibilities and liabilities as other directors.

7. Compensation to non-executive directors should be approved by the shareholders in the general meeting; restrictions were placed on grant of stock option, and it was required to make proper disclosures of compensation.

8. A whistle blower policy was to be created in a company.

Non-mandatory recommendations

Providing unqualified financial statements, training of board members, evaluation of non-executive directors' performance by a peer group comprising the entire board of directors excluding the one being evaluated.

3.3. THE DEMAND FOR DISCLOSURE

Branding and enhancing a company's reputation are part of the process of building a company's intangible assets. Managing the interface between companies and the world around them requires balancing three pressures :

1. The corporation itself desires to enhance its reputation and competitive position by putting positive information about its operations into the public domain and correcting any misinformation.

2. The information intermediaries the media and the investment analyst community want the company to provide them with information that will deliver them some comparative advantage, whether that is a scoop for the journalist or an insight for the analyst into how the company creates value.

3. The regulators, motivated by the duty to serve a perceived public good, require

certain information to be made public. The regulators specify not only the type of information they want released but also how it should be released.

It is the third of these forces that is the subject of this book. The regulators' voice is becoming a more pervasive influence over all of a company's dealings with the public. The realm of corporate privacy is contracting, as company executives are required to disclose in detail the composition of their remuneration and directors to detail their share trading.

Private rules that used to govern companies today have the force of law. Accounting standards, once agreed to as a matter of professional self regulation, are now mandatory requirements, and companies can be taken to court on matters of judgment about how the standards should apply. The rules by which companies must abide in order to gain admittance to the stock exchange are enforceable under the Corporations Act. Failure of a company to meet disclosure requirements can leave individual managers and directors open to criminal prosecution.

The annual report, which was once little more than a statement of the accounts, is today an omnibus document that is required to detail the corporate governance procedures, executive salaries, the presence of environmental management systems, the holdings of major shareholders, and a growing level of detail and discussion about the company's financial position. The disclosure over environmental and social performance is an emerging field of regulatory activity.

The half-year and full-year disclosure requirements are supplemented by the continuous disclosure of anything that may be expected to have an influence on the share price. This is the area that causes the greatest vexation as, on the margin, it can force companies to disclose information they believe would be of value to competitors or could be damaging to their position in the market.

Not only is the private domain increasingly thrown open to public scrutiny, but the way companies exchange information with the world around them is regulated as well. The days when a chief executive could go out to lunch with one of the company's big investors or a business editor and chew over the issues confronting the business are gone forever.

Intermediaries are receiving greater scrutiny. The business media has so far escaped licensing, but business journalists are required to lodge registers of their securities with regulators. Analysts have long required licences and must disclose any commercial interests they have in a company when publishing research. The operation of 'Chinese walls' between analysts and the trading operations of broking houses is under scrutiny, with the likelihood of greater regulation. In the same way, the accounting firms have come under pressure to separate their auditing from their consulting arms.

Auditors perform a statutory role to verify the public reporting of companies, and there is growing opposition to their developing any other kind of commercial relationship with their clients. Auditors also have obligations to report any malpractice, or any undue pressure placed on them by managements, to the Australian Securities and Investments Commission (ASIC). The auditing profession at large is under pressure to provide a more informative commentary on judgments that have been made in compiling the company's results.

A head of steam is developing, with ASIC asserting that it does not believe Australia's corporate sector has embraced the spirit of disclosure and signalling that it is prepared to get tough. Dissatisfaction with the standard of public financial reporting has reached a high pitch, and a new regulatory regime is being imposed on the accounting industry.

For those involved in managing a company's interaction with the world around it, the regulatory consequences of their actions are looming large.

Inside the Corporate World

Managing the mandatory demand for disclosure runs against the grain for business. Companies have long been accustomed to the idea that they can deal with the public on their own terms. They put only such information as advances their own strategic interests into the public domain. The central impetus of mandatory disclosure is to force companies to release information whether it is good news or bad, whether they want to or not.

However, companies are, in essence, internal worlds of information. The economist Ronald Coase argued that the very reason for companies to exist was to minimize the cost of transacting, or exchanging information, which is required in order to achieve complex tasks. Companies provide not merely proximity to ease the exchange of information: they provide a rule based framework for transacting. Because the rules apply only inside the corporation, there is a barrier between the company and the outside world that is intrinsic to its very existence.

Everything that takes place within a company is assumed to be secret, unless it is explicitly cleared for public release. Companies vary in their approaches to managing information, but all impose some measure of discipline on exchanges with the world around them. The delegation of communication with the outside world to a department of corporate affairs is nearly always a direct report to the CEO, and often a member of the executive committee. The internal world of information is hierarchical. Any level of management has an assumed right to access to information possessed by those at levels below it: Access to information at higher levels may be granted, but is not assumed.

The higher one ascends in an organization, the greater the scope of one's information. Strategy writer Henry Mintzberg comments, 'the manager may not know everything, but he or she typically knows more than anyone of his or her subordinates'. This is established as a universal truth, whether the organization is a multinational corporation or a street gang. The manager is at the centre of information flow. Managers, Mintzberg estimates, spend 40% of their time communicating information.

The manager has roles as monitor of the information environment, as disseminator of information throughout the organization, and as spokesperson for the organization to the external world. Managers assume the role as spokesperson precisely because they have the understanding of the totality, and of how the organization should be represented in order to further its strategic interests.

If companies are by definition internal worlds of information, it is the presence of competition that makes secrecy imperative, and that stands in the way of open and transparent corporate disclosure. In his classic work on competitive advantage, Michael Porter writes:

Information is crucial to both offensive and defensive competitive moves. Sometimes selective release of information can serve very useful purposes, in market signaling, communicating commitment and the like; but often information about plans or intentions can make it a great deal easier for competitors to formulate strategy.

He notes that companies are disclosing more and more about themselves. This is partly because of the proliferation of the business press and the increased requirements for public filings. Although some of this is legally required, a lot of what is written in annual reports, or stated in interviews and speeches, is not: 'Disclosure may stem more from concern with the stock market, managers' pride, inability to control statements by employees, or simply from lack of attention.'

Seeking competitive advantage always brooks the prospect of retaliation. It may be achieved by persuading the competition that the advantage is not threatening or, if retaliation is likely, by maximizing a lead before a competitor becomes aware. Competitors always face uncertainty about a firm's intentions and the extent of its resources. The firm, in maximizing its advantage, needs to manage the competitors' uncertainty.

It became popular throughout the 1990s to downplay the competitive nature of business, and rather to stress the potential for collaboration in broad networks and alliances, even with industry rivals. However, the reality is that contract negotiations always involve an element of finessing.

The company is not only in competition against rivals in its industry. It is also, to some extent, in competition with its suppliers, customers, staff, financiers and shareholders. Arrangements with both suppliers and customers may be subject to negotiation over contracts. In any negotiation, there is an incentive to signal to the counter party that one has less room to move than is actually the case. This also holds true for wage negotiations with staff.

Shareholders and managers are, to some extent, in competition for the same resource, and it is for this reason that disclosure of executive salaries has become mandatory. Financiers may be assured either that the security is better than in fact it is or that the company has less need of their particular service than is the reality.

Preserving secrecy is therefore an intrinsic component of corporate management. Companies limit the circulation and disclosure of information in several ways. Their hierarchical structure makes it easier to restrict information to those who 'need to know'. Companies have internal rules and procedures that must be followed. And there are informal codes of conduct, which mean you simply don't share inside information with outsiders.

A disciplined approach to the management of information relationships with the external world is required by the regulators. Even if it were possible to run a business without secrets, the regulators would not permit it. Although their general desire is that business should be as transparent as possible, the foundation stone of disclosure regulation is that the market should be informed of any price sensitive information first. This means that there must be a well organized system for identifying information that may be price sensitive within the organization, and channeling that through the officer responsible for disclosures to the share market before it leaks to the outside world.

3.4. THE BOARD – MANAGEMENT RELATIONSHIP APPROACH

The board must have a workable governance model. Such a model should address issues central to establishing and maintaining an effective process for making and administering corporate policy. A governance model should include routine approaches to the following :

- Choosing or clarifying the business or business in which the corporation will operate.
- Hiring the right people in terms of both their abilities and their values.
- Aligning the interests of the board and management with those of the shareholders.
- Developing mutually agreeable goals, policies, and standards of performance for the CEO.
- Evaluating plans to achieve agreed-upon goals.
- Remaining knowledgeable about the firm's activities and performance and evaluating the results.

- Reacting appropriately to the results by holding management accountable and rewarding or intervening as necessary.

Boards of directors creating and following this or a similar governance model should be well on their way to establishing productive partnerships with their CEOs. Boards members are advised to recall the paramount importance of hiring an effective CEO for the firm.

For example:

A major consumer goods company, PepsiCo, needed to complete significant restructuring in order to grow profitably in a more narrowly focused competitive environment. In such situations, management must recognize the need or opportunity at hand and develop a workable solution.

The results of the actions of the board and the CEO

The strategic moves to restructure the company through acquisitions and spin offs led to a dramatic improvement in the company's financial results.

On May 2, 2001, PepsiCo announced that Steven S. Reinemund was being appointed chairman of the board and CEO and Indra Nooyi was being appointed president. Between 1998 and 2001, five new directors were appointed to the board; three were CEOs of other companies, one was president of a large subsidiary of one of the largest companies in the United States, and the other was the president of PepsiCo.

The Point

As a result of these strategic initiatives, the board and management of PepsiCo achieved a major restructuring of the company during a period of transition to a new CEO. These results speak well for the environment in which the board and the CEO could work together to reposition the company in a major way. The restructuring essentially involved pruning away less effective businesses in terms of growth potential and returns on capital and leaving a smaller but richer mix of businesses for the future.

How the board should function

Given that the board has an effective person serving as CEO, that an arms length relationship exists between the board and the CEO, and that there is a workable governance model, the board can focus on its ongoing tasks. Most of these tasks may be grouped into three primary activities.

- Reviewing and influencing decisions in an advise-and-consent role.
- Reviewing and understanding the firm's results

- Determining when and how to intervene in management's affairs

The Board, the CEO, and the search for a sustainable competitive advantage

In many ways the strategic planning process in the business world is analogous to the search for the "ultimate weapon" that has characterized military initiatives throughout history. In seeking "sustainable competitive advantages" commanders have hoped to acquire access to a weapon so unique that it would tip the battle in their favour.

Thus the board must recognize the importance of strategy that creates the dual needs for the company to manage change effectively and to move decisively in implementation decisions. In fact, success of the entire process of strategic planning and execution depends on the ability of the board and the CEO to lead change and execute decisions swiftly and decisively.

Restructuring Via Acqisitions and Divestitures

The board's role in the direct management of a company's affairs takes on a more critical flavour when acquisitions or divestitures are being considered. A company's bylaws often lift the board's oversight responsibilities to a higher level when there is a move to sell a portion of the company's assets or to make major asset acquisitions through a proposed merger or by acquiring another company. While there is a general tendency on the part of the board to endorse actions recommended by the CEO, proposed mergers or acquisitions often lead to the exercising of substantial caution because the board may want to rethink or reevaluate previously adopted strategic plans.

Boards may rationally consider, in abstract terms, and acquisition that will double the size of the company or a divestiture that will remove the original product line on which the company was built. It is quite stressful, however, when the previously approved strategic initiative leads to an actual proposal for major change. Some boards are strong advocates of strategic change and lead the way for such bold moves. Other boards, though, become very reluctant participants in projects involving broad changes. These boards sometimes lose very effective CEOs who recognize the need for fundamental change and leave when the boards refuse to endorse the new direction for the company. In the final analysis, boards have the responsibility to consider all options in carrying out their duties to shareholders.

Knowing what the results are: The Oversight Responsibility

The board must, in the end, maintain an effective oversight of the CEO, the company, and its assets form the standpoint of shareholders. This means asserting itself when it must, challenging the CEO when trouble is apparent or suspected, supporting the CEO in difficult situations when it believes that support is warranted, and even altering the make up of the board, the nomination and election process if there is a perceived need for the highest level of change. It is far better for the boards to readily assume this incontrovertible responsibility than to acquiesce to a forceful CEO and learn later that the company has made a serious mistake.

The board always must be informed as to industry and general business trends; the most recent strategic initiatives of suppliers, competitors, and customers; and the strengths and weaknesses of its own organization. The board ensures this through an established process of receiving and reviewing data, with appropriate analyses provided by management, including periodic competitive analyses, industry analyses, and benchmarking studies. The board must be especially wary in its general long term financial outlook and results. Examination of these important financial issues should follow from results and projections provided by management. In particular, board members should monitor compliance with lending agreements (covenants) and the firm's bond rating.

3.5. CORPORATE GOVERNANCE IMPLICATIONS TO INDIAN BUSINESS

Corporate Governance, a phase that not long ago meant little to all, but a handful of scholars and shareholders, has now become a mainstream concern a staple of discussion in Boardrooms, academic fora and policy circles worldwide. Two events may be considered to be responsible for the heightened interest in Corporate Governance.

- Financial Crisis of 1998 in Russia; Asia and Brazil.
- Corporate Governance scandals in 2001-03 in United States, Europe which triggered off some of the largest insolvencies in the Global History.

Both these incidents brought to focus how the behaviour of corporate sectors could affect entire economies and endanger the stability of global financial systems.

In the aftermath, not only has the phrase Corporate Governance become almost a household term, but economists, policy makers and corporate world everywhere began to recognize the lurking macro-economic consequences of weak Corporate Governance systems.

Underlying Reasons for Increased Attention to Corporate Governance

- Increasing size of the firms and the role of financial intermediaries and institutional investors in fund raising. The process of capital mobilization has increasingly become one step removed from the principal owner.

- Allocation of capital has become more complex since investment choices have increased with liberalization of financial and real markets.

- Structural reforms like price deregulation and intensified competition increased Companies' vulnerability to market forces.

- Increased levels of awareness among the stakeholders viz. employees, shareholders, creditors, suppliers, customers etc. which has to lead to greater levels of accountability and responsibility.

Corporations generally work within a framework of Governance which is set by law by regulation by corporations' own constitution, by those who own and fund them and by the expectations of those they serve. This framework differs from country to country since it depends to a large extent on the history and culture of the particular country. Corporate Governance therefore represents the value framework, ethical and moral framework under which business decisions are taken. Countries and corporations are advised to start from where they are and to build on their existing structures or systems instead of borrowing these from foreign countries.

Corporate Governance therefore calls for four basic principles to be adhered to fairness, transparency, accountability and responsibility. These principles are equally relevant to privately run, public sector, state owned or widely held businesses. Private capital, in the recent times has become a prime source of funds for investment which is usually handled by intermediaries who look for equal spread of risk and return. This once again reiterates the significance of corporate governance. The more widely these principles are applied, the better are the chances for equitable and effective allocation of resources. When companies are open and transparent about their plans and purposes, they stand better chances of winning the trust of the investors. Resources flow faster into companies, which inspire trust, through their approach to Corporate Governance. Example, HDFC is one of the highly rated companies on Corporate Governance in India and this has helped to attract tremendous response to its public issue.

Similarly, best employee practices in terms of transparency in recruitment, promotions and adequate focus on career planning has put companies like Infosys and Wipro on the top of ratings related to employee management.

In other words, Corporate Governance is a continuous process. In its broadest sense it is concerned with striking balance between economic and social goals and also between individual and communal aspirations. The basic objective of the Corporate Governance, according to Adrian Cadbury is to align as nearly as possible the interests of individuals of Corporations and of society. What does each of the above get in return?

- Corporations which adopt internationally accepted governance standards are able to attract better investment and achieve their aims faster.

- The nation tends to benefit because these standards help to strengthen their economies and encourage business probability.
- Individuals benefit by a better standard of life and qualitative living through increased access to information at all levels.

Direct Implications of Corporate Governance on the Company

Increased access to external financing at lower cost. This leads to larger investment, higher growth, greater employment creation.

- The lower cost of capital results, in the long-run, in higher firm valuation. This makes investment more attractive to investors leading to higher growth and more employment.
- Good Corporate Governance is associated with a reduced risk of financial lapses which can have large economic and social costs.
- Ensuring better operational performance through better allocation of resources and prudent management helps create more wealth for the company.
- It also helps in forging better relationships with all stakeholders like employees, general public and shareholders.

Governance Catalysts within Companies

A firm's response to Corporate Governance is likely to be determined by a combination of factors like :

- Ownership and management structure (example, family controlled and managed versus family controlled and professionally managed).
- Degree of exposure to international competition for capital, customers and talent.
- Status as an independent company or part of a group.
- Nature of the Sector (Ex. high risk and new versus low risk and well established).
- Managerial Culture (orientation of senior managers towards Corporate Governance).

Corporate Governance in Developing Economies

Though issues relating to Corporate Governance are being hotly debated in the US and Europe over the last two decades, in the developing countries these have not been taken too seriously; In fact there is also a debate going on whether Corporate Governance is as important for developing countries where the national economies are dominated largely by family-owned, state owned or foreign investor owned companies that do not have shares widely traded on local stock markets; and where a number of small non-corporate firms often account for a significant proportion of local employment and output. Though the incidence of scattered financial scams has increased cautiousness among corporates, it has not acquired the seriousness that it ought to. Corporate Governance continues to be seen by many as relatively unimportant in developing countries largely because of paucity of firms with widely traded shares and employment of local people. The poor quality of local systems of Corporate Governance lies at the heart of one of the greatest challenges that most of the developing countries face today and how to transform the local systems of economic and political governance including corporate governance from systems that tend to be highly personalized and thus strongly relationship based, into systems that are more effectively rules based.

Indian Scenario

India is no exception; Corporate Governance did not receive, much attention in the initial phase of the post-1991 reform period. Since India had inherited a rules based system from Britain, there had been only steady improvement in laws relating to rules based Corporate Governance Implications to Indian Business governance. The quality of financial and non- financial disclosures, mandated by law, are stronger in India when compared to a number of developing and developed countries (World Bank, 2000). However, the recent development in the financial sector has led to increased competitiveness and greater focus on Corporate Governance. Public financial institutions are now going to the capital market to meet their capital needs, foreign institutional investors have acquired a significant stake in the Indian financial system. All this has made it necessary for the Indian Corporates to focus on greater transparency and accountability.

In India implementation of Corporate Governance has largely been based on the Kumara Mangalam Birla code, and the CII code for Corporate Governance. However, codes can only provide guidelines. Ultimately effective Corporate Governance depends on the commitment of the people in the organization. The quantity, quality and frequency of financial and managerial responsibilities toward shareholders, the quality of information that management shares with their Boards, and the commitment to run transparent companies maximizes long-term shareholder value. In Asia, India ranks third after Singapore and Hong Kong in terms of effectiveness of Corporate Governance.

Good Corporate Governance offers the promise of several benefits to Indian corporate, sector :

- A large number of foreign portfolio investors would come to India if satisfied with the transparency and disclosure standards adopted here. This is what happened in US in early 1980's and in UK in early 1990's.

- In the near future, foreign pension funds will enter India in a big way. Since these firms hold onto to their stocks for longer time, their fund managers look for better Corporate Governance.

- Indian financial institutions will not continue to support the management irrespective of performance. Instead they will convert their outstanding debt into equity, set up merger and acquisition subsidiaries and sell their shares to more dynamic entrepreneurs.

- Improve chances of Indian Financial Institutions to go for GDRs, External Commercial Borrowings and Private placements.

- Financial press will become stronger with greater access to information.

- Will pave the path for introducing full capital account convertibility for the Indian rupee.

Therefore, it may be said that Corporate Governance means relatively more to India than many other countries since it is on the steep path of economic progress, and Corporate Governance can be an important vehicle to traverse this journey. In fact good governance, not specific to any special sector is the need of the hour. Some of the issues that need to be emphasized on for Indian companies to achieve competence at global levels are :

Priorities for Regulators
- Increased concentration on enforcement rather than mere focus on models.
- Greater levels of convergence of local and international standards.
- Facilitating both financial and marketing discipline.
- Managing cross-border issues.

Priorities for Investors
- Strengthening and implementation of shareholder rights.
- Pre-emption rights, Meetings' attendance.
- Proxy voting.

Priorities for Companies
- Getting the basics right accountability structures must receive more attention.
- Focus must be on fundamental best practices (rather than whistles and bells).

Good corporate governance

- reduces risk
- stimulates performance
- improves access to capital markets
- enhances the marketability of goods and services
- improves leadership
- demonstrates transparency and
- social accountability.

Fourth, recent research has shown that countries with stronger corporate governance protections for minority shareholders also have much larger and more liquid capital markets. Comparisons of countries that base their laws on different legal traditions show that those with weak systems tend to result in most companies being controlled by dominant investors rather than a widely dispersed ownership structure. Hence, for countries that are trying to attract small investors whether domestic or foreign corporate governance matters a great deal in getting the hard currency out of potential investors' mattresses and floorboards. Collectively, these investors may be a significant source of large sums of long-term investment.

What's more, instituting corporate governance practices greatly enhances the public's faith in the integrity of the privatization process, and helps ensure that the country realizes the best return on its investments. This will, in turn, stimulate employment and economic growth.

Where does the need for corporate governance come from?

The original need for corporate governance stems from the separation of ownership and control in publicly held companies. Investors seek to invest their capital in profit-making firms so that they can enjoy these profits in the future. Yet many investors lack the time and expertise necessary to operate a firm and ensure that it provides an investment return. As a result, investors hire individuals with management expertise to run the company on a daily basis to see to it that the firm's activities enhance the company's profitability and long-term performance.

A key drawback of this arrangement is that managers and/or directors are often not owners, and thus they will not bear the brunt in terms of lost investment and lost profits if the company fails to perform. As a result, managers and/or directors may take actions that hurt the value of shareholders' investment. They, for example, may not be as vigilant as they should in overseeing the internal operations of the company, they may take excessive risks when their position is endangered or take insufficient risk when their position is secure. They may bend over backward to resist a takeover that would be in the company's best, long-term interest or entrench themselves by investing in declining industries that they are good in running, but that are not profitable. Managers may also be tempted to steal from the company by raiding the pension fund, to pay inflated transfer prices to affiliated entities, or to engage in insider trading. Such behaviour obviously harms the performance and financial viability of the firm and illustrates the need for corporate governance.

Yet, the need for corporate governance in developing, emerging and transitional economies extends far beyond resolving problems stemming from the separation of ownership and control. Developing and emerging economies are constantly confronted with issues such as the lack of property rights, the abuse of minority shareholders, contract violations, asset stripping and self-dealing.

To make matters worse, these acts often go unpunished. This is because many developing, emerging and transitional economies lack the necessary political and economic institutions in order for democracy and markets to function. Without these institutions, corporate governance measures will have little impact. Hence, in the context of developing, emerging and transitional economies, instituting corporate governance entails establishing democratic, market-based institutions as well as sound guidelines for how companies are run internally.

Each system has corporate governance challenges

Each system, the insider and the outsider, has advantages and disadvantages and thereby its own corporate governance challenges. These are briefly described below, beginning with the insider system.

Insider systems

Companies that are controlled by insiders enjoy certain advantages. Insiders have the power and the incentive to monitor management closely thereby minimizing the potential for mismanagement and fraud. Moreover, because of their significant ownership and control rights, insiders tend to keep their investment in a firm for long periods of time. As a result, insiders tend to support decisions that will enhance a firm's long-term performance as opposed to decisions designed to maximize short-term gains.

However, insider systems predispose a company to certain corporate governance failures. One is that dominant owners and/or vote holders can bully or collude with management to expropriate firm assets at the expense of minority shareholders. This is a significant risk when minority shareholders do not enjoy legal rights. Similarly, when managers are large share and/or vote holders they may use their power to influence board decisions that may directly benefit them at the company's expense. Common examples include managers that persuade boards to authorize exorbitant managerial salaries and benefits or to approve the purchase of over priced inputs from a firm in which the manager owns large shares.

Large share or vote holders have other means of damaging companies up their sleeves. One is to encourage the board to approve the purchase of a rival firm for the sole purpose of extending the company's market share and muting competition. Another is to convince the board to reject takeover offers for fear of losing control over the firm even though a takeover might improve the company's performance. This danger is exacerbated when family owned or insider controlled companies are shielded from market pressures because they are not listed on the stock market.

When banks are large share and/or vote holders in a company to which they lend, the banks may face conflicts of interest that could jeopardize the future of the bank and the company. In this scenario, banks have an obvious interest in the company's continuity, and thus, they may continue to extend credits even though the company is not creditworthy. This can also happen if insiders use their connections to public officials to obtain publicly funded bailouts or to elude bankruptcy proceedings.

In short, insiders who wield their power irresponsibly waste resources and drain company productivity levels; they also foster investor reluctance and illiquid capital markets. Shallow capital markets, in turn, deprive companies of capital and prevent investors from diversifying their risks.

Outsider systems

In contrast to insider systems, owners in outsider systems rely on independent board members to monitor managerial behaviour and keep it in check. Independent board members tend to disclose information openly and equitably, assess managerial performance objectively, and to protect shareholders' rights vigorously. As a result, outsider systems are considered more accountable and less corrupt and tend to foster liquid capital markets.

Despite these advantages, dispersed ownership structures have certain weaknesses. Dispersed owners tend to be interested in short-term profit maximization. Hence, they tend to approve policies and strategies that will yield short-term gains, but that may not necessarily promote long-term company performance. At times, this can lead to conflicts between directors and owners, and to frequent ownership changes because shareholders may divest in the hopes of reaping higher profits elsewhere both of which weaken company stability.

Small scale investors have less financial incentive to vigilantly monitor boardroom decisions and to hold directors accountable. As a result, directors who support unsound decisions may remain on the board when it is in the company's interest that they be removed.

What can be done?

It is evident that both insider and outsider systems have inherent risks. Failure to institute the appropriate mechanisms to reduce these risks jeopardizes the well being of entire economies. Corporate governance systems are designed to minimize these risks and to promote political and economic development. An effective corporate governance system relies on a combination of internal and external controls. Internal controls are arrangements within a corporation that aim to minimize risk by defining the relationships between managers, shareholders, boards of directors, and stakeholders. In order for these measures to have a meaningful effect, they must be buttressed by a variety of extra firm institutions tailored to a country's environment (referred to as external controls).

The previous point cannot be emphasized enough. Many efforts to prevent financial crises or improve firm performance by instituting corporate governance systems are doomed from the outset because they adopt internal controls without external controls and/or they failed to adapt these controls to local realities. More often than not, many policymakers and practitioners from well established market economies take the existence and well functioning of these external controls or institutions for granted and thus, overlook their importance.

This is evidenced by a quote from a recent academic survey :

"Corporate governance deals with the ways in which suppliers of finance to corporations assure themselves of getting a return on their investment. How do the suppliers of finance get managers to return some of the profits to them? How do they make sure that managers do not steal the capital they supply or invest it in bad projects? How do suppliers of finance control managers?"

From this point of view, corporate governance tends to focus on a simple model :

1. Shareholders elect directors who represent them.

2. Directors vote on key matters and adopt the majority decision.

3. Decisions are made in a transparent manner so that shareholders and others can hold directors accountable.

4. The company adopts accounting standards to generate the information necessary for directors, investors and other stakeholders to make decisions.

5. The company's policies and practices adhere to applicable national, state and local laws.

Focusing on these types of internal control processes is quite natural when the subject is corporate governance within the advanced market economies. Point number five assumes that a functioning legal system is in place. Although there are considerable differences between the Anglo-American, German, Japanese, and other systems, they all share the luxury of defining the subject of corporate governance within the context of functioning market systems and highly developed legal institutions.

Yet, many developing and emerging economies lack or are in the process of developing the most basic market institutions. Hence, corporate governance in these contexts involves a much wider range of issues. The Asian economic crisis, the continuing turmoil in Russia, and the recent experience of the Czech economy have combined to push the issue of corporate governance from the sidelines to center stage. In Asia, what began as a financial crisis is now viewed to be a crisis of corporate transparency involving relationships between government and business, between holders of debt and equity, and the legal remedies for bankruptcy and cronyism. Further, as seen in the daily papers, the lack of adequate institutions in Russia have resulted in several highly publicized cases involving allegations of asset stripping, stock register manipulation, and fraud. The Czech Republic privatization program has demonstrated the weakness of the voucher method in the absence of sound corporate governance mechanisms since it resulted in a lack of corporate restructuring and a consequent decline in competitiveness.

What these examples have in common is that they all involve the basic rules of the economy and the relationship between these rules and the way companies are governed. Solving corporate governance problems in developing and emerging economies involves going beyond a narrow view of how owners and managers of capital interrelate.

3.6. POLICIES DIRECTLY AFFECTING CORPORATE GOVERNANCE

A number of directives have already been adopted in the field of company law, independently of the White Paper. Five directives deal with company law issues in the strict sense of the word, four with accounting matters. The first directive sets general rules for the disclosure, obligations and nullity of limited companies. The formation, domestic mergers and divisions of public limited companies are the subject of the second, third and sixth company law directives. The fourth, seventh and eighth directives deal with the annual accounts, the consolidation principles and the auditor's qualifications of limited liability companies, with exemptions for small and medium-sized enterprises (SMEs).

Polices ralting to accounting & auditing are changing and amenable to different culture and systems of the countries. The following are some of the outlines described below :

Harmonisation of the Structure of Companies: Draft Company Law Directive

This draft directive relates to the structure of public limited companies (PLCs) with share capital. It defines the powers and obligations of the board. It provides that PLCs shall be structured in one of two ways: 1) the two-tier system, in which the company is managed by management under the supervision of a supervisory organ; or 2) the one-tier system in which the company is managed and controlled by a single board of directors and in which the actions of the executive members are supervised by the non-executives. The draft contains provisions obliging companies to allow a form of employee participation and control of the board, depending on the sort of company and the number of employees. Other provisions concern the holding of general meetings of shareholders proxy voting and the drawing up and auditing of annual reports.

Harmonization of Accounting Standards

Accounting systems are a reflection of corporate control systems. Whereas the American and English system, dependent as it is on the provision of capital from outsiders, insists on transparency and information, the German system is more closed.

European Guidelines for Good Practices

For the broader areas of protecting shareholder rights and ensuring corporate accountability, there is an urgent need to improve standards within the European Union. We would not propose to tackle these through constraining legislation, but rather by adopting an intermediate approach of self-regulation.

Some EU countries have developed strong shareholder oriented traditions. In Germany, for example, a bank is required to inform depositors of how it plans to vote on various issues on the agenda of the AGM and to ask for instructions on how to exercise the proxy. If no instructions are given, the bank can exercise the power according to its stated intention. In the UK, listed companies are required to state in their annual report whether they comply with the Code of Best Practice of the Cadbury report. Even before Cadbury Code was drafted, many UK companies displayed a strong shareholder orientation, as exemplified by one British company blue-chip company that dispatched some 300,000 copies of its annual report with proxy forms. Such practices, which to our knowledge rarely occur in other member states, should be further developed and extended to the single European market.

It is in the interest of a corporation to develop a shareholder friendly attitude. It might encourage the share price and the stock market capitalisation, and hence the success of the firm. This also applies to a company's responsibilities towards all its stakeholders, including employees, suppliers and clients and the communities in which it operates.

In this sense, we would propose the Guidelines of Good Practice, which establish corporate governance standards that should be observed by all listed corporations. Other (non-listed) companies, especially those that are of special public interest, should attempt to meet the requirements to the extent possible.

3.7. CORPORATE GOVERNANCE: ISSUES AND PROBLEMS

Consider the following issues:

1. How independent does a Board of Directors needs to be to enforce accountability?

In the early days of the corporation, members of the founding family constituted the shareholders, the managers, and the board of directors. As corporations became larger and needed more capital, the number of outside owners increased. As a result, professional management became necessary. While there have always been challenges to governing well, this separation of management from ownership might be said to have given birth to modern corporate governance in the West. It soon became apparent that some mechanism would be necessary to ensure that the new managers acted in the owners' interest.

An independent board of directors is supposed to serve this purpose. If board members are beholden to the CEO for their appointments, they may not have the independence needed to confront him or her when the company is in difficulty. If the CEO is also the chairman of the board as is the case in as many as ninety-five percent of American companies then the CEO is often responsible for evaluating himself. Shareholders and employees are justifiably skeptical about such an arrangement.

Some major shareholders, are now calling for an independent chairman of the board. This person would oversee the evaluation of the CEO and work with him or her to set agendas. However, this approach has its own problems. The demand for an independent chairman makes some shareholders uneasy, because it puts them in the position of telling the company how to do its business. Unless the shareholders are especially diligent in monitoring not only a company's performance but also its circumstances, they are not really in a position to run the company well. In addition, creating an independent chairman means there are two chiefs running the company. That arrangement rarely works smoothly.

And that brings me to one of my major themes in this overview: there is no magical mechanism by which to guarantee accountability. We may simply have to rely upon courageous board members to do their duty and to be willing to take the CEO and other executives to task in a responsible manner.

2. To whom should management be accountable?

Some have argued that management has a primary duty to increase shareholder value. For holders of this view, owners are the key constituency because they provide the capital necessary for a company to exist. While employees have unions to look after their interests and creditors have contractual protections, shareholders must rely upon management to safeguard their investment. Others have contended that management's primary duty is to employees and to other stakeholders. On this view, employees provide skills without which the company could not operate. Defenders of the stakeholder view also note that there is no right to exploit employees, not even if doing so were to increase shareholder value.

Any sharp contrast between the stakeholder and shareholder models is bound to be misleading. Strengthening the protections available for stockholders serve other stakeholders' interests. Conversely, if care is taken to ensure that employees understand the business and feel that their skills and judgment are respected, it is more likely that the business will thrive and that stockholders many of whom are employees will prosper. The editor of Governance magazine rightly notes that governance is "not a zero sum game." As Walter Shipley, the CEO of Chase Manhattan Bank, has observed, "A company can't consistently not pay attention to its employees, to its customers, to the communities in which it operates, and achieve its responsibilities to its shareholders. Maybe for a period of time it can, but eventually it will come up and catch you." Rather than spending so much time debating the theoretical merits of the shareholder vs stakeholder model, corporations and ethicists alike might be better advised to spend more time thinking about what types of practices improve corporate performance. By "improving corporate performance," I do not mean simply enhancing the bottom line. I mean fulfilling all of the relevant expectations which society has for a corporation and which must be acknowledged and honoured if the corporation is to perceived as credible and responsible.

It is worth keeping in mind that theory can take us only so far, especially in our rapidly changing world. Let us suppose we were all to agree that management is most effective in the long run when it seeks to maximize shareholder value. Well, which value do we mean short-term or long-term? In Japan, managers historically have seen themselves as primarily responsible to the class of permanent shareholders – i.e., large Japanese corporations that hold stock in each other. One might argue, by analogy, that American managers have more of a duty to longer-term shareholders than to day-traders. After all, it is the more permanent shareholders who will work with a company during bad times as well as good ones. What happens, though, if permanent shareholders cease to exist, as is now happening in Japan, where the system of crossholding of shares is falling apart? The US tax system does not encourage long-term investing. And with the rise of Internet trading, many investors hold stock for only a few hours or minutes. Is it realistic to expect companies to have a duty to such short-term shareholders?

3. Who should be on the board?

The board should be composed of a range of people with different perspectives and experiences. Having directors from other industries is just as essential as having women and minorities on the board. Nor should one minimize the importance of the latter. I doubt that Dow Corning would have had to file for bankruptcy if it had more women in senior positions.

Board members should be experienced not just in business but also in managing with integrity. Although directors need not have worked in business, perhaps a majority should have done so, especially in the case of for-profit boards. Business has a distinctive mission that those with business experience are more apt to recognize and understand. As the joke goes, the Republicans see the glass as half-empty, the Democrats as half-full. But the businessperson sees a glass that is twice as big as it needs to be. A board should have several members with international experience, given that today's businesses operate in a global context. Needless to say, board members must also be willing to do their duty, to take a longer-term perspective, and to strive to listen well.

4. How should investors go about enforcing accountability?

In the past, shareholders have tried to induce management to change direction by either selling shares or by attempting to develop a long-term relationship and rapport with management. The strategies of exit and of developing a voice each have ethical strengths and weaknesses. Investors with many shares to sell can get management's attention, especially if executives' pay is tied to stock options that decline in value if a sell-off makes the stock plunge. However, this strategy is of little use to smaller investors whose actions have little effect on stock price. Moreover, in this age of frantic day-trading over the Internet, it is not clear that selling stock sends any kind of reasonable or interpretable message to management.

Another concern results from the tendency of some investors to believe that the stock market is destined to go up. Investors allegedly require a much smaller risk premium, because they fear market downturns less than they used to. A dip in prices is seen as an opportunity to buy more stock at a lower price. Plus, if investors are going to hold the stock over the long-term anyway, bad corporate earnings or short-term price volatility does not spook them. In this view, stock prices are rising not because P/F multiples are good, but because the risk premiums that investors once demanded and that used to hold stock prices down are disappearing.

As a result, it might not always be advisable to rely on investors to take a walk when they are unhappy. Should we look, then, to investors to enforce accountability by developing long-term relationships with corporate management? This strategy of renouncing exit in favour of voice appears to be ethically attractive. Sometimes it is too easy to criticize from the sidelines. Making an effort to understand management's position and to work with corporate executives could be said to show more respect. And, in some cases, it may be the only responsible strategy. Would one really want investors simply to bailout of industries central to our national interest (e.g., defense contractors)? Yet, here again, there is a problem. It investors become tooaprohensive feeling (collywobbles) with management, they may lose objectivity. As the Long-Term Capital led go Fund debacle showed only too clearly, Americans are not immune to crony capitalism. Long-term relationships could lead investors to cut executives too much slack. How, too, does one develop a "relationship" with day-traders, who are not interested in any such thing? What do we do if the institutional market begins to look more and more like the frenzied retail market? Furthermore, there are structural biases toward short-term perspectives. For example, an investor gets the right to vote the stock the minute that he or she buys it.

Developing a relationship with management may prove quite difficult for some investors. Private fund managers usually have substantial staffs. Staff members can regularly meet with the senior management of the companies in which these private managers invest. Public funds, by contrast, rely upon funding from the state legislatures. The staffs of these funds are considerably smaller. Even if the public fund wants to have regular and sustained contact with management, it may find it difficult to do so. Acquiring a voice sounds like an ethically sound strategy, but some thought needs to be given to the practical mechanics of communication.

5. Who will watch the watchers?

This issue has tended to be overlooked. Those investors who attempt to hold management accountable should themselves be held accountable. The opinions of some institutional investors are surely more informed than those of others.

The institutional investors who claim to represent preach corporate governance to others, but do not always practice it themselves. As a result, they impute interests to those whom they represent. These imputations may be more or less just and fair. Institutional funds typically push to maximize returns, arguing that their beneficiaries desire the largest possible returns. Yet several studies have shown that individuals will forgo some return in order to support companies that are especially family- or environmentally-friendly. Although institutional investors claim to speak for their constituencies when they demand maximal returns, they might not be fairly representing the desires of those constituencies.

There is no reason to assume that large institutional investors are more ethically pure than senior management. Indeed, a few in the business ethics community are increasingly concerned about the rise of investor capitalism. Aggressive institutional investors may force more layoffs than senior management ever would have initiated on its own. Some public pension funds pride themselves on their sole concern to add value to their beneficiaries' pension contributions and accumulations. They avoid issues of social responsibility, which they derogatorily dismiss.

As Federal Reserve Chairman Alan Greenspan has observed during the recent debate over the privatization of Social Security funds, special vigilance is necessary when investment decisions get bound up with politics. Every decision is political to some extent - even the decision not to pursue political agendas. It is possible or desirable to completely separate political and economic decisions.

6. How will accountability be enforced in a world in which small and large investors embrace index funds?

The government has been rightly concerned about what might be called "back-door socialism." This retirement system has more than three million participants, and over half of the assets in private plans in the United States are in the 100 largest ERISA plans. So this system obviously has a lot of clout. The government was worried about the following soft scenario. Suppose that when President Kennedy got mad at the steel companies, he had directed the public pension fund to buy US Steel and to sell Bethlehem Steel. No one wanted that sort of power to be lodged with the government. Yet, it would not be fair to keep federal retirees out of the equity market. So the indexing requirement was born.

Other public funds have found indexing attractive as well. Indexed funds are cheaper to run than are actively managed funds. Indexing requires fewer people. Since many public funds are short-staffed, indexing is quite attractive.

However, this practice obviously poses certain problems for accountability. First, those who Index have decided to invest for the long haul. Since they are going to be satisfied with returns that mirror the market, they are, in some measure, giving up the right and the duty to trade. Selling individual securities ceases to be a way of enforcing accountability. To put the point slightly differently: there is a real danger that passive investment means passivity of concern.

7. How should a company align the interests of all of its employees with that of the company?

For the last decade or so, stock options have been the preferred tool for effecting such an alignment. This strategy involves well-known difficulties. A rising market rewards all managers, even the under performing ones, Executives with stock options are rewarded when the stock rises, even if the increase is far below that registered by competitors. The options are asymmetrical with shareholders' interests. The executive is rewarded on the upside and pays no penalty on the downside. Moreover, managers find ways to manipulate these options, re-pricing them to keep the options in the money.

And what does one do about lower-level employees? One might require them to invest in company stock or options on that stock. However, such a requirement raises ethical issues. Often such employees have less discretionary cash. Is it fair to ask them to choose between their job and supporting an aging parent? Senior management is in a position to make the decisions affecting the stock, while most junior employees lack such power. Requiring the latter to invest when they have little real control smacks of extortion, and it may put their retirement at risk if money they would have diversified into an index fund now must go into the company's stock. In addition, employees may become risk-averse, afraid to take any action that might reduce the value of their stock or stock options.

Nor is it clear that alignment is either just or desirable. Of course, companies want loyal employees. Management sometimes talks as if it wants employees to work together like the aligned tires of an automobile. But one needs dissenters in organizations, people who refuse to become aligned. Groupthink is a perfect case of total alignment. True loyalty virtuous allegiance presupposes the possibility of dissent. The pursuit of alignment might bring about an increase in shareholder value, but it still might not be in the long-term interest of the corporation, its employees, or its shareholders, if the alignment is achieved through repression.

8. Are we relying too much upon rules to encourage good governance?

Governing well ultimately means acting in a trustworthy fashion. No company will succeed in the long run if it is not trusted by its customers, employees, suppliers, advisors, shareholders, and other important stakeholders.

Traditionally, governments and investors have relied heavily on rules and laws to ensure such trustworthiness. However, mere compliance with rules will never guarantee trustworthiness. Trust exists precisely because we cannot spell out in advance all of the conditions that must exist in order for us to believe that those whom we trust wish us well. Trust is a virtue of uncertainty. In fact, we can think of many cases where a party becomes more trustworthy by breaking a rule. Rules will never substitute for executive discretion.

The Business Roundtable has observed correctly that no particular policy or best practice is "a substitute for, and does not itself assure, good governance." It is not desirable that one should give up completely on refining processes and establishing checks and balances. I will say more about that point shortly. A checklist or rule-based approach cannot guarantee integrity, and it runs the risk of making people complacent. Shareholders may think a board is acting responsibly if it spends a great deal of energy adopting best practices. But such a board may still fail to improve the governance of the corporation. Indeed, energy spent in this direction may prevent the board from offering real help to management. The structures and forms may look impressive, but the substance may be utterly lacking. Shareholders might learn more if directors who are standing for election had to meet with the owners and field questions.

Relying on a checklist approach poses a second danger. When boards accept the claim that they must have certain structures in place in order to be effective, it becomes tempting to make such forms a legal requirement. As these forms pass into law, boards begin to focus not on acting ethically but on instituting mechanisms for avoiding legal liability. Boardroom deliberations begin to resemble an elaborate charade, with members mouthing politically correct platitudes. Fear of lawsuits is every bit as corrosive of sound moral judgment as is vice.

Third, prudence is always needed in order to govern well. For example, we might pass a rule requiring board members to gather certain data. Whether this data is of any value, though, will depend on its quality. Did the board rely exclusively upon statistics and information provided by management, or did it seek out data from other sources e.g., third-party evaluations of the company? How reliable and comprehensive were these alternative sources? Unless the board members think about these sorts of issues, rules will be of little value.

What matters in the final analysis is corporate performance. In one fascinating study, a researcher tallied the number of corporate governance mechanisms that corporations had in place, then plotted these measures against actual corporate performance.

Surprisingly they discovered an inverse relationship between the mechanisms and performance. Form certainly matters, but not to the exclusion of performance.

9. What does it mean to govern well when the target is a moving one?

The ethical standards that we apply to companies are continually shifting. The first large corporation the Hast India Company did not receive much scrutiny from anyone. Many have argued that it was a government unto itself. Today's companies, by contrast, receive almost continuous scrutiny. During the 1980's, companies investing in South Africa were asked to adopt and comply with the Sullivan principles. Today, the international community is busy forging a new set of rules concerning bribery. Consumers routinely boycott companies who are perceived to be exploiting child labourers or torturing animals. The Internet has made it easier than ever to share this information.

The target keeps moving not because we are all hypocrites, but because the circumstances that we confront are changing, and because we are continually learning from and through our actions. Perhaps we need to commit to kaizen, the process of continual improvement. Managers should experiment with new structures. They might establish lead directors, constitute audit committees composed entirely of outside directors, or index the stock options used to compensate executives. I am all for innovation, but we should take care not to become irrationally exuberant. There is no magical way to enhance management's ability to govern well or to ensure a responsible board. Corporate governance is an art, not a science. We should hope for much but expect to be disappointed. Every innovation has its downside as well as its upside.

If a board member is a good friend of the CEO or an officer of a major customer, that member might simply say what the CEO wants to hear. On the other hand, it is simple-minded to assume that a friend or customer must always be ineffective as a board member. In some cases, such a party may actually have more leverage than an outsider would. The CEO may take criticism from a trusted friend and accept advice that he or she would never take from an outsider.

Or consider board size. When a board becomes too small, it runs the risk of tunnel vision. It will not have the range of backgrounds, experiences, and perspectives that it needs in order to function prudently. On the other hand, the board can become too large, which can lead to unfocussed discussions. As W Taylor Reeveley has observed, too many members "can shelter poor attendance, lack of preparation, avoidance of difficult issues, and failing to do anything significant amid the heaving mass of the board." So should we pick a magical number say, twelve to serve on the board? To do so would probably be a mistake because, again, some corporations may need larger boards. It has been argued that not-for-profit boards generally need to be larger than for-profit boards for two reasons. First, non-profit institutions typically serve a wider range of constituencies than do other corporations. A university should include educational, legal, business, investment, and governmental perspectives. This diversity needs to be reflected on the board of the institution. Second, non-profits need to include major donors on their board. Such reflections suggest that Aristotle is right good judgment always must look to the particular.

Change is inevitable, but there is something to be said for caution when tampering with the status quo. Sometimes a known evil may be better than a possible host of unknown evils. During the 1980's, Congress considered changing the definition of an "insider", as that term is used in securities law. After much debate, it decided to let the current definition stand. Although that definition posed many problems, those problems were well-known and well-understood. Any new definition would muddy the waters and create a whole new batch of loopholes. Therefore, it was better to live with the status quo. As I said at the beginning, experimentation needs to be prudent and selective. And we must be willing to learn from our mistakes.

Corporate Governance: The Key Issues

The primary goal of a corporation is to maximize shareholders' wealth in a legal and ethical manner.

There are three players involved in this game first, the shareholders who have trusted the company and, consequently, invested their capital either through an Initial Public Offering or through the secondary market. Secondly, the management that runs the company, which is accountable to the directors; and thirdly, the directors, who, in turn, are answerable only to the shareholders. They are not responsible to the so-called "owner influences" or to the "management". Together, these three are expected to govern the corporation ethically and legally.

Furthermore, there are three key assumptions that hold true in all well-managed corporations across the globe. First and foremost, the ability of a corporation to sell its products and/or services and earn a market rate of return on capital is a good indicator of the efficiency and effectiveness in the market place. Secondly, the sole interest of the owners (the body that comprises of each and every shareholder of the company) is in optimizing the return on their investment. Thirdly, the most effective discipline for under-performing management is the possibility of the owners exiting the corporation and the resultant opportunity for alternate management and ownership. This is an immensely important factor.

If we want to adhere to the highest principles of corporate governance, we must treat these three assumptions as axioms.

Thus, in essence, corporate governance translates into conducting the affairs of a company in a manner that ensures fairness to customers, employees, investors, vendors, the government, and to the society as a whole.

However, often, we come across examples of corporate behavior that is not in tune with this expectation. Consider the following :

- In 1985, a family-owned public corporation located in Northern India; in which the family owned 9 percent of the equity, refused to register a 14 percent stake; bought by a Non-Resident Indian (NRI). They even had the backing of the government of the day, which had changed the law to say that no more than 5 percent of the shares could be held by an NRI.

- A highly respected multinational corporation situated in Eastern India has allegedly fixed its books to show high exports, high profits, etc. But, of course, this is on subsidy why the term is considered alleged.

- As many as 250 of the 390 companies that got listed between 1991 and 1996 have completely disappeared. It is almost impossible to discover what happened to these corporations, as there are no annual ports or any other form of published information about them.

Because of lack of due diligence in its systems, the companies did not discharge their its obligations to society in general.

There is a clear pattern that one can discern in all this. An individual or a group of individuals in these corporations have taken decisions that benefit only a small selected group, to the detriment of the large body of shareholders. This is against what is universally known as the Principles of Corporate Governance.

Moreover, definition of corporate governance transcends the narrow, strict context of fairness to investors. We have to ensure that there is fairness and transparency in transactions with all the stakeholders, which include customers, employees, investors, vendors, the government, and the society at large. Corporate governance is about building confidence and increasing the trust of the stakeholders in the way the company manages its affairs. It is about bringing efficiency and effectiveness through the use of fair and transparent means.

One might question why there are such rare occasions of high standards of corporate governance in this country. I have tried to answer that oft-asked question through the reasons enumerated as follows :

The first reason is the feudal mind-set that exists in India. Unfortunately in this country, thanks to this feudal mind-set that has existed from times immemorial, the progeny has controlled the throne. From the days of the proverbial throne, we have come to the days of corporate throne. There is a tacit assumption that the son or the daughter of a "owner" has the birthright to assume positions such as Director of the company as well as Chairman and Managing Director etc. The competence of the scion in question to manage these high positions is a less important criterion for selection.

The second reason is the manifold restrictions set by the Government of India. Because we have a mind so set in the bureaucracy and in the political leadership in this country, it is necessary to frame rules to deter and punish even 5 percent of the people from committing gross violation of principles. Consequently, we have a plethora of rules, which makes doing anything worthwhile extremely difficult; and, thus, people start looking for short cuts.

The third reason, is the lack of concern for society. Aristotle said, "it is not the same thing to be a good man and a good citizen." In India, today, a good man is one who is nice to his friends and who is very nice to his relatives. A pious man who builds a temple is respected by society. We need to change this definition to say that unless one is nice to the large body of the unseen people, the large body that makes up this society, he/she is not a good person. In other words, we have to equate being a good person to being a good citizen. It is unlikely that we will have much respect towards the society or follow laws of the society unless this change in the mind-set comes in.

The fourth reason is the sense of insecurity that prevails amongst the very people who are supposed to inspire a house of confidence about the company among the stakeholder population.

The last reason is, of course, greed and ego. Most of us behave as if there is going to be no tomorrow as if the profits, the company, and the markets would vanish. In order to show that we are the greatest and the smartest managers, we tend to take short cuts and violate laws; and, therefore, we end up violating fundamental principles. My view is that if a management has a mind-set that says, "it is better to have a larger share of a smaller diminishing pie, than a small share of a larger pie," it is very likely that this management will almost certainly violate the fundamental principles of corporate governance. This is because there is a sense of insecurity and a lack of confidence in that company.

In today's context, it is important to realize that corporate governance is no longer a luxury but a necessity. There are various reasons as to why I think so:

The second reason pertains to 'employees'. Thanks to liberalization and global competition, our people have global opportunities and consequently global aspirations.

The next reason arises from the needs of the customers. Because of tremendous competition in the market place, customers have choices.

Fourthly, in these days of mega-collaboration, unless there is transparency and fairness, our vendor partner will not work with our corporations.

The next reason is the government. The government has shown tremendous faith in us and reduced taxes and duties.

The next issue is the constituency of society. The gap between the haves and the have-nots is slowly increasing.

What do the investors want from the owners? Investors want them to be trustworthy and to utilize the money for the purpose for which it is allocated.

The role of the government in the constitution of the Board is another important factor in determining corporate governance. The first requirement is the need to select external directors based on their specialization. Second, we have to define the structure of the Board to ensure its independence. This would involve the following steps :

- Appointing a Nominations Committee that consists of only external directors for bringing in other directors and promoting people within the company to the Board. The members of the Nominations Committee as well as the shareholders would also be able to remove the directors, if required. As long as the committee consists of external directors, they would be in a position to bring high quality directors.

- Appointing an Audit Committee which consists of purely external directors. The Committee must have full authority to appoint both external and internal auditors. The committee should have the powers to call any of the internal directors and seek clarifications. Also, the management or internal directors should not be present when the committee meets external directors.

In order to get the best out of external directors, they should be compensated adequately. Also, they must have stock options. The internal directors and, perhaps, members of the management who are not on the Board must have the freedom to define performance indicators for the external directors. They must allocate clear responsibility of a certain value addition for each of the external directors and then define performance indicators in consultation with them. Thus, we are likely to get high quality, independent, proactive, incentivized, and fully interested external board that is likely to ensure a high level of corporate governance. Of course, the government has to define the quality of information that the management has to share with the Board it has to define the time, frequency, and the quality of the information to be shared with the shareholders, and also the level of transparency that is expected in all financial and non-financial reporting.

To conclude, if some of these views are implemented in full faith, this country will go a long way in furthering the cause of corporate governance, improving the lot of small shareholders, enhancing the trust of the shareholders, and attracting very high quality investment from across the globe.

Another key benefit of devoting time and effort to an appropriate corporate governance structure is that effective governance will encourage the use of board expertise in ways that will maximise each director's contribution. An environment in which both the board and management set the corporate strategic direction and that allows the board to monitor performance over time without impeding the management of day-to-day operations should also limit the possibility of failure and lead to increased shareholder satisfaction.

An appropriate corporate governance framework that looks to involve the board in the strategic planning process, delineates clear board and management power sharing arrangements, establishes processes for the timely reporting and review of information, not forgetting to allow effective and responsive actions to be made thereon, will lead to improved understanding of the respective roles of the board and management. This will in turn lead to the introduction of appropriate governance processes and procedures under which management is free to manage, while the board is free to monitor, enquire and counsel.

Composition of the Board (Issues)

The following are key issues that may be considered to ensure that the board is well equipped to discharge its duties:

- The ratio of executive to non-executive directors including whether or not the chairman should be a non-executive director;
- Desired qualifications and experience of directors and prospective directors;
- Format and timing of board meetings, agenda items including availability of discussion material to enable informed discussion by directors;
- Procedures for dealing with conflicts of interest, dissent and resignation.

Board and Management Power Sharing

It is important that the board :

- Appoints a suitably qualified chief executive officer and executive team; and
- Considers the remuneration appropriate for the appointments.

Shareholder and Regulatory Expectations, Management of Business Risk

- The board is responsible for identifying any significant business risks and ensuring that processes are in place to adequately manage those risks. Considerations in this area might be :

- A detailed analysis of stakeholder needs and strategies to meet these needs;
- Awareness by directors of their legal duties and responsibilities;
- Processes required to identify business risks as well as potential opportunities, which may involve the establishment of committees, such as Finance committee and treasury committee; and
- Processes required to determine shareholder expectations and other obligations, for example the establishment of an environmental issue committee and an occupational health and safety committee.

Aligning Expectations and Risks with Objectives of Management

After determining shareholders' expectations and management risks, the board needs to ensure that its findings are aligned with management's activities. This may involve :

- The development of a strategic plan, approved by the board;
- Production of operating plans and budgets, which align with the strategic plan:
- Establishment and maintenance of internal control systems, including financial, operational and compliance controls and risk management; and
- Procedures available for directors to obtain independent professional advice at the company's expense.

Directors: The Four Most Common Mistakes Made

Duty of care and the duty of loyalty are the two components on which the director's duty towards corporate shareholders is analyzed. Addressing duty of loyalty issues is not an everyday event to the directors. Directors take duty of care very seriously, but often fail to follow guidelines, which makes them to commit common mistakes. The author primarily looks at the following four most common mistakes made by directors: Failure to document reasons for decisions; Failing to "qualify" the experts upon which the director relies; Failing to take the time for preparation and reflection that someone less sophisticated would take; and Buying into the fallacy that, if the board can't do everything, it shouldn't do anything. Addressing these common mistakes is very important for corporation's interest.

The majority of directors take their duty of care quite seriously. They attend all meetings, ask questions and call in experts when acting in areas outside of their expertise. However, despite employing these safeguards, directors continue to lake unnecessary chances by failing to follow some simple guidelines. The following are the most common mistakes made by directors :

1. Failure to document reasons for decisions: In most corporations, the vast majority of decisions are made based upon a reasonable recommendation of management. When directors' decisions are made, each director's views are memorialized in a recorded vote. Officers should be held to no lesser standard. Board books should contain management's recommendation and the basic analytical data it relied on when making its recommendation. Board books should always be drafted for two audiences: the board and the potential plaintiffs. If a voting outcome is uncertain, the pros and cons of alternatives should be summarized so that any decision made by the board is supported by some analysis.

2. Failing to "qualify" the experts upon which the director relies: All directors are entitled to rely on experts but that reliance must be reasonable. To ensure reasonable reliance, directors must have a basis for believing that matters treated by the expert are within his or her professional or expert competence. In an age of web pages and desktop published brochures, there is no excuse for failing to provide each director with a file demonstrating the expertise of the investment banker, lawyer, compensation consultant or accountant being retained and their familiarity with the unique aspects of the project which they are addressing.

3. Failing to take the time for preparation and reflection that someone less sophisticated would take: Both common sense and court decisions caution that due care is not exercised in an atmosphere characterized by a massive information download followed by an immediate decision. In spite of this, directors are consistently asked to trust experts who spend only a few days evaluating an issue central to the whole future of the corporation.

If a decision is challenged based on failure to exercise due care, the court or jury making the determination will probably not have the same ability to assemble, analyze and distribute data as does the board. Such a decision-maker might perceive speed not as efficiency (which it often is) but as carelessness. Is it any wonder then, that the short time involved in preparing for the deliberative process is prominently mentioned by courts finding a breach of the duty of care?

4. Buying into the fallacy that, if the board can't do everything, it shouldn't do anything: Expert opinions are costly. Boards predisposed to seeking an expert opinion are often dissuaded by the costs. Often, this is an understandable exercise of stewardship over corporate funds. However, a decision that the usual kind of protective opinion is not cost-justified should not end the board's inquiry.

Effective Boards: Making the Dynamics Work

The days of rubber-stamping management's decisions, collecting fees and adjourning for lunch are over at least for most boards. Some are models of effective governance, doing the job as intended. Unlike the failures, which attract widespread attention, good boards function quietly and efficiently, serving shareholder-constituents in pursuit of growing long-term shareholder value.

Why are some boards highly effective at carrying out their responsibilities and others not? Why are some stuck in the past? It's difficult if not impossible to do the right thing if a board is dysfunctional. The best boards have adopted a culture and operating style enabling them and management to operate effectively together. Some have not.

This article is about the good and the bad of board functionality. It is based on the board dynamics chapter in the recently released study, Corporate Governance and the Board What Works Best, compiled by PricewaterhouseCoopers and published by 'The Institute of Internal Auditors Research Foundation'. The report identifies the board's responsibilities and how best to carry them out, avoiding pitfalls along the way.

The Bad

Boards can be susceptible to a wide range of dysfunctional practices :

- The chief executive sees the board as a necessary evil, a burden on management;
- Management keeps directors in the dark, providing as little information as possible;
- The chief executive sets rigid agendas and runs tightly regimented, formal meetings;
- Directors violate norms of boardroom debate by aggressively challenging corporate leadership, running the risk of finding themselves isolated and possibly replaced;
- Virtually all information comes from management, making meaningful assessment of strategic direction and operating plans difficult at best.

Ineffective boards often are the result of failures of board culture. They operate on the basis of unwritten and unspoken rules for a new director wanting to make things better, inertia and resistance can be an incredible challenge.

The Good

What do effective boards do? Simply put, the opposite of the above. And changing a "bad" board culture to a "good" one is not easy. Focusing on the following helps.

Board Composition Independence

A board that's beholden to management cannot be effective. This doesn't mean a board hand-picked by a CEO is doomed to failure. But the board cannot at the same time be subservient to the chief executive and do what needs to be done on behalf of shareholders.

Boards in the United States, most of which combine the role of chair and chief executive, take any of a number of actions to ensure effective functioning :

Independent majority Maintaining a majority of directors with no ties to the company or its leader can serve as a healthy counterbalance to the CEO/chairperson;

Robust nominating committee Bringing in the right board talent is critical and a nominating committee comprised of strong, independent directors can make the difference;

Strong corporate governance committee This committee of independent directors can provide leadership to the board on critical governance activities;

Lead director One independent director is designated as the conduit bringing ideas, feedback and direction to the CEO/chairperson.

These are not mutually exclusive, with some boards employing all these measures. But generally the first three are the most widely accepted. Most boards don't support the concept of a lead director, on the basis that it can inappropriately dilute the CEO's power, lead to compromise rather than decisiveness, and result in two public spokespersons and related confusion.

The objectives of a lead director can be achieved in a way that's more widely embraced. Frequently one independent director emerges naturally who is looked to as a non-executive leader. Omitting the formal designation seems to avoid the negativity. Another tactic boards find useful is to hold executive sessions of outside directors only.

Characteristics

We all know the characteristics that directors should have including integrity, ability to think strategically, intuition, vision, capacity to make effective decisions, good interactive skills and ability to handle conflict. What it comes down to is having the appropriate skill sets and strength of character to do the right thing for the company and its shareholders, even if it's not always popular.

Board Size

For years, many boards were too large. Large boards can limit discussion of key issues. One long-time director says, "Once you get beyond 10 or 12 people on a board, you don't discuss you wait for your turn to speak." This sentiment is widely shared, and size has come down, with boards of seven to twelve members more common (the average size of public Fortune 1000 companies is now 11, according to the Directorship DataBank®. But there have been some signs of reversal. Boards are looking for expertise in technology, e-business, marketing and global business, adding members where needed. Of course, one size does not fit all. The right size brings the requisite knowledge, abilities and skills to the table in a group small enough to act cohesively.

How a Board Functions Board Meetings

Many directors have indicated disappointment with how board meetings are conducted. They say meetings are too rigid and allow little time for discussion of important issues, with insufficient advance material. They say there's too much reporting and too little discussion. The agenda has the wrong topics, and their fellow directors don't prepare sufficiently.

Many board chairpersons structure and run effective board meetings. This is not rocket science. They avoid the "bad" practices identified above and embrace the "good," and the board works well. Much has been written on conducting effective meetings, which need not be repeated here.

While meeting effectiveness is important to board effectiveness, activity occurring outside the boardroom is as, if not more, critical. Relationships cannot develop solely in a boardroom. Many board leaders recognize this and provide for director interaction outside meetings. Frequent exchanges by phone among directors are seen as a plus. Trust needs to develop among directors and with management, and is vital to effective group dynamics.

Charter

Best practice calls for a written charter, outlining responsibilities, structure, membership criteria and process.

Specific board requirements may be set forth in legislation, stock exchange listing requirements, court rulings and corporate bylaws. Board charters should reflect these rules. The company's legal counsel typically ensures that legal requirements are met in developing the charter. Charters typically address board responsibilities, composition, director selection, process, leadership, director compensation, meetings, procedures, board performance, committees and relationships.

To he useful, the charter needs to be referenced periodically to measure conformance. Even more important, directors should recognize that while the charter should be a periodic reference point, it should not serve as a day-to-day driver of activity or as a constraint to doing what needs to be done.

Committees

Most boards use committees to fulfill their mandate. But astute directors know that the presence of an effective committee doesn't allow the board to abdicate its ultimate responsibility. Effective boards monitor committee activities, guarding against a silo mentality by ensuring sharing of information, cross-committee memberships and bringing major decisions to the full board. They also don't allow issues to fall between the cracks.

Evaluating the Board

Recent surveys show that fewer than 20 percent of boards evaluate themselves as a board or the performance of individual directors. Why don't most boards evaluate themselves? Directors, by definition, are highly accomplished, successful individuals, and might not have been subjected to formal evaluations in years.

Despite an inherent distaste, many boards employing a self assessment process found it has made individual directors and the board much more effective. Directors have an accurate perspective on their peers' performance and contribution. If done well, the process is constructive and motivates positive change.

Best practice is of each director assessing his/her own performance, and that of other directors and the board as a whole. Feedback should be compiled confidentially and provided only to the individual director and a lead-type director or governance committee. Appropriate's's outside consultants to work with the board to determine rating categories, develop evaluation forms, receive and consolidate results and communicate individual and group feedback.

Orienting New Members

Orientation of new directors is vital. Many directors believe it takes much too long for new board members to contribute commonly mentioning three years. Most say new director orientation programs are not effective.

Orientation needs to provide better briefing on the company and its businesses and industries, organization, people, strategies, key issues and risks. Some boards provide orientation manuals, arrange for visits to operating sites or provide formal support, including board mentors.

Commitment

Directors see a clear link between effective boards and directors who make a significant commitment of time and energy. Attendance at every board and relevant committee meeting is a start. But it's not nearly enough. Seasoned directors say :

- Directors can't be disengaged between meetings. They need to spend time at the company, with customers and suppliers to get to know the business well enough to add value.

- Meeting four times a year is simply not enough to deal with the complexities of business today.

Also clear is that director compensation needs to be revisited as demands on time increase. Many directors and thought leaders argue that directors need to be paid more, with some companies having begun to make adjustments.

Directors, more than any other group, have an extraordinary opportunity to enhance shareholder value through their diligent and proactive oversight. Directors can make a difference, but like any worthwhile endeavour doing so isn't easy. Unless board members work well together, the company won't gain the requisite benefits. In today's world, the right board membership, with the right dynamics, can make all the difference.

3.8. CORPORATE GOVERNANCE IN INDIA

This chapter helps you to :

- Know the emergence of corporate governance in India.
- Understand the relevance of the macroeconomic policies of the state regarding corporate governance.
- Know the new economic policy and the relevance of the 73rd and 74th constitutional amendments.
- Understand the contributions of CII and ASSOCHAM/other bodies.
- Understand the strategic issues involved.

The Indian Scene

Bombay Stock Exchange (BSE) is over 150 years old. It is one of the very old stock exchanges in the world. India has the legal framework which governs the companies with appropriate administrative and regulatory structures. Around 600,000 companies are registered under the Companies Act. 14% of these companies are public limited and are listed on the exchange. It is estimated that more than 13,500 companies are listed and out of these, around 10,000 are listed in the National Stock Exchange (NSE) and Bombay Stock Exchange (BSE). Paid-up capital of the public limited companies is estimated to be more than 2100 billion rupees. Market capitalization of the companies listed on BSE ranges from 23% on India's GDP to more than 60% at the peak of the bull run.

By and large, private sector industries have fared well in catching up with internationalization of the country's corporate bodies in one or the other forms, viz., GDR or ADR issues or private venture capital funds or international equity listing on NASDAQ New York Stock Exchange or on strategic alliances of overseas companies. Being exposed to the international business ambience, private organizations backed by regulatory bodies have taken initiatives to define codes on corporate governance. The Confederation of Indian Industries (CII) has done pioneering work in this field. It formulated its code on governance during the year 1998. Likewise, we have the Kumaramangalam Birla Committee Report 2000 which also focused on the need for code on corporate governance. Further, there are committees under the Ministry of Company Affairs and the Securities and Exchange Board of India (SEBI) which also brought out reports to supplement the emerging need for public policy on corporate governance. Then, the Standing Conference on Public Enterprises (SCOPE) came out with guidelines on the subject for public enterprises. For the banking sector, the consultative group of RBI (Reserve Bank of India) gave recommendations for strengthening the boards of banks, in particular, in related areas of selection of directors, their training and contract. SEBI also amended Section 49 (dealing with listing agreement with companies) to include conditions which will reinforce governance in corporate bodies. Besides these developments, the self-regulatory section of the Institute of Chartered Accountants of India and the Company Secretaries took initiatives to raise the professional standards of the corporate bodies at par with the changing international levels. Besides, the Indian Company Law has been amended and is now under revision for meeting the exigencies of business and industry in the contemporary competitive globalized market ambience, where effective corporate governance is inevitable for success. All these initiatives establish the emergent desire as well as the need at policy making and institutional levels to make improvements in the structure and the system for effective governance in corporate organizations. In addition, corporate bodies having international presence have voluntarily taken initiatives not only to meet the basics in good governance but have done excellently, and even exceeded the international expectations on disclosure and accounting reporting.

It is observed that initiatives in the field of training and development as well as capacity building for the board level appointments have been far from satisfactory. Whereas upon ot6her area which pertains to observing the provisions of the Companies Law and is related to general management. These are the misconceptions on corporate governance. In the wake of the emerging cut throat competition, characterized by cost cutting and price wars, it is needless to point out that effective corporate governance is an absolute necessity. Corporate governance has assumed significance the world over as a subject of study, which mainly covers the following topics :

- Economic theory
- Financial aspects
- Dynamics of human behaviour
- Communication
- Legal provisions

Training at the top level has always been a casualty because those who reach the top position have the misconception that they have acquired all the knowledge by virtue of the long and rich experience. By the time they move to the top position, they believe that they do not need any more inputs from training. Unfortunately, they are not able to appreciate that training is needed at each level of responsibility demanding new skills. It is very difficult to motivate the top echelon of the management to go on training.

Indian Institute of Management, Bangalore and the Management Development Institute, Gurgaon, organize courses for the top management on corporate governance. Some professional bodies also hold sessions and discuss burning topics of corporate governance in their annual workshops/meets and conferences. In 2003, the FICCI organized a one-day program on corporate governance in Grand Hotel, Delhi, where, about 1500 delegates participated in discussion sessions.

The Government of India has set a National Foundation for Corporate Governance. The Foundation has drawn up plan to organize training for trainers' course. Participants will be drawn from select groups of trainers from prestigious institutes in the country. The body will also organize training for the directors and initiate research on corporate governance and corporate governance advocacy.

In India, management consultants are aware that integrity has collapsed in corporates and there is an urgent need to streamline corporate management in the wake of large-scale scams. The globalization, among other requirements, calls for a clean business environment for investments. Countries which do not meet the required standards of clean environment will lose the opportunities for capital investments. In this background, Indian corporates are looking for potential candidates for positions at the board level.

But, the demand for trained and competent personnel is not being met by the poor supply. Large PSUs in the country are finding it hard to identify the right caliber persons for meeting SEBI requirement to fill up 50% of board appointments as independent directors.

It is pertinent to note that as per its rules, the Stock Exchange of Malaysia, the regulating authority, has made training of directors mandatory. Certification of training and re-training of the directors is absolutely necessary. In fact, certification of directors training schemes has been started in more than thirty countries.

Corporate governance has assumed further significance in the wake of the failures of companies in the UK, namely, Barings Bank, BCCI, and Maxwell Group, etc. These incidents were followed by the popularly known Asian Crisis in 1997 which witnessed total collapse of the local currencies and total mess up of the corporate bodies. The entire region was under fiscal stress. But, India was fortunate not to have been adversely affected by the episode, for reasons of its stable legal system and systematic sound administrative/political system and fiscal management.

The fact that corporate governance gained further importance is also attributed to the regulatory bodies and professional institutions. The International Monetary Fund (IMF) brought out its principles on transparency and the OECD (Organization for Economic Cooperation and Development) has set forth its codes on governance, which are taken as global standards. The group of thirty member countries has adopted the revised version of the code in April 2004.

Private companies like Wipro, ICICI Bank, Infosys and the PSUs like ONGC, are making efforts to internationalize their businesses. These organizations will have to observe the codes of corporate governance in the respective countries. Moreover, since we live in a global village, India is a party to initiate efforts for promotion of good governance. Good governance makes good business. Some Indian companies, like ICICI Bank, realising the need for corporate governance have voluntarily defined their code on governance. A major effort was initiated by SCOPE (Standing Conference on Public Enterprises) in 1997, which had organized a top level meeting of the CEOs and directors. Representatives from Ministries and the CAG (Comptroller and Auditor General of India) and Central Vigilance Commission also participated in the session. SCOPE prepared the report and recommended reforms in corporate governance in public enterprises (PEs). The report also underlined the need for interface between PEs and the government. At this point of time, another significant contribution was made by the Administrative Staff College of India (ASCI), Hyderabad. ASCI had issued a special journal on corporate governance which was edited for the ASCI Journal of Management. The special issue highlighted the dilemmas and dynamics of corporate governance as well as the contradictions in Indian and international perspectives on the subject.

Y.R.K. Reddy, a very strong enthusiast on corporate governance, with his worldwide experience and knowledge of the subject set out to give birth to an institute on corporate governance, called the Academy of Corporate Governance, in November, 2001 at Hyderabad. The academy has published its monthly electronic journal on corporate Government.

- creation of a movement in India on corporate governance, to share, generate, promote knowledge and competency in corporate governance and improve transparency, reinforce the concepts of accountability and competitive performance, which will bring socio-economic benefits to the society.

The New Economic Policy

July, 1991 is memorable with the birth of the new economic policy in India. Issues related to banking and capital market reforms came into focus. The Narsimham Committee Report on financial system (1992), has recommended large measures for improvement in the banking and finance sectors. Some of the major recommendations include measures of provisioning, improved project monitoring, exposure ceiling, improved appraisal, etc. In the early reform period, the emphasis was on recapitalization of the banks. In this regard, a beginning was made by allowing them to raise funds.

Banks and financial institutions started working as passive observers on the companies and their defaults on loans. Divestment at this time was not evident and significant legal reforms and policy for improvement in the quality of corporate governance, were still awaited.

It has been observed that due to the imperfect financial market coupled with weak regulatory and institutional capacity in the country, a number of big companies in corporate sector in India which were poor performers were open to the threat of takeovers. Of course, the threat of takeovers existed but it was thwarted in time by many roadblocks like the practice of the capital market and shortcomings in the law and regulatory systems. Moreover, bulk of the equity was held with the public finance system.

A study of the above subject has revealed the need for reclassification of the existing equity portfolios held by the financial institutions so that they can plan to dispose of a part of the equity in order to get improved quality of portfolio, and also to use a part of the proceeds for the creation of takeover funds. Meanwhile, the Companies Act 1956 is under review which should make rules for restructuring the corporate bodies.

It is expected that the task forces and industrial bodies appointed by the government will give recommendations for reforms which are overdue. While a series of reforms in the capital market rules have taken place, ample scope exists for in-depth reforms in the financial sector. Policy on liberalization is already in vogue for promotion of the foreign institutional investments (FIIs) and the FDIs (foreign direct investments). The new rules and regulations on takeovers need to be scrutinized.

Contribution of Confederation of Indian Industries (CII)

The need for effective corporate governance cannot be overemphasized for overall corporate effectiveness. While the debate on the subject in the country at different institutional levels/forums continues, deregulation and the need for higher quality of internal governance in organizations have been underlined in the backdrop of the focus given to faster industrial growth.

Michael Gillibrand opines that corporate governance is not a panacea... which will lead to development; but it can contribute to improvement in efficiency and complement the macroeconomic policies (of the state). As such, corporate governance will not succeed in the absence of corresponding macroeconomic as well as public governance reforms, which should nurture policy ambience for reform at macroeconomic level.

As we are aware that the government of India has undertaken programs under the five-year plans for structural adjustments/reforms (under constitutional amendments, the 73rd and 74th) in all spheres of social life. The main aim behind the programs is to effect improvement in macroeconomic governance. Corporate governance has become a significant aspect in the social and economic life of the society. Let us briefly explain the union government's initiatives on structural reforms/adjustments.

India's 8th plan (1992-97) advocated a strategic departure from the erstwhile centralized planning to a more market oriented approach. Accordingly, financing of metropolitan development has to be from local resources and self-financing in nature. The plan also encouraged public agencies to mobilize financial resources for construction, operation and maintenance of drinking water projects, improve revenue recovery mechanism and enforcement machinery, undertake computerization of operations, involve communities in activities, etc.

The new economic policy (1991) together with the 8th plan and the 74th amendment to the Constitution of India sought to change the direction in terms of be reclassified and will incorporate modifications in classifying the companies. The modifications will include raising the paid-up capital for constitution of a company and financial as well as non-financial disclosures, mergers, corporate restructuring, monitoring, accounts and auditing, enforcement and winding up of the business. The new rules will underline the usage of funds. Further, it will also cover in its ambit, small and medium size companies which do not attract the public although such companies will enjoy some exemptions. The new rules will be an improvement upon the existing rules and procedures under the Companies Act.

What Needs to be Done by Corporate Bodies

Companies must observe certain systemic improvements of functioning as follows :

- Formulate well-considered/planned corporate strategy and also set the procedure for implementation and evaluation of the practices.

- Decentralize perspective planning/strategic policy formulation/implementation.

- Top management and the board must play significant role in the information sharing and to this end, involve participation of all across the board.

- Introduce training and development interventions for internal education on all issues relative to governance and roles of people at each level.

- Introduce adult education at board level for empowering the internal organization and encourage the community of people for assimilation and absorption of learning.

Initiate steps for transfer of knowledge across the board once the foregoing measures are in position. The top management team/CEO have the onerous task of managing the operations and the board has to concentrate on prudent control of the company, keeping the overall picture of the organization in mind. However, in case the board starts devoting its valuable time and energies on the operational aspects of the organization along with the management, the prudent control function of the company may suffer.

The board must be fully aware of its distinctive roles, responsibilities and also limits. The board and the top management need to keep strategic interface with the external constituents like the state, the regulatory bodies, communities and the environment. Corporate bodies must clearly articulate in its policies and activities, their prime duty towards protecting the society and conserving the environment against the organization's hazardous operations. They use the natural resources of the community where they carry out the operations. Corporate bodies are no doubt, a source of strength for the society, because they provide opportunities for employment, they compensate handsomely for the jobs done and thereby, contribute to the society's purchasing power. However, corporate bodies cannot be absolved from their responsibilities towards the society/community and this aspect must always be borne in mind in case of any policy change on strategic issues of the business. In other words, the objectives set by the corporate bodies in their policies towards the community must always be achieved.

Even after the new economic policy is implemented, there will always be a good efforts which the state policy relative to macroeconomic dimensions will still need to be eercised.

Economic liberalization has opened doors to FDIs (foreign direct investments) and FIIs (Foreign Institutional Investments) which will bring many opportunities to enter into mergers and acquisitions, tie-ups of technical and economic nature. The new environment has generated opportunities for growth and expansion to Indian companies.

This will accord challenging opportunities to our companies to prepare themselves for cross cultural learning, so that they are able to excel in the market both at home and overseas. However, in order to seize the opportunity, Indian companies must develop innovative strategies to benefit from it. The existing institutions need to be reinforced and new ones brought into perspective to have a realistic appraisal of the long-term implications of the decisions of the corporates.

Corporate governance is one of the major parts of overall governance in the country. Governance as defined by World Bank is an exercise of authority/power of the government for development in social and economic fields. The dimensions of central governance can be outlined as follows :

- public sector management
- civil services reform
- economic, political and social accountability
- design/nature of institutions, legal and constitutional provisions, decentralization and policy making
- transparency in rules, procedures and regulations.

A corporation is an established institution and an integral part of the society. It just cannot wait for the overall governance to improve. Companies with effective internal governance mechanisms always produce the best results. Hindustan Lever, an Indian MNC, is an illustrative example. Coca Cola India has recently bagged the award for best governance. Further, we have the House of Tatas, Wipro, and Infosys, etc., who have defined codes on governance which are also practised. No doubt, these organizations too, are affected by the macroeconomic policies of the country, and yet they manage the operations and the stakeholders' interests in a responsible way.

All the best managed companies have one thing in common: they have their own well defined codes, duly shared and internalized across the board and the codes are well practised by all.

Consequences of Bad Governance

Consequences of poor governance can be seen not only at the company level but also at macro/systemic levels. Poor governance is reflected by the following indicators :

At Company level:

- low confidence in stakeholders and financiers due to which investors are not ready to risk their capital for investment in the company
- poor quality of management which is reflected in overall poor results.

At the macro level, poor corporate governance leads to :

- stagnation and slow growth of capital market due to public being not ready to risk their money
- stagnant, stunted individual growth
- poor employment generation
- low GDP (gross domestic product) growth
- low efforts for alleviation of poverty
- low human development indicator

Poor corporate governance normally coexists with loss of integrity, incidences of high corruption. Perceived indices of transparency can act as proxy measures.

- poor governance is an impediment for inflow of international capital
- poor governance leads to hazards of regional financial crisis triggered by the collapse of the domestic currency.
- FIIs withdraw money invested due to perceived loss of faith in the capital market.

As we can see from the above that corporate governance is a part of the macroeconomic system of a country and as such, by and large, corporate governance cannot succeed in the absence of corresponding macroeconomic and public reforms. Good business needs hassle free environment, strong legal system and at macro level, the right structure where business can flourish.

3.9. CHALLENGES FOR CORPORATE GOVERNANCE IN THE DEVELOPING WORLD

The developing world has also faced its own corporate governance challenges. For instance, in Russia a substantive share of the profits of an oil company was siphoned off by its controlling shareholder, leaving the company in debt to its creditors, employees and the state. In the Czech Republic, thousands of small shareholders lost their investments as 'tunneling' schemes by insiders stripped privatized companies of their assets. The economic crises in East Asia and other regions have demonstrated how macro-economic difficulties can be exacerbated by a systemic failure of corporate governance stemming from weak legal and regulatory systems, inconsistent accounting and auditing standards, poor banking practices, thin and unregulated capital markets, ineffective oversight by corporate boards of directors, and little regard for the rights of minority shareholders. Unfortunately, the brunt of the impact has been shouldered by the poor, setting back social and economic gains by as much as a generation in some countries.

Globalization and the Relevance of Corporate Governance

Increasingly, for developing and transition economies, a healthy and competitive corporate sector is fundamental for sustained and shared growth sustained in that it withstands economic shocks, shared in that it delivers benefits to all of society. Slow economic growth remains a major cause of poverty in man and low-income in countries, but the record also shows that a focus on growth alone is not enough. Poverty persists in part because the benefits of growth are distributed unevenly and because poor governance diminishes growth's potential impact on poverty (World Bank 1999). Countries are coming to realize that just as public governance (public administration, including service delivery, regulations, and tax administration) is important in the public sector, so corporate governance is important in the private sector.

Moreover, public governance can have a major impact (positive or negative, depending on its quality and effectiveness) on private corporate behaviour. Countries also realise that good governance of corporations is a source of competitive advantage and critical to economic and social progress. With globalisation, firms must tap domestic and international capital markets in quantities and ways that would have been inconceivable even a decade ago. Increasingly, individual investors funds, banks, and other financial institutions base their decisions not only on a company's outlook, but also on its reputation and its governance. It is this growing need to access financial resources, domestic and foreign, and to harness the power of the private sector for economic and social progress that has brought corporate governance into prominence the world over.

Sound corporate governance is important not only to attract long-term patient foreign capital, but more especially to broaden and deepen local capital markets by, attracting local investors individual and institutional. Unlike international investors, who can diversify their risk, domestic investors are often captive to the system and face greater risks, particularly in an environment that is opaque and does not protect the rights of minority shareholders. As a group, however, domestic investors frequently constitute a large potential pool of stable long-term resources critical to development. If local capital markets are to grow, corporate governance standards will need to improve to give investors the protection required to encourage them to provide capital.

Many developing and transition economies lack the supporting institutions and human resources so critical to sound corporate governance. The challenge for them is to adapt systems of corporate governance to their own corporate structures and implementation capacities, public and private, to create a culture of enforcement and compliance. They need to do so in a manner that is credible and well understood both internally and across borders – and they need to do it far more quickly than did developed countries before them. Because effective corporate governance can promote enterprise and ensure accountability, it is an essential foundation of the global financial architecture and central to the World Bank Group's mission to fight poverty.

Corporate governance has only recently, emerged as discipline in its own right, although the strands of political economy it embraces stretch back through centuries. The importance of the subject is widely recognised, but the terminology and analytical tools are still emerging. The burgeoning literature on corporate governance has largely neglected in transition economies. This report develops a framework for corporate governance-reform based largely on the operational experience of the World Bank Group and practitioners in the field. This framework is used to identify the major elements and processes of reform required in emerging market economies and the contribution that the World Bank Group, together with its partners, can make to the objective of promoting enterprise and accountability.

Balancing Divergent Interests

What makes corporate governance necessary? Put simply, the interests of those who have effective control over a firm can differ from the interests of those who supply the firm with external finance. The problem, commonly referred to as a principal-agent problem, grows out of the separation of ownership and control and of corporate outsiders and insiders. In the absence of the protections that good governance supplies, asymmetries of information and difficulties of monitoring mean that capital providers who lack control over the corporation will find it risky and costly to protect themselves from the opportunistic behaviour of managers or controlling shareholders.

Without meaningful protection for external capital providers, those who control the corporation can use their position to misappropriate economic benefits, often at the expense of the long-term performance and value of the enterprise. Where poor corporate governance is the norm, the problem extends beyond under performance in the corporate sector to greater vulnerability of the financial system, since it is difficult for local capital providers (banks and institutional investors) to avoid governance risks. Lack of meaningful protection for capital providers makes it harder for firms to get financing on favourable terms.

Just what constitutes corporate governance is still a topic of debate. From a corporation's perspective, the emerging consensus is that corporate governance is about maximizing value subject to meeting the corporation's financial and other legal and contractual obligations. This inclusive definition stresses the need for boards of directors to balance the interests of shareholders with those of other stakeholders employees, customers, suppliers, investors, communities in order to achieve long-term sustained value for the corporation.

From a public policy perspective, corporate governance is about nurturing enterprises while ensuring accountability in the exercise of power and patronage by firms. The role of public policy is to provide firms with the incentives and discipline to minimize the divergence between private and social returns and to protect the interests of stakeholders.

A Corporate Governance-Framework: The Internal and External Architecture

These two definitions from public and private perspectives provide a framework for corporate governance (figure) that reflects an interplay between internal incentives (which define the relationships among the key players in the corporation) and external forces (notably policy, legal, regulatory and market) that together govern the behaviour and performance of the firm.

The Internal Architecture Defines the Relationships Among Key Players in the Corporation

In its narrowest sense, corporate governance can be viewed as a set of arrangements internal to the corporation that defines the relationships between managers and shareholders. The shareholders may be public or private, concentrated or dispersed. These arrangements may be embedded in company law, securities law, listing requirement

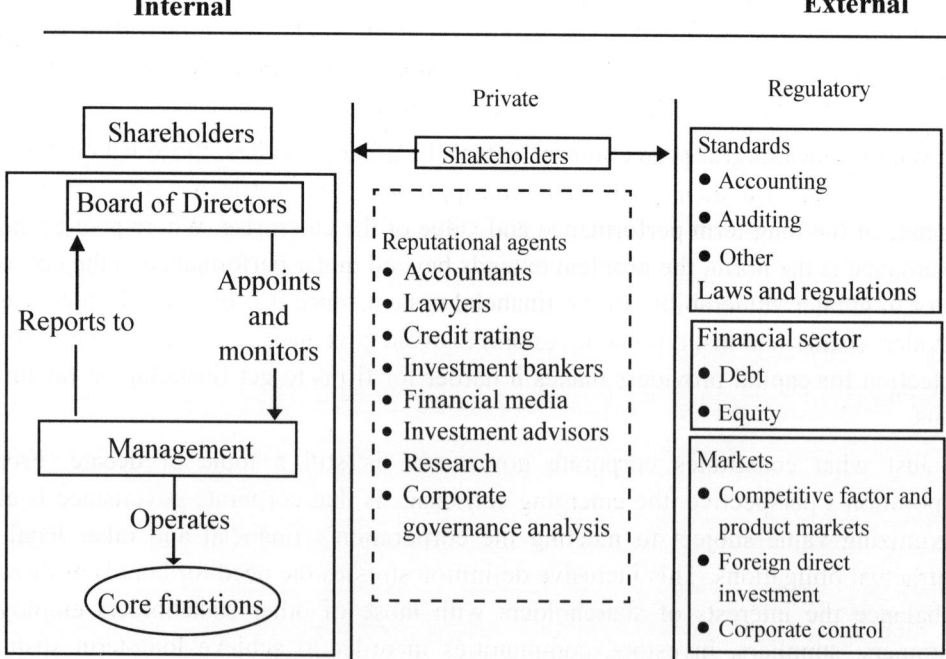

Fig. 3.1: Modern corporations are disciplined by Internal and external factors

and the like or negotiated among the key players in governing documents of the corporation, such as the corporate charter, by-laws, and shareholder agreements.

At the center of this system is the board of directors. Its overriding responsibility is to ensure the long-term viability of the firm and to provide oversight of management. In many countries, the board is responsible for approving the company's strategy and major decisions and for hiring, monitoring, and replacing the management. In some countries the board has fiduciary responsibility for ensuring compliance with laws and regulations, including accounting and financial reporting requirements. For a going concern, the board is answerable to shareholders and in some systems to employees and creditors. Its task is to protect the interests of the company. When the company runs into financial difficulty, the duty of the board shifts to the company's creditors; the primary duty of the director is to the company rather than to shareholders.

The governance problems that need to be addressed vary according to the ownership challenge in the corporate sector. At one end of the spectrum is the publicly traded company with widely dispersed shareholdings. There, the challenge is for outside shareholders to control the performance of managers. Since managers dominate, the key governance mechanism is the rules for selecting directors, who need to have enough independence to ensure that they will properly monitor managers' performance. At the other end of the spectrum is the closely held company with a controlling shareholder and a minority of outside shareholders, where the manager acts at the dictate of the controlling shareholder. There, the primary governance issue is how outside shareholders can prevent the controlling shareholder from extracting excess benefits through self-dealing or disregard the economic rights of minority shareholders; Common protections include limits on self-dealing by insiders, anti-dilution provisions, and appraisal or withdrawal rights for minority shareholders. Where a publicly traded corporation is dominated by a controlling shareholder, additional governance mechanisms may include voting rights, provision for outsider representation on the board, and takeover rules limiting the 'control premium' that insiders can appropriate.

External Rules Provide a Level Playing Field and Keep Players in Line

These internal mechanisms for corporate governance are strengthened by external laws, rules, and institutions that provide a level, competitive playing field and discipline the behaviour of insiders, whether managers or shareholders. In developed market economies, these policies and institutions minimize the divergence between social and private returns and reduce costly agency problems, primarily through greater transparency, compliance mechanisms and monitoring by regulatory and self-regulatory bodies. Notable among the institutions that discipline corporations are the legal framework for competition policy, the legal machinery for enforcing shareholders' rights, systems for accounting and auditing, a well-regulated financial system, the bankruptcy system and the market for corporate control.

Transparent, Efficient, and Liquid Equity and Bond Markets

Efficient securities markets send price signals rapidly, rewarding or penalising insiders through changes in the value of their interests in the company or in the company's access to capital. The system of rewards and penalties is severely diluted, however, if markets are not transparent, investments are costly to exit, or, in the case of institutional investors, if the investors themselves are poorly governed.

Firms' Performance is Monitored and Spurred by Reputational Agents and Activist Shareholders

Developed markets increasingly feature a dense network of reputational agents who significantly reduce monitoring costs. They include accounting and auditing professionals, lawyers, investment bankers and analysts, credit rating agencies, consumer activists, environmentalists, and media. Keeping an eye on corporate performance and insider behaviour, these reputational agents can exert pressure on companies to disclose relevant information, improve human capital, recognise the interests of outsiders, and otherwise behave as good corporate citizens: Some can also put pressure on government through their influence over public opinion.

Dearth of Models

These internal and external features have come together in different ways to create a range of corporate governance systems that reflect specific market structures, legal systems, traditions, regulations, and cultural and societal values.

Effects of Globalization

Despite the diversity of corporate governance systems, the globalization of markets is producing a degree of convergence in actual operations and governance practices. Countries and firms compete on the price and quality of their goods and services (which has led to a convergence of cost structures and firm organisation that in turn has spilled over into firm behaviour and decision-making). They compete for financial resources in global capital markets. Increasingly, they also compete on their regimes for corporate governance.

Uniform Standards and Gaining Currency

Similarly, governments, which retain priority in protecting savers, investors, suppliers, and the broader interests of the economy, are increasingly requiring that corporations operate in a fair, transparent, and accountable manner. Numerous public and private bodies have responded by establishing standards and norms related to important aspects of corporate governance. Among them are the International Accounting Standards Committee, the Bank for International Settlements (BIS, for banking supervision and prudential regulation), the International Organisation of Securities Commissions, the World Trade Organisation, and the International Labour Organisation.

Agreement on Basic Principles for Corporate Governance is Spreading

Through a consultative process involving OECD members and observers, the private sector, international organisations, and a range of stakeholders, the OECD has distilled from diverse national practices a set of Principles of Corporate Governance. The principles deal mainly with internal mechanisms for directing the relationships of managers, directors, shareholders, and other stakeholders. They are also intended primarily for listed companies that function within an effective legal and regulatory environment with adequate competition.

Challenge of Corporate Governance in Emerging Markets

In advanced market economies the rich and complex governance system (of policy, laws, regulations, public institutions, self-regulated professional bodies, and managerial ethos) has evolved over centuries. In emerging markets, however, many elements of this mosaic are absent or countries are ill-equipped to address the corporate governance challenges they face. These challenges are all the more daunting because of the complexity of the ownership structure in the corporate sector, interlocking relationships between government and the financial sector, weak legal and judicial systems, absent or underdeveloped institutions, and scarce human resource capabilities.

The Range of Corporate Structures makes the Problems more Complex

Ownership patterns across developed, developing, and transition economies are extremely varied. Among successful developed economies, both dispersed and concentrated shareholdings, have provided an efficient base for growth and capital accumulation as long as there has been a well, functioning legal and regulatory framework, active oversight by reputational agents, and adequate institutional and professional infrastructure.

Transition economies face a different problem. Much of their corporate sector consists of 'instant corporations' created through mass privatization programs implemented without the legal and institutional structures necessary to operate in a competitive market economy. With diffuse ownership, this has sometimes allowed insiders to strip assets and leave little value for minority shareholders. In both systems there is a need to build institutions and professional capacity.

These corporate structures complicate the problems associated with asymmetries of information, imperfect monitoring and opportunistic behaviour and make corporate governance reform more complex.

Less Competitive Markets and Weaker Institutions make the Solutions more Difficult

In emerging market economies, the business environment lacks many of the elements needed for a competitive market and a culture of enforcement and compliance. Inadequate competition policies entrench large dominant firms, prevent new entry and discourage entrepreneurship.

Focus on the Fundamentals/Basics

Though reform is difficult, many countries have taken some of the necessary steps and a few have taken most of them, improving their institutions and human resources. Those that have strayed the course have seen impressive gains in corporate governance and economic performance. But even in this group, reform has been a long, uneven, and sometimes fragile process of successes and reversals. And some important institutions are just beginning to emerge, such as reputational agents and active shareholders.

Reforms have proved most effective when they have focused on fundamentals and have combined a complementary mixture of laws consistently enforced and incentives for firms to take voluntary actions. They have emphasised a comprehensive strengthening of external sources of discipline and internal incentives to improve corporate governance, especially by making corporate boards more effective and competent to exercise their duties of oversight and control over management.

Resistance from Powerful Interest Groups

Reform of corporate governance systems is politically difficult. Vested interests within firms generally oppose greater transparency and disclosure of both financial and non-financial information, arguing that the requirements are costly to comply with and put them at a disadvantage relative to local or foreign competitors. These immediate drawbacks, they claim, outweigh the potential longer-term benefits of higher share values and lower financing costs that can come with greater transparency. Worried about diluting their privileged position in the company's decision making, insiders often oppose such substantive corporate governance requirements as one-share one-vote, cumulative voting, public tender offers, and independent directors. Giving greater power to minority shareholders is often opposed on the grounds that it could lead to foreign control of local firms, ignoring the benefits that they could bring. Large firms tend to have considerable political influence and access to the public media, opening the door for bribery and corruption. In developing countries, transition economies regulators or supervisors rarely have the political, human, and financial resources to prevail against the determined opposition of these vested interests.

Tough disclosure requirements and substantive changes in corporate governance are sometimes also opposed by members of exchanges (brokers, dealers, banks), who fear a loss of revenue if the measures discourage firms from listing. The threatened loss of privileged access to information can also provoke resistance to reform, particularly in smaller economies where ownership and control of industrial companies may overlap.

The Solution: Ownership with due Diligence

The challenge for developing countries is to take the next steps toward sound corporate governance before another crisis erupts. The important initial steps already taken will not become fully effective until the supporting institutions and implementation capacity evolve and adjust to new monitoring and regulatory needs. The culture of state intervention and policy influence by large conglomerates will have to adapt to a global environment that puts a large premium on a culture of compliance and enforcement.

Effecting this change of culture will require a combination of regulatory reform and voluntary private action in a sustained process of consensus and capacity building involving all the players. Each country will have to find its own formula by assessing its strengths and weaknesses, setting priorities and sequencing reforms, creating strong institutions and developing the necessary human capital. The winning formula has to be adapted to the corporate structure and to the implementation capacity in the private and public sectors. It has to provide both the incentives and the discipline for the private sector to adopt and consistently practice sound principles of corporate governance. It also needs to encourage a broadening and deepening of local ownership that will enable firms to compete more effectively in world markets often by adhering to best practices and rules set by global markets.

Strategies of the World Bank Group

The World Bank Group has long been active in supporting client countries in undertaking difficult structural changes requiring reforms of legal and regulatory structures, the financial sector, and enterprises, including privatisation of state-owned enterprises. These programs have addressed many issues central to corporate governance: creating competitive markets, establishing regulatory and supervisory capability in banking and capital markets, introducing greater transparency, adopting international accounting and auditing standards, and strengthening the competence and independence of boards of directors. Further, a scarcity of qualified professionals often poses the most daunting challenge to effective reform, the Bank has also financed technical assistance operations in support of institutional development and capacity building in many areas affecting corporate governance, including auditing and accounting standards, legal and judicial systems, financial sectors, and capital markets.

Sharing Knowledge and Best Practices

- Developing human capacity and building institutions to sustain and expand corporate governance practices.

- Addressing corporate governance issues that go beyond a specific country.

Time is short. Crises highlight challenges and offer opportunities for governments and the private sector to change behaviour and the rules of the game. But while reforms are most often initiated in the wake of crisis, they should not be viewed in the context of a short-term anti-crisis package. Change will take a concerted effort in building consensus and sharing experience, expertise and resources among all players. Above all, the private sector must see that implementing reform is in its own best interest. Likewise reform of the public sector is central to an effective partnership. Because reforms are likely to yield results only over the medium to long run sustainability and comprehensiveness in design and staying power during implementation are critical.

3.10. WHY ALL THE FUSS ABOUT CORPORATE GOVERNANCE?

In order to answer this question, we need to delve into what exactly corporate governance is. If we review the Final Report, the term 'Corporate Governance' refers to the 'processes and structure by which the business and affairs of the company are directed and managed, in order to enhance long-term shareholder value through enhancing corporate performance and accountability, whilst taking into account the interests of other stakeholders'.

The Greenbury Code for Corporate Governance

- Limit the size of the board so that each director can contribute, and avoid coalitions. Separate the roles of CEO and Chairman to avoid potential conflicts of interests

- Avoid inside directors on the committees so that executives do not audit, evaluate and reward themselves

- Ensure a majority of outside directors so that tough questions are asked. Require directors to resign upon retirement, or upon change in employment or responsibilities
- Limit the number of other boards of directors on which directors can serve
- Place the whole board up for re-election each year to maintain a mix of skills
- Impose term limits to introduce fresh and potentiality critical viewpoints while avoiding groupthink
- Establish a set of qualifications for directors, and use them to screen new candidates
- Impose a retirement age to maintain a mix of skill, energy, enthusiasm and commitment
- Develop guidelines for the use of committees to ensure that basic tasks are fulfilled and complex topics are explored in sufficient depth
- Rotate directors through the various committees to ensure a mix of views
- Ensure that outside directors, as a group, meet alone on a specific number of occasions every year
- Choose a lead director to prevent insiders from dominating the agenda.
- Ensure unrestricted access for board to management so that information is not filtered
- Establish additional modes of information flow to ensure sufficient information
- Insist on regular attendance of board meetings by all directors
- Establish an orientation program so that new directors can contribute quickly
- Develop effective recruitment and evaluation processes for the board
- Ensure that the management reports regularly to the board on succession planning.

Now this is an interesting concept, as it implies that more than just a company's shareholders are to be taken as important stakeholders in an organization. One of the best examples of a company that clearly does consider all of its stakeholders is the US firm Johnson & Johnson. They have a shared system of values termed the 'Credo'. The Credo is only about a page long and acts as a guide as to how the company plans to interact with its stakeholders. This includes a policy of bearing their fair share of taxes and being responsible to the communities in which they live and work, as well as the world community as a whole. All entities within the Johnson & Johnson group follow the Credo and it is currently available in 36 languages to enable them to do so.

At times, financial performance shows how successful a company has been in the past, whereas the 'measures that matter' may highlight how successful a company will be in the future. That is to say, non-financial measures may well be a better indicator for long-term investment selection than financial performance.

Obviously, one of the key non-financial measures is the quality of a company's management and some of the determination of that quality can be taken from the way in which a company looks to govern itself. These factors, combined with increased global competition, are placing good corporate governance practices high on the agenda of many Singapore board and senior executive meetings.

So, given that we understand what is meant by 'corporate governance' and have some idea of its importance, where does one go to find the panacea, the corporate governance structure that fits all? Well, there is no such beast. Given the current plethora of different business models that companies follow, even within the same industry and market segment, it would be impossible to establish a standardised corporate governance methodology that could adequately address governance issues in all companies.

Therefore, the shareholders, who have elected a board of directors to be responsible for corporate governance and managing the affairs of the business, should ensure that the board has effective involvement in strategic policy, risk management and performance assessment. The directors, who are accountable to the shareholders for their actions, should determine the appropriate corporate governance practices applicable to the company given its business model, and then ensure that the necessary procedures to ensure that the governance objectives can be met are put in place. The prospective investors, here demand transperances.

Further, communication of corporate governance matters should not just be limited to annual reports. Additionally, communication of such information should not only be made to a selected group of institutional investors but rather to all shareholders. When disclosing such information, companies should be as descriptive, detailed, forthcoming and meaningful as they can.

Another key benefit of devoting time and effort to an appropriate corporate governance structure is that effective governance will encourage the use of board expertise in ways that will maximise each director's contribution. An environment in which both the board and management set the corporate strategic direction and that allows the board to monitor performance over time without impeding the management of day-to-day operations, should also limit the possibility of failure and lead to increased shareholder satisfaction.

In summary, an appropriate corporate governance framework that looks to involve the board in the strategic planning process, delineates clear board and management power sharing arrangements, establishes processes for the timely reporting and review of information, not forgetting to allow effective and responsive actions to be made thereon, will lead to improved understanding of the respective roles of the board and management. This will in turn lead to the introduction of appropriate governance processes and procedures under which management is free to manage, while the board is free to monitor, enquire and counsel.

4

CORPORATE SOCIAL RESPONSIBILITY (CSR)

INTRODUCTION

CSR "analyses economic, legal, moral, social and physical aspects of environment".

—*Barnard* (1938)

Corporate Social Responsibility (or CSR as we will call it) is a concept which has become dominant in business environment. Every corporation has a policy concerning CSR and produces a report annually detailing its activity. And of course each of us claims to be able to recognise corporate activity which is socially responsible and activity which is not socially responsible. There are two interesting points about this: firstly we do not necessarily agree with each other about what is socially responsible; and although we claim to recognise what it is or is not when we are asked to define it then we find this impossibly difficult. Thus the number of different definitions is huge and in this chapter we will look at some of these.

DEFINITIONS OF CSR

The broadest definition of corporate social responsibility is concerned with what is or should be the relationship between global corporations, governments of countries and individual citizens. More locally the definition is concerned with the relationship between a corporation and the local society in which it resides or operates. Another definition is concerned with the relationship between a corporation and its stakeholders.

For us, all of these definitions are pertinent and each represents a dimension of the issue. It has its ethical base. However this debate is represented in that it is concerned with some sort of social contract between corporations and society.

This social contract implies some form of altruistic behaviour the converse of selfishness whereas self-interest connotes selfishness. Self-interest is central to the Utilitarian perspective championed by such people as Bentham, Locke and J.S.Mill. The latter, for example, is generally considered to have advocated as morally right the pursuit of the greatest happiness for the greatest number although the Utilitarian philosophy is actually much more based on selfishness than this something to which we will return later. Similarly Adam Smith's free-market economics, is predicated on competing self-interest.

These influential ideas put interest of the individual above interest of the collective. The central tenet of social responsibility however is the social contract between all the stakeholders to society, which is an essential requirement of civil society.

There is however no agreed definition of CSR so this raises the question as to what exactly can be considered to be corporate social responsibility. According to the EU Commission (2002).

> "... CSR is a concept whereby companies integrate social and environmental concerns in their business operations and in their interaction with their stakeholders on a voluntary basis."

Corporations are part of society

A growing number of writers however have recognised that the activities of an organisation impact upon the external environment and have suggested that one of the roles of accounting should be to report upon the impact of an organisation in this respect. Such a suggestion first arose in the 1970's and a concern with a wider view of company performance is taken by some writers who evince concern with the social performance of a business, as a member of society at large.

Indeed the desirability of considering the social performance of a business has not always however been accepted and has been the subject of extensive debate. Thus Hetherington (1973) states :

> "There is no reason to think that shareholders are willing to tolerate an amount of corporate nonprofit activity which appreciably reduces either dividends or the market performance of the stock."

Conversely, writing at a similar time, Dahl (1972) states :

> "every large corporation should be thought of as a social enterprise; that is an entity whose existence and decisions can be justified insofar as they serve public or social purposes."

Similarly Carroll (1979), one of the early CSR theorists, states that :

> "business encompasses the economic, legal, ethical and discretionary expectations that society has of organization at a given point in time".

More recently this was echoed by Balabanis, Phillips and Lyall (1998), who declared that :

> "in the modeen commercial era, companies and their managers are subjected to well publicized pressure to play an increasingly active role in [the welfare of] society."

Profit is all that matters

Some writers have taken the view that a corporation should not be concerned with social responsibility and you are certain to come across the statement from Milton Friedman, made in 1970:

> "there is one and only one social responsibility of business - to use its resources and engage in activities designed to increase its profits so long as it stays within the rules of the game, which is to say, engages in open and free competition without deception or fraud".

Equally some people are more cynical in their view of corporate activity. So Drucker (1984) had the opinion that:

> "business turns a social problem into economic opportunity and economic benefit, into productive capacity, into human competence, into well-paid jobs, and into wealth".

CSR is conditional

While Robertson and Nicholson (1996) thought that:

> "a certain amount of rhetoric may be inevitable in the area of social responsibility. Managers may even believe that making statements about social responsibility insulates the firm from the necessity of taking socially responsible action."

> "whether or not business should undertake CSR, depends upon the economic perspective of the firm that is adopted".

So we can see that CSR is a contested topic and it is by no means certain that everybody thinks that it is important or relevant to modem business.

The principles of CSR

Because of the uncertainty surrounding the nature of CSR activity it is difficult to define CSR and to be certain about any such activity. It is therefore imperative to be able to identify such activity and we take the view that there are three basic principles which together comprise all CSR activity. These are:

- Sustainability
- Accountability
- Transparency

Sustainability will be considered in detail in terms of action in future while accountability and transparency will be considered in terms of responsibility to give efforts to action.

Sustainability

This is concerned with the effect of action taken in the present upon the options available in the future. If resources are utilized in the present are no longer available for use in the future, and this is of particular concern if the resources are not renewable.

Sustainability therefore implies that society must use no more of a resource than can be regenerated.

Viewing an organization as part of a wider social and economic system implies that these effects must be taken into account, not just for the measurement of costs and value created in the present but also for the future of the business itself.

Accountability

This is concerned with an organization recognizing that its actions affect the external environment, and therefore assuming responsibility for the effects of its actions. This concept therefore implies a quantification of the effects of actions taken, both internal to the organization and externally. More specifically the concept implies a reporting of those quantifications to all parties affected by those actions. This implies a reporting to external stakeholders of the effects of actions taken by the organization and how they are affecting those stakeholders.

This concept therefore implies a recognition that the organization is part of a wider societal network and has responsibilities to all of that network rather than just to the owners of the organization. Alongside this acceptance of responsibility therefore must be a recognition that those external stakeholders have the power to affect the way in which those actions of the organization are taken and a role in deciding whether or not such actions can be justified, and if so at what cost to the organization and to other stakeholders.

Transparency

Transparency, as a principle, means that the external impact of the actions of the organization can be ascertained from that organization's reporting and pertinent facts are not disguised within that reporting. Thus all the effects of the actions of the organization, including external impacts, should be apparent to all from using the information provided by the organization's reporting mechanisms. Transparency is of particular importance to external users of such information as these users lack the background details and knowledge available to internal users of such information. Transparency therefore can be seen to follow from the other two principles and equally can be seen to be a part of the process of recognition of responsibility on the part of the organization for the external effects of its actions and equally part of the process of transferring power to external stakeholders.

Environmental issues and their effects and implications

When an organization undertakes an activity which impacts upon the external environment and is affected by environment in ways which are not reflected in practice accounting of that organization. The environment can be affected either positively, through for example a landscaping project, or negatively, through for example the creation of heaps of waste from a mining operation.

These actions of an organization impose costs and benefits upon the external environment. These costs and benefits are imposed by the organization without consultation, and in reality form part of the operational activities of the organization. These actions are however excluded from traditional accounting of the firm, and by implication from its area of responsibility. Thus we can say that such costs and benefits have been externalized. The concept of externality therefore is concerned with the way in which these costs and benefits are externalized from the organization and imposed upon others.

Issues of socially responsible behaviour are not of course new and examples can be found from throughout the world and at least from the earliest days of the Industrial Revolution and the concomitant founding of large business entities (Crowther 2002) and the divorce between ownership and management - or the divorcing of risk from rewards (Crowther 2004). According to the European Commission CSR is about undertaking voluntary activity which demonstrates a concern for stakeholders.

Examples of such externalization include

- Environmental degradation through such things as polluted - and dead - rivers or through increased traffic casts costs upon the local community through reduced quality of life;

- Causing pollution imposes costs upon society at large;
- Waste disposal problems impose costs upon whoever is tasked with such disposal;
- Removing staff from shops imposes costs upon customers who must queue for service;
- Just in time manufacturing imposes costs upon suppliers by transferring stockholding costs to them.

In an increasingly global market then one favourite way of externalizing costs is through transfer of those costs to a third world country.

Stakeholder theory points to :

The argument for Stakeholder Theory is based upon the assertion that maximizing wealth for shareholders fails to maximize wealth for society and all its members and that only a concern with managing all stakeholder interests is inseperable must be considered in the decision making process of the organization. The theory states that there are 3 reasons why this should happen:

- It is the morally and ethically correct way to behave
- Doing so actually also benefits the shareholders
- It reflects what actually happens in an organization

According to this theory, stakeholder management, or corporate social responsibility, is not an end in itself but is simply seen as a means for improving economic performance. This assumption is often implicit although it is clearly stated by Atkinson, Waterhouse and Wells (1997) and is actually inconsistent with the ethical reasons for adopting stakeholder theory. Instead of stakeholder management improving economic, or financial, performance therefore it is argued that a broader aim of corporate social performance should be used (Jones and Wicks, 1999).

Social Benefits

- Enhanced company or product image - this in itself can lead to increased sales
- Health and safety benefits
- Ease of attracting investment and lowered cost of such investment
- Better community relationships - this can lead to easier and quicker approval of plans through the planning process
- Improved relationship with regulators, where relevant
- Improved morale among workers, leading to higher productivity, lower staff turnover and consequently lower recruitment and training costs

- General improved image and relationship with stakeholders

The above social benefits entail some accounting problems:

The steps involved in the incorporation of environmental accounting into the risk evaluation system can therefore be summarized as follow :

- Identify environmental implications in term of costs and benefits
- Quantify those costs and incorporate qualitative data regarding less tangible benefits
- Use appropriate financial indicators
- Set an appropriate time horizon which allows environmental effects to be fully realized.

4.1. WHY IS CSR IMPORTANT?

CSR is important because it influences all aspects of a company's operations. Increasingly, consumers want to buy products from companies they trust; suppliers want to form business partnership with companies they can rely on; employees want to work for companies they respect large investment funds want to support firms that they see as socially responsible; and nonprofit NGOs want to work together with companies seeking practical solutions to common goals. Satisfying each of these stakeholder groups (and others) allows companies to maximize their commitment to their owners (their ultimate stakeholders), who benefit most when all of these groups' needs are being met. As Carly Fiorina, former chair and chief executive officer of Hewlett-Packard, noted :

I honestly believe that the winning companies of this century will be those who prove with their actions that they can be profitable and increase social value-companies that both do well and do good... Increasingly, shareowners, customers, partners and employees are going to vote with their feet rewarding those companies that fuel social change through business. This is simply the new reality of business one that we should and must embrace.

CSR is increasingly crucial to success because it gives companies a mission and strategy around which multiple constituents can rally. The businesses most likely to succeed in today's rapidly evolving global environment will be those best able to balance the often conflicting interests of their multiple stakeholders. Lifestyle brand firms, in particular, need to live the deals they convey to their consumers.

A Rational Argument for CSR

CSR in businesses is seeking to maximize their performance by minimizing restrictions on operations. In today's globalizing world, where individuals and activist organizations feel empowered to enact change, CSR represents a means of anticipating and reflecting societal concerns to minimize operational and financial limitations on business.

The rational argument for CSR is summarized by the iron law of social responsibility, which states that in a free society discretionary abuse of societal responsibilities leads, eventually, to mandated solutions. Restated, in a democratic society, power is taken away from those who abuse it. The history of uprisings from Cromwell in England, to the American and French Revolutions to the overthrow of the Shah of Iran or Saddam Hussein in Iraq underscores the conclusion that abusers seed their own destruction.

Acting proactively in a socially responsible manner is a rational business response particularly so in light of the overwhelming anecdotal evidence that discretionary abuses lead to a loss of decision-making freedoms and financial repercussions for profit organizations.

An Economic Argument for CSR

Summing the moral and rational arguments for CSR leads to an economic argument. To incorporate CSR into operations offers a potential point of differentiation and competitive market advantage upon which future success can be built, besides avoiding moral, legal, and other sanctions.

CSR is an argument of economic self interest for business. CSR adds value because it allows companies to reflect the needs and concerns of their various stakeholder groups. By doing so, a company is more likely to retain its societal legitimacy, and maximize its financial viability, over the long term. Simply put, CSR is a way of matching corporate operations with societal values at a time when these parameters can change rapidly.

CSR influences all aspects of a business's day-to-day operations. Everything an organization does interacts with one or more of its stakeholder groups. Companies today need to build a watertight image with respect to all stakeholders. Whether as an employer, producer, buyer, supplier, or as an investment, increasingly the attractiveness and success of a company today is linked to the strength of its image and its brand(s). Concerning investments, for example, about 10% of all investment in the United States is classified as socially earmarked for development.

At a deeper level, societies rest upon a cultural heritage that grows out of a confluence of Refugio, mores, and folkways. This heritage gives rise to a belief system that defines the industries of socially and morally acceptable behaviours by people and organizations. Though are always codified into 40gma or laws, the cultural heritage leads to an evolving definition of social justice, human rights, and environmental stewardship, the violation of which is deemed morally wrong and socially irresponsible. To violate these implicit moral boundaries can lead to a loss of legitimacy.

A Moral Argument for CSR

CSR broadly represents the relationship between a company and the principles expected by the wider society within which it operates. It assumes businesses recognize that for profit entities do not exist in a vacuum and that a large part of their success comes as much from actions that are congruent with societal values as from factors internal to the company.

Charles Handy makes a convincing moral argument for businesses going beyond the goals of maximizing profit and satisfying shareholders above all other stakeholders: The purpose of a business. . . is not to make a profit only. It is to make a profit so that the business can do something more or better. That "something" becomes the real justification for the business. . . . It is a moral issue. To mistake the means for the end is to be turned in on oneself, which Saint Augustine called one of the greatest sins. . . . It is salutary to ask about any organization, "If it did not exist, would we invent it?" "Only if it could do something better or more useful than anyone else" would have to be the answer, and profit would be the means to that larger end.

At one level, the moral argument for CSR reflects a give-and-take approach, based on a meshing of the firm's values and those of society. Society makes business possible and provides it directly or indirectly with what for profits need to succeed, ranging from education and healthy workers to a safe and stable physical and legal infrastructure, not to mention a consumer market for their products. Because society's contributions make businesses possible, those businesses have an obligation to society to operate in ways that are deemed socially responsible and beneficial. And, because businesses operate within the larger context of society, society has the right and the power to define expectations for those who operate within its boundaries :

This changing nature of investment in companies and the evolving relationship between company and owners has seen the importance of shareholders and their value for businesses. Manager now have to concentrate on their time on the considerations of quarterly results, dividend levels, and share price in order to keep demanding shareholders happy. This perspective often comes at the cost of long-term strategic considerations of the company and its business interests.

Many observers see this development as the fundamental purpose of a company issuing shares, as well as how those shares are later traded on the stock market.

Businesses still have a duty to provide a return for investors. It is central to their economic mission; but the idea that shareholders have the best interests of the firm at heart no longer necessarily holds true. In today's business environment, a broader stakeholder perspective will provide the stability necessary for managers to chart the best course for the company so that it remains a viable entity over the medium to long term. This is in the interests of a company's investors rather than those of its speculators.

CSR represents an argument for a firm's economic interests, where satisfying stakeholder needs becomes central to retaining societal legitimacy (and, therefore, financial viability) over the long term. Much debate (and criticism) in CSR springs from well-meaning parties, arguing the same "facts" from different perspectives, breaking down along partisan and ideological lines.

The Cultural and Contextual Assumptions

Different societies define the relationship between business and society in different ways. Unique expectations spring from many factors, with wealthy societies having greater resources and perhaps, more demanding expectations that emerge from the greater options wealth brings. The reasoning is straightforward: In poor democracies, the general social well being is focused on the necessities of life: food, shelter, transportation, education, medicine, social order, jobs, and the like. Governmental or self-imposed CSR restrictions add costs that poor societies can ill afford. As societies advance, however, expectations change and the general social well being is redefined. This ongoing redefinition and evolution of societal expectations causes the CSR response also to evolve.

Differences in CSR expectations among rich and poor societies are a matter of priorities. For example, the need for transportation evolves into a need for non-polluting forms of transportation as society becomes more affluent. Though poor societies value clean air just as advanced ones do, there are other competing needs that may take priority, one of which will be the need for low-cost transportation. As a society prospers economically, new expectations compel producers to make vehicles that pollute less a shift in emphasis. In time, these expectations may evolve from a discretionary to a mandatory (legal) requirement.

CSR has become an increasingly relevant topic in recent decades in corporate boardrooms, business school classrooms, and family living rooms.

On the one hand, these ever changing standards and expectations compound the complexity faced by corporate decision makers. Worse, those standards vary from society to society, even among cultures within a given society. Faced with a kaleidoscopic background of evolving standards, business decision makers must consider a variety of factors on the way to implementation. For example, in the early history of the United States, the Alien Tort Claims Act "was originally intended to reassure Europe that the fledgling U.S. wouldn't harbour pirates or assassins. It permits foreigners to sue in U.S. courts for violations of the law of "nations".

Concerns over domestic and international income disparity, gender issues, discrimination, human rights, spirituality and workplace religiosity, technological impacts on indigenous populations, and other issues all affect societal well-being. Unless firms take actions directly affecting stakeholders in these areas, however, the study of these topics might -fall under ethics, public policy, sociology, or developmental economics courses, which are better suited to explore these complex and socially important topics in greater depth.

What CSR is and is Not

CSR embraces the range of economic, legal, ethical, and discretionary actions that affect the economic performance of the firm. A significant part of a firm's CSR, therefore, is complying with the legal or regulatory requirements faced in day-to-day operations. To break these regulations is to break the law, which is not socially responsible. Clearly, adhering to the law is an important component of any ethical organization. But, legal compliance is merely a minimum condition of CSR. Strategic CSR focuses more on the ethical and discretionary concerns that are less precisely defined and for which there is often no clear societal consensus.

Ideally, leaders should address stakeholder concerns in ways that carry strategic benefits for the firm. CSR is not about saving the whamor ending poverty or other worthwhile goals that are unrelated to a firm's operations and are better left to government or NGOs.

The acid test of good corporate philanthropy is whether the desired social change is so beneficial to the company that the organization would pursue the change even if no one never knew about it.

Beyond the desired changes are the approaches employed to achieve those changes. Too often, the end (shareholder wealth, for example) has been used to justify the means (polluting the environment). Strategic CSR is concerned with both the ends of economic viability and the means of being socially responsible.

The connection between these two concepts is an important focus for strategic CSR, which sets it apart from other social responsibility areas. This distinction becomes apparent when discussing an issue such as ethics, which is concerned about the honesty, judgment, and integrity with which various stakeholders are treated. There is no debate: Ethical behavior is a prerequisite assumption for strategic CSR. It is hard to see how a firm's actions could be both socially responsible and unethical. Ethics, however, is not the central focus for strategic CSR, except insofar as constituents are affected or the

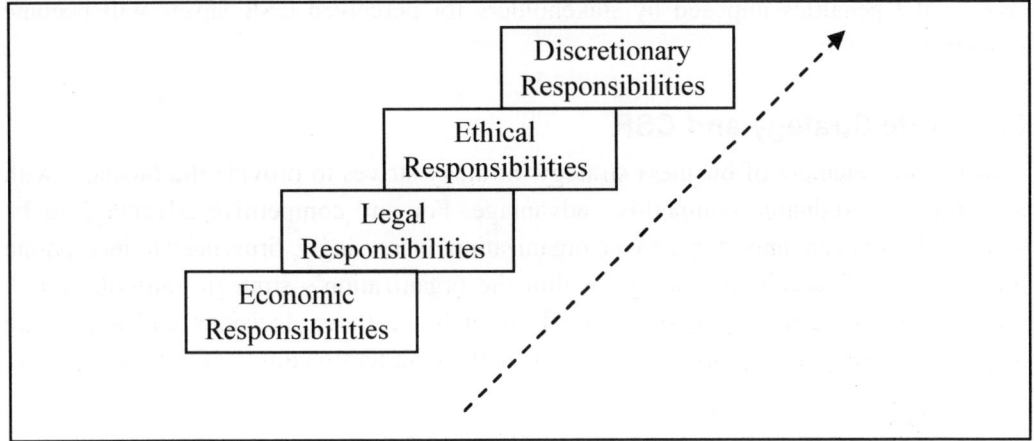

Fig. 4.1: The Hierarchy of Corporate Social Responsibility

larger society defines the firm's evolving. It is a much broader concept than business ethics, for example (see Figure). And, because these issues affect all aspects of strategy, as well as farm a key element of day-to-day operating activities, CSR's cutting edge can be, controversial, especially among those stake-holders whose interests are not considered primary by decision makers.

Corporate Social Responsibility Hierarchy

Archive Carroll, University of Georgia, was one of the first academics to make a distinction between different kinds of organizational responsibilities. He referred to this distinction as a firm's "pyramid of corporate social responsibility".

Fundamentally, a firm's responsibility is to produce an acceptable return on its owners' invest. An important component of this, within a law-based society, is a duty to act within the legal framework drawn up by the government and judiciary. Taken one step further, a firm has an ethical responsibility to do no harm to its stakeholders and within its operating environment. Finally, firms have a discretionary responsibility, which represents more proactive, strategic behaviors that can the firm and society, or both.

Once of the central theses of this book is that what was ethical, or even discretionary in Carroll's model, is becoming increasingly necessary today due to the changing environment within which busi-ness operate. As such, ethical responsibilities are more likely to stand on a par with economic and legal responsibilities as foundational for business success. In order to fulfill its fundamental economic obligations to owners in today's globalizing and wired world, a firm should incorporate a broader and broader stakeholder perspective within its strategic outlook. As societal expectations of the firm rise, so the penalties imposed by stakeholders for perceived CSR lapses will become prohibitive.

Corporate Strategy and CSR

CSR is a key element of business strategy. Strategy strives to provide the business with a source of sustainable competitive advantage. For any competitive advantage to be sustainable, strategic importance to the organization. Increasingly, firms need to incorporate the concerns of stakeholder groups within the organization's strategic outlook or risk losing societal legitimacy. CSR helps firms embrace these decisions and adjust the internal strategic planning process to maximize the long-term viability of the organization.

4.2. IMPACT OF SOCIAL RESPONSIBILITY ON CORPORATE GOVERNANCE

If there is a Corproate Social Responsibility, How does a Firm Evidence its Satisfaction of that Responsibility?

This topic is all the more important these days as we move from a unique occurrence where a firm shares information with regard to social and environmental issues to a quasi-regulatory environment that arose in the mid-1990s where firms produce reports pursuant to generally accepted principles.

Even in focusing on social reporting, there is no one structure or group of topics for a corporate social report. Though some elements are mandatory pursuant to other regulations, such as corporate charitable contributions, pension fund adequacy, employee share ownership schemes and employment data.

Firms are engaging in this voluntary reporting process for a variety of reasons. The benefits to a transparent organization include a positive impact on reputation, enhance shareholder relations, clearer and more transparent corporate governance and greater trust within the investment community.

Formal efforts at standardized corporate responsibility reporting began in the early 1990s. In 1991, seven companies had published sustainability reports; at that time, however, much of the reports' focus was on the environment. The reporting trend has since transformed itself, addressing not only environmental issues, but also economic and social performance.

Accountability, credibility and transparency are now all a necessity in the company's social reporting procedure.

A broad compendium of global initiatives, principles and standards designed to stimulate change and to promote good corporate citizenship and encourage innovative solutions and partnerships, has emerged several organizations have created processes or standardized reporting structures to assist organizations in quantifying their social reporting as well as in creating benchmark data against which they can gauge their activities and decisions. Though many of these are industry specific, others apply cross-industry. These initiatives refer to :

Global Reporting Initiative

Global Sullivan Principles

Social Accountability

UN Global Compact

OECD Guidelines for Multinational Enterprises Conventions

ISO 14000

These voluntary initiatives are considered credible and aulhentic because of their association with reputable international organizations and agencies, despite lhe absence of formal regulatory schemes. Each of the following initiatives shares a common mission: to promote an economic environment where smart, sustainable development and good corporate citizenship coexist.

The Global Sullivan Principles of Corporate Social Responsibility

The "Global Sullivan Principles, "which several multinational companies have endorsed within the last year, were announced at the United Nations in November 1999 by Baptist Minister Leon H. Sullivan from the US. In 1997, Rev. Sullivan drafted a similar corporate code of conduct for US companies operating in South Africa in an effort to abolish that country's apartheid policies. In his words, the overarching objective of these principles is "to support economic, social and political justice by companies where they do business," including respect for human rights and equal work opportunities for all peoples.

The Principles

As a company which endorses the Global Sullivan Principles we will respect the law. and as a responsible member of society we will apply these principles with integrity consistent with the legitimate role of business. We will develop and implement company policies, procedures, training and internal reporting structures to ensure commitment to these principles throughout our organization. We believe the application of these principles will achieve greater tolerance and better understanding among peoples, and advance the culture of peace.

Accordingly, we will :

- Express our support for universal human rights and. particularly, those of our employees, the communities within which we operate, and parties with whom we do business.

- Promote equal opportunity for our employees at all levels of the company with respect to issues such as colour, race, gender, age, ethnicity or religious beliefs, and operate without unacceptable worker treatment such as the exploitation of children, physical punishment, female abuse, involuntary servitude, or other forms of abuse.

- Respect our employees' voluntary freedom of association.

- Compensate our employees to enable them to meet at least their basic needs and provide opportunity to improve their skill and capability to raise their social and economic opportunities.

- Provide a safe and healthy workplace; protect human health and the environment; and promote sustainable development.

- Promote fair competition including respect for intellectual and other property rights, and not offer, pay or accept bribes.

- Work with governments and communities in which we do business to improve the quality of life in those communities their educational, cultural, economic and social well-being and seek to provide training and opportunities for workers from disadvantaged backgrounds.

- Promote the application of these principles by those with whom we do business.

We will be transparent in our implementation of these principles and provide information which demonstrates publicly our commitment to them.

Importance of CSR in Corporate Governance

The requirement from corporates has moved beyond just getting shareholder value. The stakeholder theory is now an integral part of corporate governance. Socially responsible companies are lauded for their involvement in the welfare of all the stakeholders, the community, and the environment. There is a general acceptance that the government alone cannot manage the multifarious needs of the modern globalized society. Public-private partnerships have to be the order of the day to balance the interest of stakeholders with the profit requirements of the shareholders.

Carroll (1979) has succinctly summarized the concept of responsibility as given below :

Economic responsibility: The company has to be profit-oriented and market-driven.

Legal responsibility: Since society gives the sanction to the business to operate, it is the duty of business to obey the laws and regulations laid down by society.

Ethical responsibility: The company has to go beyond the law and honour the trust and expectations of society. The company should also be extremely culture-sensitive to provide the right services.

Discretionary (or philanthropic) responsibility: Undertake voluntary activities and expenses, keeping the greater good of society in mind.

Many countries have created company laws, which incorporate CSR as a formal duty of the company. For example, The Companies Act of 2006 of the UK formally includes CSR as a responsibility to be undertaken by companies. In India, the Narayana Murthy Committee also recommends CSR as an integral part of corporate governance.

Corporate governance is a powerful tool, which can ensure that CSR permeates throughout the company. It should operate at the internal, intermediate, and outer circles. However, we must remember that every company would have to find and frame its own framework for implementation of CSR because every company is unique, and similarly, every culture in which it is operating is distinct. Today, environmental pollution has become a major concern and many companies are including their efforts to improve the situation in sustainable reports that they are issuing along with CSR reports.

A brief list of obligations catering to important areas related to some important stakeholder is given below.

List of obligations to society, investors, and employees

Obligations to Society

1. National interest should take priority
2. Political non-alignment
3. Legal compliances
4. Rule of law
5. Honest and ethical conduct
6. Corporate citizenship
7. Ethical behaviour
8. Social concerns
9. Corporate social responsibility

10. Environment-friendly
11. Healthy and safe working conditions
12. Trusteeship
13. Accountability
14. Effectiveness and efficiency
15. Timely responsiveness
16. Uphold brand of the country

Obligation to Investors

1. Towards shareholders
2. Measures promoting informed shareholder participation
3. Transparency
4. Financial reporting and records

Obligation to Employees

1. Fair employment practices
2. Equal opportunities employer
3. Encouraging whistle-blowing
4. Humane-treatment
5. Participation
6. Empowerment
7. Equity and inclusiveness
8. Participative and collaborative environment

Source: Adapted from Corporate Governance, Principles, Policies, and Practices by A.C. Fernando (2006).

The Social Impact

We all know that long-term success is an outcome of interdependence and the people we are dealing with have their unique emotions and feelings. The business fraternity needs to realize that the law and order situation is also dependent on the economics of the country, its roots are often embedded in a terrible sense of deprivation felt by the local community. Unless the business community contributes to the basic development needs, its very survival would be threatened and it is in its own interest to participate in the 'nation building' effort. Another famous business tycoon of India, Shri Ramakrishna Bajaj (1970), grandson of Jamnalal Bajaj, also expressed the same view when he said, 'The business community is an essential ingredient of our democratic society and it has a duty not only to create wealth but also to promote the ethical and social goals of the community) Unless it fulfills both these functions and thereby plays its due role as a responsible section, it will not be able to ensure its own survival'. Yet, we as managers behave as if we are alone and only our needs matter. The commonly heard cliche is that we have no choice. It is this seeming lack of choice that is the real killer of humanity in organizations that are undergoing massive and speedy change. People suffer from the insecurity that they are not in control of anything and try to control whatever they can, in any way they can. Problems are assuming huge proportions and gigantic shapes. They can only be solved when looked at in the broader light of global interaction.

Sir Julian Huxley (1957) has rightly pointed out that 'human evolution is not biological but psychological'. The spirit of cut-throat competition only for wealth has led us to accumulation of more and more wealth at the cost of ethics. This has wrought social havoc at every level leading to fear, purposelessness, drug addiction, anxiety, neurosis, unethical life, and untimely death. As early as 1939, Arnold J. Toynbee questioned the efficiency of scientific management when he wrote 'what was the extent of sacrifices of personal freedom that workers would be prepared to make for the sake of increasing size of the cake of which they were each demanding a larger slice? How

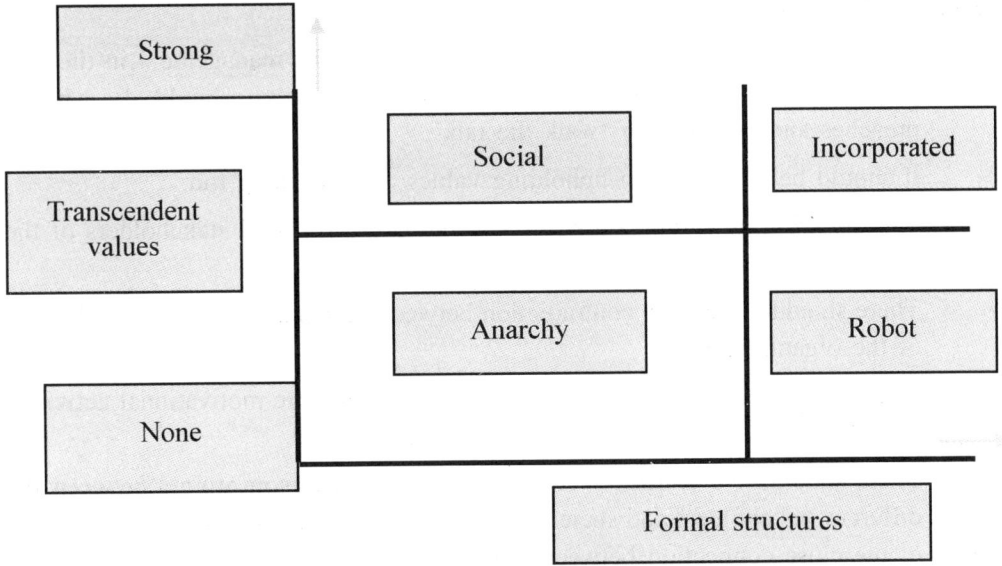

Fig. 4.2: Matrix of transcendent and formal organization

far would the urban industrial workers go in submitting to scientific management?... Western man had brought himself into concentration on a sensationally successful endeavour to increase his material well being. If he wants to find salvation, he would find it only in sharing the result of his material achievement with the less materially successful majority of the human race.' Toynbee called him the western man, but is it not true for all our self-seeking materially motivated professionals? 'Consensus management' with vertical, horizontal, and diagonal consensus among workers and management, is becoming common. To achieve this, an individual has to move beyond himself and create a family within the corporation. The idea of family carries with it the concept of basic human values of love, integrity, honesty, selflessness, etc. Therefore, we are again back to the principle of being ethical, which leads to good business.

Corporate social performance encompasses an organization's commitment to behave in an economically and environmentally sustainable manner, while honouring the interest of the direct stakeholders and benefitting the greater community at large. The success of this performance depends totally on what people really think about business and that entirely is based upon how people behave within a corporate.

A brief analysis of the various combinations in the matrix given will help us understand that a transcendent organization is an incorporated organization.

The management heads should remember that 'CSR is a process, not an event'. There should be a continuous effort to keep the process moving in a positive direction and generate collective support within the organization in support of CSR. For this purpose, a few simple steps are needed :

1. The organization should reflect the confidence of its management in the role play. This means the management of an organization should do what it preaches and promises or 'walk the talk'.

2. It should be consistent in upholding values over the long run.

3. The organization should show respect and courtesy to all stakeholders of the organization.

4. There should not be any contradiction between verbal and non-verbal activities of the organization.

5. There should be a positive and sincere attitude and the motivational activities should be fresh and interesting.

6. There should be a constant endeavour to find a common ground between the different stakeholders and society at large, based upon emotional understanding of the close connection between them.

7. Organizations should stay focused in proving that good ethics means good business in the long run.

The quickest path to success is to decide on the process and put down the goals, objectives, and action plan, so that the focus is visible and transparent.

4.3. CORPORATE SOCIAL RESPONSIBILITY IN THE ERA OF ECONOMIC LIBERALIZATION

Economic Liberalization

Corporate social responsibility (CSR), also called corporate responsibility, corporate citizenship, and responsible business is a concept whereby organizations consider the interests of society by taking responsibility for the impact of their activities on customers, suppliers, employees, shareholders, communities and other stakeholders, as well as the environment. This obligation is seen to extend beyond the statutory obligation to comply with legislation and sees organizations voluntarily taking further steps to improve the quality of life for employees and their families as well as for the local community and society at large.

In the increasingly conscience-focused marketplaces of the 21st century, the demand for more ethical business processes and actions (known as ethicism) is increasing. Simultaneously, pressure is applied on industry to improve business ethics through new public initiatives and laws (e.g. higher UK road tax for higher emission vehicles).

Business ethics can be both a normative and a descriptive discipline. As a corporate practice and a career specialization, the field is primarily normative. In academia, descriptive approaches are also taken. The range and quantity of business ethical issues reflects the degree to which business is perceived to be at odds with non-economic social values. Historically, interest in business ethics accelerated dramatically during the 1980s and 1990s, both within major corporations and within academia. For example, today most major corporate websites lay emphasis on commitment to promoting non-economic social values under a variety of headings (e.g. ethics codes, social responsibility charters). In some cases, corporations have rebranded their core values in the light of business ethical considerations (e.g. BP's "beyond petroleum" environmental tilt).

The term CSR itself came in to common use in the early 1970s although it was seldom abbreviated. The term stakeholder, meaning those impacted by an organization's activities, was used to describe corporate owners beyond shareholders from around 1989 (Marlin & Marlin, 2003).

The era of Economic liberalisation focuses on :

Auditing and reporting

To demonstrate good business citizenship, firms can report in accordance with a number of CSR reporting guidelines or standards, including :

- Account Ability's AA1000 standard, based on John Elkington's triple bottom line (3BL) reporting

- Global Reporting Initiative's Sustainability Reporting Guidelines

- Verile's Monitoring Guidelines

- Social Accountability International's SA8000 standard

- Green Globe Certification / Standard

- The ISO 14000 environmental management standard

- The United Nations Global Compact promotes companies reporting in the format of a Communication on Progress (COP). A COP report describes the company's implementation of the Compact's ten universal principles.

- The United Nations Intergovernmental Working Group of Experts on International Standards of Accounting and Reporting (ISAR) provides voluntary technical guidance on eco-efficiency indicators, corporate responsibility reporting and corporate governance disclosure.

Some nations require CSR reporting, though agreement on meaningful measurements of social and environmental performance is difficult. Many companies now produce externally audited annual reports that cover Sustainable Development and CSR issues ("Triple Bottom Line Reports"), but the reports vary widely in format, style, and evaluation methodology (even within the same industry). Critics dismiss these reports as lip service, citing examples such as Enron's yearly "Corporate Responsibility Annual Report" and tobacco corporations' social reports.

4.4. BANKS AND CORPORATES

Banks, in a broad sense, are institutions whose business is handling other people's money. A joint stock bank, also known as commercial bank, is a company whose business is banking. These are more particularly institutions that deal directly with the general public, as opposed to the merchant banks and other institutions more concerned with trade and industry. These banks specialize in business connected with bills of exchange, especially the acceptance of foreign bills. A merchant banker is thus a financial intermediary who helps in transferring capital from those who possess it to those who need it. Merchant banking includes a wide range of activities such as management of customers' securities, portfolio management, project counseling and appraisal, underwriting of shares and debentures, loan syndication, acting as banker for refund orders, handling interest and dividend warrants, etc. Thus, a merchant banker renders a host of services to corporates and promotes industrial development in the country. Further, there are also investment banks which acquire shares in limited companies on their own account, and not merely as agents for their customers. Sometimes, banks are set up to handle specialized functions for particular industries such as the Industrial Development Bank of India (IDBI), National Bank for Agricultural and Rural Development (NABARD) and Export-Import Bank (Exim Bank).

Banks are thus a critical component of any economy. They provide financing for commercial enterprises, basic financial services to a broad segment of the population and access to payment systems. In addition, some banks are expected to make credit and liquidity available in difficult market conditions. The importance of banks to national economies is underscored by the fact that banking is virtually universally a regulated industry and that banks have access to government safety nets. It is of crucial importance, therefore, that banks have strong corporate governance. -

There has been a great deal of attention given recently to the issue of corporate governance in various national and international forums. In particular, the OECD has issued a set of corporate governance standards and guidelines to help governments "in their efforts to evaluate and improve the legal, institutional and regulatory framework for corporate governance in their countries, and to provide guidance and suggestions for stock exchanges, investors, corporations, and other parties that have a role in the process of developing good corporate governance".

Why Corporate Governance in Banks?

If we examine the need for improving corporate governance in banks, two reasons stand out: (i) Banks exist because they are willing to take on and manage risks. Besides, with the rapid pace of financial innovation and globalization, the face of banking business is undergoing a sea-change. Banking business is becoming more complex and diversified. Risk taking and management in a less regulated competitive market will have to be done in such a way that investors' confidence is not eroded. (ii) Even in a regulated set-up, as it was in India prior to 1991, some big banks in the public sector and a few in the private sector had incurred substantial losses. This, along with the massive failures of non-banking financial Companies (NBFCs), had adversely impacted investors' confidence.

Moreover, protecting the interests of depositors becomes a matter of paramount importance to banks. In other corporates, this is not and need not be so for two reasons: (i) The depositors collectively entrust a very large sum of their hard-earned money to the care of banks. It is found that in India, the depositor's contribution was well over 15.5

times the shareholders' stake in banks as early as in March 2001. This is bound to be much more now. (ii) The depositors are very large in number and are scattered and have little say in the administration of banks. In other corporates, big lenders do exercise the right to direct the management. In any case, the lenders' stake in them might not exceed 2 or 3 times the owners' stake.

Banks deal in people's funds and should, therefore, act as trustees of the depositors. Regulators the world over have recognized the vulnerability of depositors to the whims of managerial misadventures in banks and, therefore, have been regulating banks more tightly than other corproates.

To sum up, the objective of governance in banks should first be protection of depositors' interests and then be to "optimise" the shareholders' interests. All other considerations would fall in place once these two are achieved.

As part of its ongoing efforts to address supervisory issues, the Basel Committee on Banking Supervision (BCBS) has been active in drawing from the collective supervisory experience of its members and other supervisors in issuing supervisory guidance to foster safe and sound banking practices. The committee was set up to reinforce the importance of the OECD principles for banks, to draw attention to corporate governance issues addressed by previous committees, and to present some new topics related to corporate governance for banks and their supervisors to consider.

Banking supervision cannot function effectively if sound corporate governance is not in place and, consequently, banking supervisors have a strong interest in ensuring that there is effective corporate governance at every banking organisation. Supervisory experience underscores the necessity of having the appropriate levels of accountability and checks and balances within each bank. Put plainly, sound corporate governance makes the work of supervisors infinitely easier. Sound corporate governance can contribute to a collaborative working relationship between bank management and bank supervisors.

Recent sound practice papers issued by the Basel Committee underscore the need for banks to set strategies for their operations and establish accountability for executing these strategies. In addition, transparency of information related to existing conditions, decisions and actions is integrally related to accountability in that it gives market participants sufficient information with which to judge the management of a bank.

Sound Corporate Governance Practices in Banks

As mentioned earlier, supervisors have a keen interest in determining that banks have sound corporate governance. The practices to be viewed as critical elements of any corporate governance process are :

1. Establishing strategic objectives and a set of corporate values that are communicated throughout the banking organization: It is difficult to conduct the activities of an organization when there are no strategic objectives or guiding corporate values. Therefore, the board should establish strategies that will direct the ongoing activities of the bank. It should also take the lead in establishing the "tone at the top" and approving corporate values for itself, senior management and other employees. The values should recognize the critical importance of having timely and frank discussions of problems. In particular, it is important that the values prohibit corruption and bribery in corporate activities, both in internal dealings and external transactions.

The board of directors should ensure that the senior management implements policies that prohibit (or strictly limit) activities and relationships that diminish the quality of corporate governance, such as :

- Conflicts of interest.

- Lending to officers and employees and other forms of self-dealing (e.g., internal lending should be limited to lending consistent with market terms and to certain types of loans, and reports of insider lending should be provided to the board, and be subject to review by internal and external auditors).

- Providing preferential treatment to related parties and other favoured entities (e.g., lending on highly favourable terms, covering trading losses, waiving commissions). Processes should be established that allow the board to monitor compliance with these policies and ensure that deviations are reported to an appropriate level of management.

2. Setting and enforcing clear lines of responsibility and accountability throughout the organization: Few public policy issues have moved from the wings to center stage as quickly and decisively as corporate governance. Virtually every major industrialized country as well as the Organisation for Economic Cooperation and Development and the World Bank has made efforts in recent years to refine their views on how large industrial corporations should be organized and governed. Academics in both law and economics have also been intensely focused on corporate governance. Oddly enough, despite the general focus on this topic, very little attention has been paid to the corporate governance of banks. This is particularly strange in light of the fact that a significant amount of attention has been paid to the role that the banks themselves play in the governance of other sorts of firms. In this paper, we explain the role that corporate governance plays in corporate performance and argue that commercial banks pose unique corporate governance problems for managers and regulators, as well as for claimants on the firms' cash flows such as investors and depositors.

The intellectual debate in corporate governance has focused on two very different issues. The first concerns whether corporate governance should focus exclusively on protecting the interests of equity claimants in the corporation, or whether corporate governance should instead expand its focus to deal with the problems of other groups, called "stakeholders" or non-shareholder constituencies. The second issue of importance to corporate governance scholars begins with the assumption that corporate governance should concern itself exclusively with the challenge of protecting equity claimants, and attempts to specify ways in which the corporation can better safeguard those interests.

The Anglo-American model of corporate governance differs from the Franco-German model of corporate governance in its treatment of both issues. The Anglo-American model takes the view that the exclusive focus of corporate governance should be to maximize shareholder value. To the extent that shareholder wealth maximization conflicts with the interests of other corporate constituencies, those other interests should be ignored, unless management is legally required to take those other interests into account. The Franco-German approach to corporate governance, by contrast, considers corporations to be "industrial partnerships" in which the interests of long-term stakeholders, particularly banks and employee groups should be accorded at least the same amount of respect as those of shareholders. The Anglo-American model of corporate governance also differs from the Franco-German model in its choice of preferred solutions to the core problems of governance. Specifically, the market for corporate control lies at the heart of the Anglo-American system of corporate governance, while the salutary role of non-shareholder constituencies, particularly banks and workers, is central to the Franco-German governance model.

At the outset, we note that it is strange that paradigms of corporate governance differ on the basis of national boundaries rather than on the basis of the indigenous characteristics of the firms being governed. Of course, the extent to which either the Anglo-American model or the Franco-German model of corporate governance exists as more than theoretical constructs is a matter of debate. There is doubt about the extent to which the European system really protects the interests of non-shareholder constituencies, just as there is debate over whether the interests of U.S. management are as closely aligned with those of shareholders as is generally claimed. However, differences in corporate governance systems do exist. Moreover, the distinctions between these two paradigms of corporate governance are quite useful in framing the analysis in this paper.

We begin with an overview of the topic of corporate governance and proceed to a discussion of the particular corporate governance problems of banks. We embrace the view that a corporation is best defined as a complex web or "nexus" of contractual relationships among the various claimants to the cash flows of the enterprise. The defining principle of American corporate governance is that an implicit term of the contract between shareholders and the firm is that the duty of managers and directors is to maximize firm value for shareholders. The legal manifestations of these contracts are the fiduciary duties of care and loyalty that officers and directors owe to shareholders.

Although we support the general principle that fiduciary duties should be owed exclusively to shareholders, we believe that the scope of the duties and obligations of corporate officers and directors should be expanded in the special case of banks. Specifically, directors and officers of banks should be charged with a heightened duty to ensure the safety and soundness of these enterprises. Their duties should not run exclusively to shareholders. Thus, we support a hybrid approach to corporate governance in which most firms are governed according to the U.S. model, while banks are governed according to a variant of the Franco-German paradigm. Our variant calls for bank directors to expand the scope of their fiduciary duties beyond shareholders to include creditors. In particular, we call on bank directors to take solvency risk explicitly and systematically into account when making decisions, or else face personal liability for failure to do so.

4.5. CORPORATE GOVERNANCE IN BANKING SECTOR

Corporate governance systems depend upon a set of institutions (laws, regulations, contracts, and norms) that enable Bank to operate as the central element of a competitive market economy. These institutions ensure that the internal government procedures adopted by the Banks are enforced and that management is responsible to owners (shareholders) and other stakeholders.

The key point in this definition is that the public and private sectors Banks work together to develop a set of rules that are binding on all and which establish the ways in which have to govern themselves.

Where to start?

A useful first step in creating or reforming the corporate governance system is to look at the principles laid out by the OECD and adopted by the governments which are members of the OECD itself.

The Rights of Shareholders

These include a set of rights including secure ownership of their shares, the right to full disclosure of information, voting rights, participation in decisions on sale or modification of corporate assets including mergers and new share issues. The guidelines go on to specify a host of other issues connected to the basic concern of protecting the value of the corporation.

The Equitable Treatment of Shareholders

Here the OECD is concerned with protecting minority shareholders' rights by setting up systems that keep insiders, including managers and directors, from taking advantage of their roles. Insider trading, for example, is explicitly prohibited and directors should disclose any material interests regarding transactions.

The Role of Stakeholders in Corporate Governance

The OECD recognizes that there are other stakeholders in companies in addition to shareholders. Banks, bond holders and workers, for example are important stakeholders in the way in which companies perform and make decisions. The OECD guidelines lay out several general provisions for protecting stakeholder interests.

Disclosure and Transparency

The OECD also lays out a number of provisions for the disclosure and communication of key facts about the company ranging from financial details to governance structures including the board of directors and their remuneration. The guidelines also specify that annual audits should be performed by independent auditors in accordance with high quality standards.

The Responsibilities of the Board

The guidelines provide a great deal of detail about the functions of the board in protecting the company, its shareholders, and its stakeholders. These include concerns about corporate strategy, risk, executive compensation and performance, as well as accounting and reporting systems.

A well-regulated banking sector

A healthy banking system is an absolute prerequisite for a well-functioning stock market and corporate sector. The banking sector provides the necessary capital and liquidity for corporate transactions and growth. Good governance within the banking system is especially important in developing countries where banks provide most of the finance. Moreover, financial market liberalization has exposed banks to more fluctuations and to new credit risks.

As evidenced by the Asian and Russian crises, poorly governed banking systems and massive capital flight can seriously damage national economies. A framework that fosters a flexible yet safe and sound financial system is, therefore, crucial.

What is needed in this context are sound prudential requirements and effective banking supervision practices. The Bank for International Settlements provides some useful sets of standards and best practices that can be adapted to particular national systems to establish a sound, level playing field. The proposed New Capital Adequacy Framework offers more flexible and calibrated techniques for assessing capital adequacy and risks to align regulatory capital requirements with underlying risks.

The proposed framework is based on three pillars: minimum capital requirements; supervisory review of an institution's internal assessment process and capital adequacy; and effective use of disclosure to strengthen market discipline as a complement to supervisory efforts. (Detailed information about the standards is available online at the Bank for International Settlements' website) Together the pillars offer important mechanisms for ensuring that banks are managing capital responsibly and efficiently, and that they are financially viable. Moreover, each pillar fosters better governance within the corporate sector.

The first pillar, minimum capital requirements, provides banks and supervisors with a range of tools to accurately assess different types of risks so that a bank has an adequate amount of capital to cover these risks.

Determining the accuracy of capital adequacy requirements is useful only if such requirements are upheld. To this end, each bank has a set of policies and procedures to ensure adequate capital requirements, in particular, and sound bank management, in general. Such measures include undertaking credit risks; monitoring and disciplining large borrowers effectively; and adhering to stringent auditing procedures. In the end, it is the results of these internal processes that count and these depend on two factors.

The first is effective corporate governance of borrowers, usually firms. Banks need accurate information about a firm's condition in order to assess risks appropriately. This demands that a company has well-documented and thoroughly audited books that are made available to banks. In other words, firms need to have a well functioning corporate governance system in place. Banks that lend to companies under false pretences or to firms that engage in fraud can suffer greatly when a firm is unable to repay its obligations. (Banks that recently lent to Enron are examples.) Hence, banks are increasingly making sound corporate governance practices a lending prerequisite.

The usefulness of internal bank processes also hinges on effective monitoring mechanisms that ensure compliance. The second pillar, the supervisory review process, is designed to do just that. Based on a set of standards, supervisors review a bank's internal processes in order to determine to extent the which these measures assess the bank's capital adequacy needs relative to a thorough evaluation of risks.

Besides the new standards that are part of the proposed framework, the Basel Committee on Banking Supervision developed a comprehensive blueprint for an effective supervisory system entitled, "Core Principles for Effective Banking Supervision." The "Core Principles Methodology" explains how to implement and assess the principles.

The third pillar of the new framework bolsters the first two by strengthening disclosure requirements and thereby enhancing market discipline. The only way that market participants can evaluate the soundness of their dealings with banks is if they can understand and have access to banks' risk profiles and capital adequacy positions in a timely manner. Regular disclosure of this information will discipline banks because market participants will flock to banks that have sound practices and are financially viable. Market participants will avoid banks that take excessive risks without adequate capital provisions, and possibly those that do not undertake enough risk in order to remain competitive.

Disclosing banks' risk assessments can also improve corporate governance. Banks' company risk rankings provide important information about a corporation's financial viability. Shareholders can use this information to press management for changes or to discipline management by shifting their capital elsewhere.

Similarly, disclosing information about banks' ownership structures and relationships with other firms or the public sector fosters good governance of banks and corporations, and helps prevent moral hazard and financial meltdowns. Many developing countries experienced financial crises that stemmed from undisclosed transactions that were not conducted at arm's length.

Examples include the frequent and substantial direct or connected lending by banks to firms in the bank's business group that were not creditworthy. When this happens on a large scale, the impact can be as great as any other economic shock. In short, links between the government, banks and corporations should be at least disclosed so that shareholders and board members can respond accordingly, and at best, served. Similarly, there is increasing discussion about whether or not developing countries should require that commercial and investment banking activities be separated.

Banking sector flourishes under sound securities, markets

Efficient securities markets discipline insiders by sending price signals rapidly and allowing investors to liquidate their investment quickly and inexpensively. This affects the value of a company's shares and a company's access to capital. A well functioning securities market requires :

- Laws governing how corporate equity and debt securities are issued and traded, and stipulating the responsibilities and liabilities of securities issuers and market intermediaries (brokers, accounting firms and investment advisers) that are based on transparency and fairness. In particular, laws and regulations governing pension funds and allowing for open-ended mutual funds are extremely important;

- Stock-exchange listing requirements should be based on transparency and stringent disclosure standards independent share registries would be useful in this regard;
- Laws protecting minority shareholders' rights (See the OECD Principles of Corporate Governance); and
- A government body such as a securities commission that has independent and qualified regulators empowered to regulate corporate securities transactions and to enforce securities laws.

The existence of competitive markets as external control on companies has greater impact as smooth recurring of Banks.

The existence of competitive markets is an important external control on companies forcing them to be efficient and productive lest they lose market share or go under. The lack of competitive markets discourages entrepreneurship, fosters management entrenchment and corruption and lowers productivity. For this reason, it is crucial that laws and regulations establish a commercial environment that is fair yet competitive. To this end, governments can :

- Remove barriers to entry
- Enact competition and anti-trust laws
- Eliminate protectionist barriers including the protection of monopolies
- Eliminate preferential treatment schemes such as subsidies, quotas, tax exemptions, to name a few
- Establish fair trade priorities
- Remove restrictions on foreign direct investment and foreign exchange
- Reduce the cost of setting-up and running a formal business

The World Trade Organization (www.wto.org) and the International Labour Organisation (www.ilo.org) provide standards that are useful in creating a competitive and equitable commercial environment.

The Banking sector can have important role to play in take-over markets. The illustration is given below :

Another vital element of a competitive commercial environment is the existence of a market for corporate control. Such a market disciplines insiders and encourages them to improve firm performance or risk losing control over it, or face bankruptcy. This means that firms or investors can, under certain conditions, take control of an underperforming firm in the hopes that, by running it themselves, they will create additional value. In this case, specific laws and implementation rules (such as those governing proxy contests and takeovers) need to be clear and specific (so that management can't delay or derail a takeover attempt).

Establishing effective takeover markets can incur strong resistance even within well-developed economies. This is particularly the case when companies are faced with hostile takeovers or with takeover bids from foreign companies or foreign nationals. For the past twelve years, the European Commission has been struggling to adopt a Europe-wide takeover code that would pave the way for cross-border mergers and acquisitions. An attempt last year was fiercely opposed by German companies who argued that the failure to ban the use of shares with special voting rights including the right to veto a takeover put companies without such rights at a disadvantage. Moreover, such rights could be used to shield management from accountability. A new code that suspends special voting rights during takeovers and stipulates fair share acquisition prices is being finalized and has the initial endorsement of German leaders and the European internal market commissioner.

Establishing orderly and transparent takeover markets is crucial so that mergers and acquisitions can be and are undertaken for economically justifiable reasons in a manner that is fair to all stakeholders. Valid takeover contests and well-implemented mergers and acquisitions usually strengthen corporate governance by improving a firm's internal management and thereby providing more economic benefits for outsiders and creditors than if the firm continued to underperform under the former management.

4.6. CORPORATE GOVERNANCE IN INDIAN BANK

Corporate Governance is concerned with the systems and processes for ensuring proper accountability, probity and openness in the conduct of an organization's business. Thus, it is the process under which the organizations try to hold the balance between economic and social goals and between individual and communal goals. In a nutshell, we can say that the corporate governance framework strives to the efficient use of resources and equally to require accountability for the stewardship of those resources. The basic aim of Corporate Governance is to align as nearly as possible the interests of individuals, corporations and society.

Corporate Governance has three important features, and these are :

(a) Transparency in operations and decision-making.

(b) Accountability for the decisions taken

(c) Accountability for the stakeholders

For example, it the duty of the Board members to ensure that in the case of shareholders, the investments and the return on investment is safeguarded. This means that the managements of the company have to ensure that the decisions taken by them actually create wealth and do not destroy wealth. In case the net earnings are less than the cost of capital it is considered as net destruction of wealth and can not be considered as good governance.

Indian Banks Some Sound Practices for Corporate Governance

According to the Organization for Economic Cooperation & Development (OECD), some of the sound corporate governance practices for Banks in India include :-

(a) The Board of a bank should be broad-based with induction of non executive directors of sufficient caliber and number for their independent views to carry the desired weight in the Board's decisions.

(b) The Board is responsible to establish certain strategic objectives and a set of corporate values for the senior management, employees and the Board members themselves.

(c) The Board should set and enforce clear systems & procedures, lines of responsibility and accountability throughout the bank.

(d) The Board should ensure that senior managers exercise supervisory role with respect to line managers in specific business and activities with great sense of propriety.

(e) The Board should recognize the importance of the audit process, communicate this importance throughout the bank and ensure effective utilization of the work by internal and external auditors.

Recent Steps Taken by Banks in India for Corporate Governance

(a) Induction of non executive members on the Boards

(b) Constitution of various Committees like Management Committee, Audit Committee, Investor's Grievances Committee, ALM Committee etc.

(c) Gradual implementation of prudential norms as prescribed by RBI

(d) Introduction of Citizen's Charter in Banks

(e) Implementation of "Know Your Customer" concept.

The primary responsibility for good governance lies with the Board of Directors and the senior management of the Bank.

The concept of corporate governance, which emerged as a response to corporate failure, and widespread dissatisfaction with the way many corporates function, has become one of the wide and deep discussions across the globe recently. It primarily hinges on complete transparency, integrity and accountability of the management. There is also an increasingly greater focus on investor protection and public interest. Corporate governance is concerned with the values, vision and visibility. It is about the value orientation of the organisation, ethical norms for its performance, the direction of development and social accomplishment of the organisation and the visibility of its performance and practices.

Indian Banking Industry

Indian banking has around 200 years of history and has undergone many transformations since independence. But, Liberalisation, Privatisation, Globalisation and Information Technology are currently changing the Indian banking radically.

Earlier, banking was virtually a monopoly of the public sector banks with full protection from the State. But the process of reforms in the Indian banking system has thrown them out to more liberal and free market forces. Now the banks, more particularly the public sector ones, feel the real heat of the competition. The interest rate cuts, dwindling margins and more number of players to serve a reduced number of bankable clients have all added to the worries of the banks. The customer has finally come to hold the center stage and all banking products are tailor-made to suit his tastes and preferences. This sudden change in the banking environment has bereaved the banks of all their comforts and many of them are finding it extremely difficult to cope with the change.

Need for Corporate Governance in Banks

- Since banks are important players in the Indian financial system, special focus on the Corporate Governance in the banking sector becomes critical.

- The Reserve Bank of India, as a regulator, has the responsibility on the nature of Corporate Governance in the banking sector.

- To the extent that banks have systemic implications, Corporate Governance in the banks is of critical importance.

- Given the dominance of public ownership in the banking system in India, corporate practices in the banking sector would also set the standards for Corporate Governance in the private sector.

- With a view to reducing the possible fiscal burden of recapitalising the PSBs, attention towards Corporate Governance in the banking sector assumes added importance.

Pre requisites for Good Governance

There are some pre-requisites for good corporate governance. They are :

- A proper system consisting of clearly defined and adequate structure of roles, authority and responsibility.
- Vision, principles and norms which indicate development path, normative considerations and guidelines and norms for performance.
- A proper system for guiding, monitoring, reporting and control.

Recommendations of the Birla Committee

The report of the Committee on Corporate Governance, set up by the Securities and Exchange board of India, under the Chairmanship of Kumar Mangalam Birla, is the first formal and comprehensive attempt to evolve a Code of Corporate Governance, in the context of prevailing conditions of governance in Indian companies, as well as the state of capital markets. The committee has identified the three key constituents of corporate governance.

Why care about Corporate Governance?

Corporate governance matters for development in the following ways :

1. Increased access to financing investment, growth, employment
2. Lower cost of capital and higher valuation of investment, growth
3. Better operational performance a better allocation of resources, better management, creates wealth
4. Less risk at the firm and country level fewer defaults, fewer financial crises.
5. Better relationship with stakeholders improved environment, social/ labour
6. All of these relationships matter for growth, employment, poverty reduction.
7. Empirical evidence has documented these relationships.
8. Quite strong relationship

But so far mainly documented for non-financial corporations that are listed on stock exchanges.

Better corporate governance translates into somewhat higher returns on assets. But much better higher returns on investment, relative to cost of capital.

What do we know about CG of banks?

*   So far, little evidence on the standard CG-questions and more complex issues of CG and regulation/supervision.
*   Some have documented effects of bank ownership
    *   LSV /BCL: banking systems with more state-ownership: less stable, less efficient, worse credit allocation
    *   More foreign banks: more stable, efficient, competitive effects
*   Few so far investigated bank governance
    *   Many studies on effects of laws & regulations for corporations
    *   But few on banks, except for recent evidence from Caprio, Laeven, and Levine

Bank ownership: possible ownership and control patterns

*   Widely-held, not controlled by any single owner
*   Controlling owner
    *   Family (individual)
    *   State
    *   Widely-held (non-financial) corporation
    *   Widely-held financial institution
    *   Other (trust, foundation, which may be "shell")
*   With small or large deviations of control rights from ownership (cash-flow) rights

Difference between ownership and control

*   Controlling owner vs. widely-held bank
    *   Controlling owner if direct + indirect control> (say) 10%
    *   Widely-held if no entity owns> 10% directly + indirectly
*   Ultimate owners versus direct owners
*   If any major shareholders are (FINF) corporations, then find their major shareholders. Continue until ultimate owners
    *   **Example:** Shareholder has x% of indirect control over bank A if she controls directly firm C that, in turn, controls firm B, which directly controls x% of bank A. Control chain can be long
*   Controlling owner - if any - will be the one with the maximum direct + indirect control

(1) Banks are generally not widely-held

(2) Family ownership of banks is very important, and so is the state ownership

(3) Cross-country differences are large, though

> o In 14 of 44 countries, the controlling owner has more than 50% of voting shares. But in Australia, Canada, Ireland, UK, and US, either NO bank has a controlling owner or the average is less than 2%.

(4) Legal protection of shareholder is associated with more widely-held banks, i.e., with better legal protection less need/desire for close control

Effects of control on firms and banks

- Some firm and bank owners can be better than others
- Insiders may expropriate firm resources
  - o Expropriation = theft, transfer pricing, asset stripping, nepotism, and "perquisites" that benefit insiders
- Ownership & shareholder protection laws ® value
  - o Cash-flow rights up ® expropriation less ® valuation higher
  - o Shareholder protection laws better ® valuations up
  - o Interactions between ownership & protection ® valuation
- Are banks different?
  - o Laws insufficient with powerful, complex, opaque banks?
  - o Regulations: laws superfluous or superceded?
  - o Role of ownership less important?

Valuation and shareholder rights

- Valuation
  - o Market-to-Book Value
- Rights: shareholder rights (0-6)
  - (1) Mail proxy votes
  - (2) Not required to deposit shares
  - (3) Proportional representation of minorities on board allowed
  - (4) Oppressed minorities mechanism

(5) When shareholders have preemptive rights, they can only be waived by a shareholders meeting

Supervision and regulation powers

* Official
  * o Power to change internal organization, management, directors, etc
  * o Power to supercede rights of shareholders to intervene
  * o Power, get reports from, and take legal action against auditor
* Restrict
  * o Regulatory restrictions on banks (i) securities, (ii) insurance, (iii) real estate, and (iv) owning non-financial firms
* Capital
  * o Regulatory restrictions on source of funds, BIS minimum, risk-based, are loan, security.
* Independence
  * o Degree of supervisory independence from government and banks

Valuation effects of bank ownership and equity rights

* When cash-flow rights of controlling owner higher and equity rights stronger, bank valuation higher. Effects can be large :
  * o A one-standard deviation increase in shareholder protection laws (1.25) raises market-to-book by 0.28, or 21 percent of mean
  * o A one-standard deviation increase in cash flow rights (0.27) raises market-to-book by 0.42, or 31 percent of mean
* More cash-flow rights can even offset some of negative effects of weak equity rights
* Suggests strong owners, both in share and in their rights, can help corporate governance of banks
* Surprising, perhaps, quality of supervision and the degree of regulation does not robustly influence valuations

Implications for CG of banks

* Bank ownership

- o Be very careful on state ownership: negatively related to valuation, stability and efficiency
- o Consider inviting foreign banks
- Bank governance, regulation and supervision
  - o Strong private owners necessary, but they need to have their own capital at stake
  - o Better shareholder protection laws can improve functioning of banks
  - o Supervision/regulation less effective in monitoring banks

4.7. CORPORATE MISCONDUCT AND MISGOVERNANCE

The term 'corporate misconduct or misgovernance' refers to 'frauds committed by corporate entities to wilfully erode shareholder value'. The corporate misconduct stretches beyond malpractices in accounting, reporting, operations and misconduct. Such improper corporate behaviour and aberrations are the product of (i) a culture of corporate greed and (ii) opportunism and rationalisation. The 1990s have been mainly responsible for instigating the corporates to indulge in various types of misconduct. During this period, investor expectations were higher and corporate boards exploited these expectations to their benefit. During this period, equity based executive compensation had become the trend. Outside professionals such as auditors, lawyers, consultants who are expected to function as a part of cheeks and balances system began to emphasise closer client relationships. Contrary to popular belief, the decadence in corporate conduct is not a recent phenomenon. Rather corporate scandals have existed across continents for the past several years. Many of the corporate codes and regulatory bodies have been created to deal with various instances of misconduct. In this chapter, we have considered the reasons for corporate frauds, the concerns for and the ways to prevent corporate corruption, and the need to promote business ethics besides a few notorious instances of corporate scandals and misconduct.

Corporate Misconduct and Frauds

In recent times, fraud in the business arena has become a matter of serious concern because of its capacity to cause business collapses as seen in the western industrialized countries and particularly the Anglo-American societies. In the U.S.A., business crime spread rapidly in the post 1970s as a consequence of widespread abuses of office during Richard Nixon's presidency. Money-laundering became more prevalent and a variety of frauds occurred in real estate and insurance. In the 1980s, crimes got extended to take within its purview the mainstream business. Reagan Presidency reinforced the view that "what is good for business is good for America. Imbued by this ideal, he implemented a policy of deregulations. This resulted in insolvencies, extraordinary recklessness in banks, and savings and loan institutions. Flagrant misconduct such as insider dealing in securities, distortions in capital market and increased debt burden on companies became the order of the day.

Similar has been the story in Britain where the financial sector had been marred by a series of corporate scandals in the 1980s. Hutton has pointed out towards serious frauds worth billions of pounds against thousands of individual investors at Lloyd's insurance market. Another instance has been the hostile take over of Distillers by Guinness in 1986 by using massive share support scheme that violated the Companies Act. The Bank of Credit and Commerce International (BCCI),' the world's, seventh largest bank with 400 branches in 70 countries had to be closed down by the Bank of England in 1991 for proven cases of money-laundering, bribery, corruption, evasion of foreign exchange regulations, falsification of accounts, misappropriation of depositors' money, blackmail and massive fraud.

The Reasons for Corporate Misconduct

In the advanced capitalist societies, there prevails a culture of competition which spurs and motivates rule breaking. As the power of business concern expands, so does the potential for abuse of power. The occurrence of structural changes such as globalisation of economy, advancements in technology, industrial concentration, deregulation etc., have all provided opportunities for deviance. As a consequence, organisational power, trust and impression management have begun to be used as a tool in deviant activity and also as a camouflage against discovery, enforcement and prosecution. The organisation is used to rationalise and justify the misconduct. The unified and coherent facade of a business organisation serves as a political arena where reality is constantly redefined. Managers themselves are shaped as well as shape the business environment through the use of various strategies.

Basically, corporate frauds are a manifestation of the failure of corporate governance mechanism. The factors responsible for such failure and for the possibility of misuse of corporate power have been described below:

Excess of Management Power

The Anglo-American model of corporate governance is based on separation of ownership and control. As pointed out by Berle and Means, in this model, the management group comes to possess enormous powers. The concentration of management power has been legitimised on the grounds of orthodox notions and ideologies such as the primacy of contract; the reliance on self regulation, emphasis on profit maximisation, and the belief in market's correction discipline. All these reasons are adduced to justify the appropriateness of the current corporate governance structure. However, such concentration of power is liable to be abused.

Ballooning Executive Pays

Excessive executive pays are regarded as an instance of corruption and corporate scandals. Excessive executive remuneration are a way to siphon off corporate funds through seemingly legitimate means. Conyon et. al. have given evidence to show that the salaries of senior managers have risen faster than average earnings.

Misuse of Executive Power

Corporate corruption occurs due to misuse of power by corporate bigways. The present scheme of corporate governance places enormous powers in the hands of corporate management with the possibility of its misuse. The various ways adopted to abuse this power include the following:

   (i) Expropriation of investors' funds by simply taking out the money and putting it elsewhere from where it may be easily taken out,

   (ii) Transfer pricing mechanism,

   (iii) Rise in perquisites of higher level officers,

   (iv) Irrational expansion of the firm,

   (v) Adoption of anti-takeover measures.

Lack of public understanding

Corporate corruption could be deep if fubern far the malaise. Very few in the public realize that activities of large corporations affect a large number of people. The activities of large corporations kill or injure far mare persons due to unsafe working conditions, violation of consumer law: environmental degradation etc. There is misconception in the public mind that corporate violations are usually of a non-criminal nature. Consequently, corporate misdemeanours largely go unpunished and hardly raise public scorn.

Problem of externality

There are several reasons which encourage corporations to overlook their social responsibilities. Markets are not perfect. Due to imperfect markets, maximization of profit does not lead to maximization of social wealth. Firms maximize their profits by externalizing more of their costs i.e., transferring the costs on to third parties. These externalities occur in the farm of unemployment, adverse impact on environment, saving on costs of training etc. As to who bears these costs, different countries have answered this question in their own ways.

Bureaucratic corporate structures

In a modern corporation, the responsibility for a corporate act is often distributed among a number of parties. Corporate acts are brought about by several actions or omissions of many different people. For example, one team may design a car, another may test it, and the third build it. One group may defraud buyers and the other silently cooperate therein. No one feels himself responsible because of large scale bureaucratic nature of the corporate organizations.

Corporate Corruption

The two basic factors of personal motivation and organizational opportunity must come together for corruption to occur. Corruption is often caused by a combination of greed and lax control mechanisms which provide the motive and the opportunity. If there is no opportunity, there can be no corruption. The attractiveness of an opportunity is strongly influenced by an individual's perception of how likely "he or she is likely to get caught, how severe will be the punishment and how big will be the payoff from indulgence in corruption. Most people who are likely to commit white collar crime such as bribery, hold positions of trust. They have easy access to other people's property and also the opportunity to use the authority or position of trust corruptly. They tend to rationalize their corrupt practices a process called "Neutralization". Through neutralization, the perpetrators convince themselves that the victim deserved it and that they were just getting even or that they are just borrowing or taking it for the family's sake.

White collar crime, fraud, corruption and the stewardship in the corporate sector had always been thought to exist in the past. But their magnitude in the present times has assumed alarming proportion and jeopardized the very existence of corporate organizations.

Corruption Prevention Strategies

To tackle the menace of corruption in any organisation, most of the anti-corruption agencies have recommended the adoption of corruption prevention strategies. The specific component of these strategies may vary but their effective implementation should lead to similar ends.

Legal Action as a Deterrent

Perhaps the greatest single factor that can keep a check on corruption by the directors is the initiation of strong legal action against the erring corporate official. An Australian

National Safety Council, Casel needs a mention to illustrate the suggested remedy. In this case, the honorary non-executive chairman Max Eise was held to be personally liable for the corporate losses of $97 million. No one suggested that Mr. Eise had acted dishonestly. Indeed he was associated with the National Safety Council in a spirit of public service and not for any financial gain. The Judge agreed that Mr. Eise had been a victim of fraud by the Managing Director Mr. Friendrich whom he described as "manipulative and deceitful and an accomplished liar." Nevertheless, Mr. Eise was found not to have performed his directorial duties adequately. The Judge, Justice Tadgell held: "As the complexity of commerce has greatly intensified... the community has of necessity come to expect more than formerly from directors... In response, Parliaments and Courts have found it necessary in legislation and litigation to refer to the demands made on directors in more exacting terms than formerly, and the standard of capability required of them has correspondingly increased."

Business ethics as an antidote to corporate frauds

Focusing on corporate frauds, corporations should promote ethical practices in all their pursuits. Ethics relate to right or wrong in the workplace and placing emphasis on doing what is right regarding products/ services. Attention to ethics in the workplace sensitizes both the top management and staff about how they should act, providing them with a strong moral compass in times of crisis and confusion.

Following benefits arise from practising ethical business :

- Cultivation of strong teamwork and productivity
- Ensuring that policies and procedures are legal and ethical
- Avoidance of criminal acts of "omission" and early detection and reporting of violations
- Promotion of strong public image of the business
- Legitimisation of managerial actions, improvement of trust in relationships between individuals and groups
- Focus on quality management, strategic planning and diversity management.

To reap the above benefits, organisations should establish ethics programme that conveys corporate values. These programmes often use codes and policies to guide decisions and behaviour. These programmes also provide guidance in the event of ethical dilemma. The code of ethics may help define acceptable behaviour, promote high standards of practice, provide a benchmark for self evaluation, establish a framework for professional behaviour and act as a mark of occupational maturity. The following guidelines could be used to establish an ethics programme:

- Establish organisational rules to manage ethics
- Schedule ongoing assessment of ethics requirements
- Establish required operating values and behaviour
- Align organizational behaviour with operating values
- Develop awareness and sensitivity of ethical values
- Integrate ethical guidelines into decision making
- Structure mechanisms to resolve ethical dilemmas
- Facilitate ongoing evaluation and updation of the programme
- Help convince employees that attention to ethics is not just a knee-jerk reaction done to get out of trouble or improve public image
- Service in public entities is a public trust requiring employees to place loyalty to the Constitution, the laws, and ethical principles above private gain
- Employees shall not hold financial interest that conflicts with conscientious performance of duty.
- An employee shall not solicit or accept any gift or other item of monetary value from any person or entity seeking official action from, doing business with, or conducting activities regulated by the employee's agency, or whose interest may be substantially affected by the performance or non-performance of the employee's duties.
- Employees shall put forth honest effort in the performance of their duties
- Employees shall act impartially and not give preferential treatment to any private organisation or individual
- Employees shall protect and conserve the entity's property and shall not use it for other than authorised activities
- Employees shall disclose waste, fraud, abuse and corruption to appropriate authorities. Employees shall comply with all laws, rules and regulations
- Employees shall endeavour to avoid any actions creating the appearance that they are violating the law or ethical standards
- All the entity's information and records must be treated as confidential and disclosure or dissemination thereof should be within the operation of applicable rules and procedures.

Whistle Blower's protection

Whistle blowing is an attempt by a corporate employee to disclose wrongdoing in or by an organization. It may be internal i.e., reporting a corrupt activity to the higher ups in the organization. When the wrongdoing is reported to external authorities such as the Government or media, the whistle blowing is said to be external. As part of good governance, a corporate should provide channels and procedures that facilitate internal whistle blowing e.g., setting toll free telephone or appointing an ethics officer empowered to investigate into the allega-tion, publishing such programmes etc. The whistle blowers should be protected against termination, harassment or discrimination. The revised clause 49 has included non-mandatory obligation upon listed companies to formulate and disclose its whistleblower's policy in the Report on Corporate Governance.

Illustrative cases of corporate misconduct

Since the opening up of the Indian economy in 1992, there has been a spate of corporate frauds and corporate scandals. The blame has to be equally shared,by the private sector and the public enterprises including the banks. The casualty of all such scams and scandals has been the erosion of both investors' faith and wealth in the equity markets. According to Indian Express the market capitalization of listed shares on the BSE plunged from Rs. 6,92,565 crore on the last trading day of 2000 to Rs. 5,66,897 crore on December 31, 2001 washing a whopping Rs. 1,25,000 crore of investors money. On a year to year basis, the sensex lost around 710 points (1896). The loss works out to over 5.6 per cent of India's projected GDP figure of Rs. 22,20,000 crore for the year 2001-2002. There has been massive price rigging in the counters of companies like Himachal Futuristic Communications, Lupin Laboratories, Zee Tele Films Padmini Polymers, etc. The SEBI woke up rather too late and by that time, the speculators like Ketan Parekh had defrauded the whole system. It also exposed the lack of transparency in the working of the regulator and an insufficient system of checks and balances. So much so, even the mutual funds which have been proclaimed to be the best bet for small investors, suffered heavily.

The concept of mutual funds got a dressing down at the hands of UTI. The US-64 considered a holy cow, has been badly mauled. The cat came out of the bag when UTI announced on Dec. 29, 2001 that the "trust has decided that for holdings in excess of 5000 units per unit holder, the repurchae price as on May 31, 2003 will be Rs. 10 per unit or NAV, whichever is higher. This facility was also extended to unit holders of US-64. On Dec. 28, 2001, the NAV of US-64 had been declared as Rs. 5.81 per unit. This lack of accountability has ridden the UTI with all kinds of manipulative mechanisms leading to a rude shock to the investing middle classes of India who had hitherto considered UTI a great moral agency.

Enron Imbroglio

Formed from a 1987 merger between Houston Natural Gas, and Inter North two natural gas pipeline companies, Enron profited from the 1990's deregulation of gas and electricity prices. It transformed itself from power provider to energy broker. Its operations stretched across four continents. After a decade of its inception, Enron dominated the energy "spot" and future markets. In July 2001, Fortune ranked Enron as the seventh largest US Corporation by turnover based on reported revenues for the previous year. After the bust of new technology boom, Enron's shares continued to rise on the basis of its strong revenues and profitability. The company owned substantial tangible assets pipelines, power stations, reservoirs besides enjoying vast revenues from trading business. Enron was quite good in cultivating political connections. The chairman of Enron, Kenneth Lay distributed largesse to politicians. Its political clout enabled the passage of a rule change that explicitly excluded energy derivative contracts and interest rate swaps from the Government's supervision giving the company freehand to speculate in energy futures. Under Clinton administration, political donations won the company over one billion dollars in subsidized loans. Bill Clinton felicitated Kenneth Lay as a good 'Corporate citizen' on account of company's enlightened personnel policies. Senior members of George W. Bush including President's economic adviser were on Enron's pay-roll. Investigations by the U.S. Congress into Enron collapse revealed that 212 of the 248 members who sat on Senate Committees had been in receipt of money from either Enron or Arthur Anderson.

Citibank despite being aware of the laxity enjoyed by the company overlooked the same. This was done by these conglomerates by employing a complex process of financial engineering. Moreover the banks had already passed the risk to million of employees whose pension money was invested in Enron shares/bonds/ credit derivatives/ special purpose entities. Enron encouraged its own employees to become investors on a large scale. About 57 percent of Enron's 21000 workers were members of the plan.

Enron's strategy of pushing up its prices was plain fraud. With active connivance of its auditor Arthur Anderson, it used borrowed funds to jack up the stock prices. Borrowings were made through undisclosed partnerships off the balance sheet. No surprise, the company collapsed under the weight of huge repayment dues. It finally reached the stage of bankruptcy and had to file for protection under Chapter of the U.S. Bankruptcy Act on December 2, 2001. The use of creative accounting, and special purpose entities were nothing but fraudulent devices employed to defraud innocent investors.

Initiatives to Curb Corporate Misconduct

Insider Trading

By insider trading is meant 'the use by corporate insiders and their associates of confidential price sensitive information which has been obtained by them by virtue of their position so as to make a profit or avoid a loss by dealing in the securities of the relevant company'.

Notwithstanding the controversy as to who shall be regarded as "corporate insiders", it is definite that the directors are the foremost among them. The resentment against use of confidential information springs from the fact that it places the insiders in a far better position than others in the market. The insiders have the advantage of trading on certainties whereas others can at best speculate.

Grounds of Criticism of Insider Trading

Unfair

Uncertainty constitutes an integral part of business. Trading in corporate securities entails a substantial element of chance. But the odds should not be heavily cast against outsiders vis-a-vis corporate insiders. It is unfair for an insider to have an edge over an outsider merely due to former's access to company information something not attributable to any merit. The consequent advantage secured by an insider must be considered illegitimate because it has not been acquired due to any intrinsic merit. Moreover, insider information is 'property' rightfully belonging to the company. The use of such property by corporate insiders constitutes a breach of their fiduciary obligation. Professor Manne, however, has criticized the resort to the principle of fairness for denouncing insider dealings to be hypocritical and self-righteous. The theory of open market postulates that for an appropriate appraisal of securities, there must be a free interaction between buyers and sellers. Interposing by the corporate insiders, equipped as they are with price sensitive information, would necessarily impair the accuracy of open market appraisal and the supposed liquidity of securities. Unfairness is the outcome of inequality in the bargaining position of insiders and outsiders. The equality of bargaining position as a legal concept has come to be recognized as an integral part of the corpus of common law unless in future some higher principle emerges to override it and render it infructuous as a juristic tool.

Adverse Impact on Corporate Management

Directors owe the fiduciary duty of loyalty to the company and to their corporate principles. If they are allowed to indulge in unrestricted insider trading, their attention will be diverted from the pursuit of company's objectives. Moreover, if insiders are allowed to trade freely on the basis of corporate insider information, they will be tempted to manipulate corporate developments and news release so as to accommodate their personal dealings. Frequent insider trading would tend to harm company's reputation. A company whose insiders are suspected of manipulating inside information, may experience considerable difficulty in finding sufficient capital. Financial and professional institutions would be averse to becoming associated with managements of such dubious integrity.

Jolt to Investors Confidence in Stock Market

Public-confidence in the fairness of nation's capital market is an obvious necessity. Securities markets play a vital role in the provision and allocation of scarce capital resources. Public respect for stock markets will dwindle if it begins to be suspected that market unduly favours certain privileged individuals. It becomes imperative to ensure that each investor who comes to the market should have the feeling that he is subject to the same degree of risk as everyone else. Mere suspicion that abuses occur is just as harmful as the proof that abuses have taken place.

Adverse Impact on Foreign Investment

The suspicion that insider trading is prevalent on a nation's securities markets is injurious to its domestic and international goodwill. Compared to domestic investors, international investors possess far greater degree of latitude of investing their funds. It is for these reasons that many countries have made considerable efforts to arrest such abuse as 'insider trading'.

Need for regulation of Insider Trading: Of the various reasons emphasizing the need for inhibiting insider trading, the most important is the need to preserve public confidence. To allow insiders to trade on the basis of privileged information destroys the occurrence of a level playing field between the insiders as investors and the remaining shareholders. It highlights the need to regulate insider trading. In England, the need to regulate insider trading was considered by the Cohen Committee for the first time.

The Jenkins Committee endorsed the proposition that directors should not be penalized merely because they are in a better position than outsiders. But where the directors acted on "a particular piece of information" not known to the general body of shareholders, remedy should be available to the other party to the transaction. The Committee was of the view that when a director makes a personal profit and causes a loss to the company by making use of privileged information, the company should be able to seek restitution.

Jenkins Committee also opined that when a director of a company is guilty of improper use of confidential information which might be expected to materially affect the value of the securities involved in a transaction, he should be liable to compensate the person who suffers from his action.

Sanctions against Insider Trading

Judicial opinion differs as regards the form of penalty which could be deemed a sufficient deterrent against insider trading practices. Some of these sanctions are mentioned below:

Imposition of Fine

In Mount Charlette Investment Ltd. & Gale Lister Co. Ltd.", the Panel (Norbury Insulation Panel) censured Grimshawe & Co. for violating General Principle of the City Panel on Take Overs and Mergers on the ground that as a financial adviser, it had advised shareholder to do something different whilst it had itself done something entirely different in the same case.

In Johnson & Sons Ltd., an employee of a bank which was involved in the negotiations for public offer, traded through a nominee on the basis of inside information. The Panel named the individual concerned, reprimanded him and ordered him to pay the profits to an approved charity.

Imposition of fine on the erring director is considered to be a sufficient remedy although some of the penologists have doubted its effectiveness in regulating insider trading. The argument is that unless the prescribed fine is in excess of the wrong-doer's illicit profit, it would fail to act as a deterrent. To constitute a meaningful punishment, the fine must not only deprive the wrong doer of his gain but go beyond that.

Surrender of Profits

Accordingly, if a director uses information which he has obtained by virtue of his office or in the course of carrying out his functions, for dealing in shares or debentures of his company, or any other company in the same group as his company and that information is not generally available to investors (e.g., where the company has not communicated certain information to Stock Exchange), the company may require the director to account for the profits and it is immaterial that the company could not have engaged in the transaction. The liability of the director to account for the profits may be based on the following facts :

a. that the information was given to him confidentially so that he is under an equitable obligation to use it exclusively for the company's benefit

b. in considering the information as property of the company which the directors have misappropriated to their own advantage

c. on the simple ground of conflict of director's fiduciary duty with company's interests.

However, the right to recover a profit improperly made by a director belongs to the company and not to the shareholders. It would be necessary to consider two things in all cases of misuse of insider information, namely :

1. Whether the action is within the scope of duties of office or employment of the insider, and

2. Whether the knowledge, of which profitable use was made can be described as property of the company?

What is crucial is not acquisition of information but its 'use'.

Public Disgrace Through Reporting

An important aspect of reporting is that it hits companies and an individual's greatest assets, namely, their self-respect and prestige, both of which are considered to be essential for increasing the ability to produce material wealth. From the standpoint of deterrence, the significance of disclosure consists in the fact that others could see the misfortune that befalls the wrongdoers.

IMPRISONMENT IN CASE OF BLATANT MISUSE OF INSIDE INFORMATION

In extreme cases, imprisonment proves really effective.

Some illustrations of insider trading in India

On April 16, 1996, Hindustan Lever announced the merger of Brook Bond Lipton India with itself. Frenetic trading in the two scrips preceded the announcement. Once the information became public, the trading volume came down and so did the stock price.

On March 7, 1996, Ciba Geigy AG and Sandoz AG announced the merger of their worldwide operations. Trading flared up in the stock of Indian subsidiaries a couple of days before the formal announcement of merger.

On July 9, 1996, Tata Group announced buyout of Unisys Corp to form Tata Unisys Ltd. The company's stock began its upward journey from June 1996 and reached the peak on announcement and then started its journey southwards.

Insider trading has plagued the stock markets since the times New York Stock Exchange was established in 1792. The 1980's biggest US stock market scandal Ivan Boesky Dennis Levine Michael Milken Junk Bond scandal had its origin as well its sustenance in insider trading. Surprisingly, even the U.S. Securities Exchange Commission has found it difficult to tackle insider training. It however, successfully convicted Boesky for insider trading.

For every known case of insider trading, many more go unnoticed. All established cases of insider trading were those when insiders went overboard in their greed to maximise their gains. Since the issue of Insider Regulations by SEBI in 1992, there has been no proven case of insider trading. In most cases, SEBI has mostly reacted to press reports

Modus operandi of insider trading

Insider trading can be prompted by prior information of various events connected with the company. These events may include the ensuing announcement of financial results, information on bonus or right offers or more sensitive information relating to merger / amalgamation/demerger.

Proving Insider Trading

Proving insider trading is not easy because it cannot be done by anybody who may be even remotely connected with a particular company. Unless trading volume and price movements are exceptionally abnormal, it is not noticed. Even a low level employee who buys 100 shares with prior knowledge of an event concerning his company can be hauled up for insider trading but such small events would go unnoticed unless there is more than one person trading in huge quantities on such counter.

If abnormal price movements in a stock are noticed, it is not difficult to obtain trading records for the period of abnormal activity. In that case, it will not be very difficult to trace the ultimate beneficiary. This is how Ivan Boesky insider trading ring was busted in US in 1980s. But this is a post mortem exercise. By the time the regulator traces the trail and nails down the ultimate beneficiary, the culprits may have flown the coop. An insider trading transaction can be camouflaged in various ways without the trail leading to the insider. Classic method of doing this would be to form an interest group of different people who have access to price sensitive information by virtue of their position. The person having the inside information may pass it to another who buys on the former's behalf or vice versa. Loot is thereby shared without anybody tracking the culprit. This form of insider trading is called 'tippee trading'.

Similarly in the course of his everyday job, a newspaper correspondent may get across to price sensitive information ahead of everybody else except the auditors. So does the employee in the finance department who comes to know of company's performance before the outside world. There is nothing to prevent these people from using the information to their advantage. In this sense, insider trading may be said to refer to, 'any trading done by a person who has access to price-sensitive information by virtue of either his position or any contract ahead of rest of the market.'

Regulation of Insider Trading in India

With the object of preventing and curbing insider trading, the SEBI has issued the SEBI (Prohibition of Insider Trading) Regulations, 1992. The regulations render insider trading a criminal offence, in certain circumstances, punishable under the SEBI Act. Insider trading is not only a menace but also a big stumbling block in the adoption of good governance practices by corporates. For self-aggrandizement and profiteering, all norms of honesty and fair play are thrown to the winds and the image of the entire corporate sector is tarnished due to manipulative activities of a few. Hence, the regulations to curb insider trading have a special significance in the Indian context. Insider trading was first tackled in the USA. In UK, insider dealing was made a criminal offence in 1980 and there exists a separate piece of legislation, viz., the Company Securities (Insider Dealing) Act, 1985 which was re-enacted in 1993.

The Regulations do not define the term "insider trading", Instead, regulation 2(d) defines dealing in securities, "as an act of buying, selling, or agreeing to buy, sell or deal in any securities by any person either as principal or agent."

Insider Meaning and definition

Regulation 2(e) defines insider as many person who is, or was connected with the company, or is deemed to have been connected with the company and who is reasonably expected to have access, by virtue of such connection, to unpublished price sensitive information in respect of securities of the company, or who has received or has/had access to such unpublished price sensitive information."

Categories of insiders

The definition of the term 'insider' contains two phrases, namely (i) any person connected with the company, and (ii) any person deemed to have been connected with the company. Accordingly, insiders can be of two types as follows :

(i) **Primary Insiders:** They are directly connected with the company. By virtue of their proximity or nearness to the company, they have easier access to unpublished price sensitive information.

(ii) **Secondary Insiders:** They are indirect insiders. They are not as close to the company as the primary insiders but nonetheless are placed in a relationship whereby they have or are expected to have access to unpublished price sensitive information.

Meaning of 'unpublished price sensitive information'

'Unpublished price sensitive information' is any information which relates to following matters or is of concern, directly or indirectly to the company, and is not generally known or published by such company for general information but which 'if published or known, is likely to materially affect the price of securities of the company':

(i) financial results (both half yearly and annual) of the company,

(ii) intended declaration of dividends (both interim and final),

(iii) issue of shares by way of public offer, rights, bonus, etc., or buy back of securities,

(iv) any major expansion plans or execution of new projects,

(v) amalgamation, mergers and take overs,

(vi) disposal of the whole or substantially the whole of the undertaking,

(vii) such other information as may affect the earnings of the company,

(viii) significant policies, plans or operations of the company.

Prohibitions Against Insider Trading

There are three prohibitions:

(1) **Prohibition of Dealings Regulation:** Prohibits an insider from dealing in securities on his behalf, or on behalf of any other person on the basis of unpublished price sensitive information. This prohibition is applicable to dealings in the securities of listed companies.

(2) **Prohibition on communication:** The communication of any unpublished price sensitive information by an insider to another.

(3) Counselling or procuring any person to deal in securities on the basis of unpublished price sensitive information is also prohibited. The last two clauses seek to prohibit any attempt to aid or assist any person to indulge in insider trading.

Contravention and punishment

Regulation merely declares that if any insider who deals in securities, communicates any information or counsels any person for dealing in securities in contravention of provisions of regulation, he shall be guilty of insider. Any such contravention will be an offence under the SEBI Act and the offender will be punishable under section 24 of the Act : the Punishment under section 24 of the SEBI Act shall be of one year of imprisonment, or fine or both. The amount of fine has not been prescribed. SEBI can file a complaint with the previous approval of Central Government.

Investigation

Before taking any action, SEBI undertakes investigation on the basis of complaints from investors, intermediaries or any other person regarding insider trading, or suo motu upon its own knowledge or information to protect the investors. In making this investigation, SEBI has been obliged to give reasonable notice to the insider against whom investigation is sought to be ordered unless SEBI dispenses with such notice. The investigator is to submit a report to SEBI within one month of conclusion of investigation. The findings of investigation are to be conveyed to the "insider" who can make representation to SEBI.

Disclosure of information

One of the effective ways to curb insider trading is to require companies to make adequate disclosures. Sachar Committee had rightly pointed out that unfair profits can be made in share dealings by using confidential information not generally available to investing public. Therefore, disclosure of shareholdings by specified insiders in the Annual Reports and restriction on transfer of shares by specified insiders during a stipulated period prior and subsequent to closure of companies' accounting year could deal with the menace more effectively.

Summary of the 'Model Code of Conduct for Prevention of Insider Trading for Listed Companies

The following model code (Part A) has been formulated by SEBI under regulation 12(1) of the SEBI (Insider Trading) Amendment Regulations, 2002.

Appointment of Compliance Officer

A listed company shall appoint a senior employee as compliance officer who shall report to the MD or CEO. Such officer shall be responsible for setting forth policies, procedures, and monitoring the adherence to rules for preservation of 'price sensitive information'; pre-clearing of designated employees and their dependents' trades, monitoring of trades and implementation of code of conduct under the overall supervision of the Board.

Duty of employees and directors to preserve confidentiality of price sensitive information.

No employee or director shall pass price sensitive information to any person, directly or indirectly.

Handling price sensitive information on 'need to know' basis:

The price sensitive information should be disclosed only to those within the company who need the information to discharge their duty.

Limited access to confidential information

Files containing confidential information shall be kept secure. Computer files must have adequate security of login and password, etc.

Prevention of misuse of 'price sensitive information'

All directors, officers and designated employees shall be subject to following trading restrictions :

i. No trading during the closure of 'trading window', i.e., trading period. The trading window shall remain closed at the time of declaration of financial results, dividends; issue of securities, any major expansion plans or execution of projects: amalgamation, merger, takeovers, and buyback; disposal of whole or substantially whole of undertaking; any changes in policies, plans or operations.

ii. The trading window shall be opened 24 hours after the information mentioned in previous para has been made public.

iii. All directors, officers or designated employees shall make dealings in securities only in a valid trading window but not when the trading window is closed.

iv. In the case of ESOP, exercise option may be allowed during the period, the trading window is closed. But sale of shares in exercise of ESOPs shall not be allowed when trading window is closed.

Pre-clearance of trades

All directors, officers or designated employees shall seek pre-clearance of transactions above the threshold limit fixed by the company. For such clearance, an application shall be made to the Compliance Officer containing particulars about the securities to be dealt in. The concerned director shall give an undertaking to the company that he does not have access or has not received any price sensitive information. In case he has access to or receives 'price sensitive information' after signing the undertaking but before execution of transaction, he shall inform the compliance officer of change in his position and shall refrain from dealing in securities until the information becomes public.

Other restrictions

The order shall be executed within one week after the pre-clearance approval, else the pre-clearance shall have to be reobtained. The directors, etc., shall hold their investments including IPO for a minimum period of 30 days in order to be considered as being held for investment purposes. The holding period may be waived by the compliance officer in case of personal emergency.

Reporting requirements

Details of securities transactions including statement of dependent family members and holdings in securities shall be supplied by the concerned directors, etc., to the compliance officer who is obliged to maintain record of all such declarations.

Penalties for violation of code

The penalties for violation of code may be imposed by the company but such action shall not preclude SEBI from taking any action for violation of the SEBI (Prohibition of Insider Trading) Regulations, 1992.

These rules require the organization to adopt a "Chinese Wall" policy to prevent misuse of confidential information. This is to be done by separating "inside areas" which routinely have access to confidential information from "public areas", i.e., those areas which deal with sale/ marketing/investment/ advice, etc. The employees in "inside area" shall not communicate any price sensitive information to a "public area".

To monitor Chinese wall procedures and trading in client securities based on inside information, the organisation or firm shall restrict trading in certain securities and designate these as "restricted/ grey list". The securities of a listed company shall be put on such list if the organisation is handling any assignment for the listed company or is preparing appraisal report or is handling credit rating assignment and is privy to price sensitive information.

Any security which is being purchased or sold or is being considered for purchase/ sale by the organisation on behalf of the client shall be put on the restricted/ grey list.

Listed intermediaries such as credit rating agencies, AMCs, broking companies, etc., whose securities are listed in a recognised stock exchange shall comply both with Parts A and B of the Code.

4.8. BUSINESS VALUES FOR THE 21ST CENTURY

Cultural Values

Culture has a great influence on an individual's value framework. This chapter discusses the salient features of Western and Indian cultures. It also debates /devise views on cultural relativism, individual relativism, moral relativism and moral absolution.

Cultural relativism states that ethics is a function of culture. Individual relativism says that what is right or wrong depends upon the attitude of the individual. But there are many critics of this view who affirm that moral and ethical views are universal and apply to all cultures.

The twentieth century has been a melting pot of various cultures. Scientific inventions have made the world a global village. The story of this period is one of progress towards greater justice but also one of regression. It has been a time of transition from a fragmented to a global society.

The liberal modern culture is proving to be far less value-dominated and much more self-indulgent. The prevailing state of business, society and polity is influencing long-standing and cherished value systems and causing a drift. But the diversity in behaviour among cultures is only apparent and cultures do agree on certain fundamental ethical standards. A discussion on this would involve anthropology, history, theology and a host of other disciplines.

Culture and tradition of each family is a unit of the total culture of any society and nation. Confusions in the moral life outside which is on one's own inner mind have ruined family traditions. When the moral integrity of individuals in families is destroyed, the morality of the society also slowly wanes. When the general morality of society has decayed, individuals, blinded by uncontrolled passion, start acting without restraint. Lust and greed knows no logic and cares least for better evolution or better culture. When the unity of the human is shattered, and when purity of living and sanctity of thought are destroyed in society, the generation inflicts sorrow and suffering on itself.

Western and Indian Culture

Values are an integral part of any culture and the operation of a society is closely related to these values. The source of values, their transmission, and integration follow a pattern specific to a particular culture. The attitudes of people are conditioned by the value system of their culture (Feather, 1985).

Every culture is an integration of social and religious practices. Components of religion are intrinsically meant to induce particular behavioural patterns, leading to proper attitude formation for a healthy society. Customs are the habits of a community, which secure social approval for conduct and are exemplified in the life of the best people of the community. They are the unwritten laws of social tradition, distinguished from the codified tradition. They are in essence, social approval (Tejomayananda,1994).

Cultures differ in physical setting, economic development, the state of science and technology, literacy rate and in many other ways. The commonly held principle underlying various cultures is that social institution and individual behaviour should be ordered so that they lead to the greatest good for the greatest number (Huddleston, 1998).

Western culture is based upon industrialization and large-scale production. The financiers and industrialists, with the aid of modern scientific knowledge, struggle hard to discover and satisfy new desires, and to the extent an individual has fulfilled his newly created desires, is taught by the day's civilization that he is that much happier than before. Some of the key points of Western culture are :

- Self-centred: Motivated by principle of benefit
- Exploitation of nature: Exploit natural resources to satisfy desires
- Capital intensive: Industrialization and large-scale production
- Market oriented: Creating 'fresh needs' to develop and expand the market.

In spite of the best that may be available in life, a sense of incompleteness drives people to experience spiritual ways of life. Developed nations are not daring to go beyond the values of economic expansion and materialistic gains.

Indian Culture

Very little is known of the origin of the great and rich faith recorded in Bhagavat Gita, the Vedas and Upanishad tells us of their beliefs in seeking the truth and living a harmless life. The Hindu culture is based on this pursuit of truth (specially yoga and karmayoga).

Indian culture operates at two levels the individual and the group. At the individual level it talks of development of the self and at the group level of duties/responsibilities of individual towards social groups, i.e. family and community. Indian culture is oriented towards harmonious group activities. It feels the strong emotional need for vertical leaders. In summary the Hindu culture operates on the following paradigms :

- Development of self: spiritual pre-eminence
- Harmony with nature: living with nature
- Labour intensive: importance to individual sills of production
- Society oriented: importance to family and community as a unit.

These paradigms can be found in the practices of social, virtual and religious pursuit.

Ethical Values and Culture

Cultural relativism is a philosophical position which holds that ethics is a function of culture. According to this view there are no absolute principles in ethics: principles are relative in nature and hold sway within given culture. Relativism points to the fact that deferent individual on culture hold different views about what constitute moral behaviour.

Norman Bowi (1990) presents an analysis of cultural relativism and its relationship to ethical decision making in corporation. He argues that there may not be different moral standards among cultures. Bowi says that there are universal moral principles which applly to all culturees.

Individual Relativism

Individual relativism states that what is right or wrong, good or bad depends on the attitude of individual. The relativists say ethics is irrelevant since there is no universal right or wrong. It is all relative.

The same argument holds true for the advocate of cultural relativism because you need a stable world environment for global harmony. An unstable country or region is highly inimical to international peace and ethical peruses.

Moral Absolutism and Relativism

Moral absolutism is a philosophy that regards morality as absolute. That is not dependent on society or situation. In one sense this is a contrast to moral relativism. In moral absolutism, morals are held to be independent of social customs and are instead inherent in the laws of universe, the nature of humanity or some other fundamental source. Moral absolutism emphasizes on right and wrong.

Moral absolutism regards action as inherently good and bad. Like for instance would always be an action even if done to promote some other good (saving a life). In this sense it is contrast to consequentialism, the view that the morality of action is dependent on the context or consequence of their action.

Many religions have morally absolutist position regarding the system of morality. They therefore regard to such a moral system as absolute perfect an unchangeable. The philosopher Emanuel Kant was a promoter of moral absolutism.

Moral relativism believes that moral standards are nt absolute but instead emerge from social customs and other sources. A modern interpretation of this statements might be that things exist only in the context of the people who observe them.

Moral relativism stands in contrast to moral absolutism. Those who believe immoral absolutism are highly critical of moral relativism. Some have been known to equate it with outright morality and immorality.

Managing for value

Value: the Core

What is the world scenario today? Spectacular human progress, reflecting the contributions of knowledge and technology to society.

Knowledge is the distinct resource of any business or industry. From the start of recorded history to date, it has been growing exponentially. It is estimated that information is doubling every eighteen months. Technical knowledge is said to double itself every six to eight years. Eighty to ninety percent of the Scientists, Engineers, Technicians and Professionals who ever lived are alive today, and are endeavoring to unravel new truths. Reading is the path to knowledge. Knowledge is not only power, but also, the crux of competitive advantage. Even more so, in this technology age, anyone can copy other's plans, products, processes and procedures. However, no one can duplicate the company's intellectual capital knowledge and knowhow the real source of values.

Technology is knowledge, systematically applied to useful purposes. Evolving in leaps and bounds, it will only be advancing and accelerating. The evolution and successful acceptance of technology will be directly proportional to the ability of the technology to improve the quality of human life. A firm is a collection of activities that employ technology, in some form or the other, for its performance. Technological change, shortening the average life cycle of products, leads to new products, new equipment that produce them, and new tools that assist production. It also reduces the time-distances between two points on the earth, and so, increases the speed with which one transacts business. Technology has powerful effect on cost and differentiation and pervades the value chain. It is more than robots replacing people. It is aiding people to do more with same talent and time. That is why, customers and users prefer products and services with latest technology and producers favour technological change, per se.

Expanding knowledge and advancing technology are at the core of everything: everything of value.

Lessons to Learn

In the 19th century, industrial revolution demonstrated technology's power in improving human existence. In the 20th century, science, transformed into usable industrial technology, continued to hold mastery over the material world, with spectacular wealth-creating additions.

A hundred years ago, the idea of men in space, a person living with an implanted organ, or a device capable of following half a billion instructions per second, would have been considered absurd; even ridiculous. Fifty years ago, none had heard of jumbo jets, sound barrier, guided missile, reusable spacecraft, satellite communication, colour television, video cassette recorders, cellular phones, photo copiers, fax machines, answering machines, e-mail, lap-top computers, bio-genetics or bio-engineering.

Years ago, the issue was to manage technology metallurgical, chemical, engineering, electronics, nuclear, etc. It was enough to delegate it to a technical specialist. The specialist went to work and found some competitive advantage. Thus, textile technology improved spinning of cotton and weaving of textiles. Instrument technology helped better process controls. Engineering technology led to mechanization, mass production and economies of scale.

Then, it became managing the use of technology to problems. The focus shifted to choosing a specific technology from many and using it on the jobs, to secure better business results. It involved the user. If the user knew what additions were wanted, and had the resources, these were acquired and used. Mechanisation was used to speed up production and reduce labour; automation, to produce mass at low costs; and robots, to increase manufacturing flexibility. Thus, coal, gas, oil and nuclear fission became available for power generation; and electronics, pneumatics and hydraulics, for instruments in industrial controls.

It has now become managing with technology, to search and exploit new opportunities, for satisfying human wants and needs. There are many: namely, information technology, micro technology, nano technology, biotechnology, and what have you. For example, WalMart's success is in managing the use of two: computer technology and information sharing, in a creative way.

Technology decisions have integrated with business decisions. Also involved is the timing for abandoning an old technology, when its limit of exploitation is imminent; and the launching of new ones. Austrian and Japanese steel makers used a technological innovation oxygen blown converters; and US steel makers, Nucor and Lone Star continuous casting; to create improvements within the value adding chain in the age old, traditional steel making technology.

Learning from the past, we progress into the future, escalating endeavours to discover ever more effective ways of turning knowledge and technology into commercially useful new products and services of value.

Success Through Value

Success once depended on access to material. For example, a steel plant owned captive ore and coal deposits and stone quarries. Vertical integration meant uninterrupted production. Success also depended on size, like the super large plants, biggest machines, longest production runs, or most market share. It may not enable making the best product (or service). However, it could make the most at lowest costs. Cost was invariably a dominant consideration. Success also came from strategies, like making everything within the company.

It controlled availability and quality. Being the sole possessor of critical product knowledge, patents and copyrights, meant monopoly: again success.

Success came from strength in machines, the quick producer, replacing workers. The factory system views labour as a tool, for either specialization or elimination with labour saving devices. Mechanization, automation and robots became reliable replacements for humans. At first, the blue-collar jobs vanished. Then, accounting and finance jobs, taken over by batch-processing computers. Computers are now replacing white-collar workers, seeping into the ranks of middle management.

It appears as though there is no limit to replacing humans in the workplace. For, technological advances in electronically controlled machines are getting smarter and better. It would have made matters worse had not the single-handed, problem-solving specialist started giving way to extensive team approaches. The worker is coming back now as the primary means of adding value to product, process and service.

Success stems from knowledgeably applying a combination of available resources to the needs of each individual situation. Clinging to yesterday's methods of manufacture and product lines, tools, techniques, mindsets and attitudes is a prescription for suicide. Former assumptions, practices and procedures, methods and strategies are increasingly changing. Even what the industrial leaders think as the work their companies should do. This implies that individuals have to increasingly and continuously learn and relearn value adding knowledge and skills.

4.9. MEASURING GOVERNANCE NATURE OF CORPORATE EXCELLENCE

There is a growing global consensus for evolving sensitive reporting and measurement tools to capture the changes and impact of governance, democratization and human rights on development initiatives. Developing such tools is all the more important because participatory democracies with responsive, accountable and transparent governance are essential to reduce poverty.

Measuring Governance

New global standards of governance are emerging and increasingly there is an expressed need to measure governance. Citizens desiring better performance from their governments are anxious to know the extent to which the stated intentions of government are matched by results. Also, citizen's groups and businesses want information on governance to help them decide the design and target of their advocacy and lobbying efforts. To determine how good a job they are doing, and whether or not mid-course corrections are needed, government officials and decision-makers want to know more about people's perception of governance.

Similarly, governments would like to measure the governance of their neighbours to build better relations and to assess security risks. Donor agencies and other investors need such knowledge to base sound investment decisions. As official development assistance declines, and is replaced by investment from private investors, pension funds and the like, development financing is likely to be more and more attracted to countries with good governance. Many international agencies may want to know governance practice in countries to understand their compliance to international conventions and treaties. International ratings agencies such as Standard and Poor, Moody's, International Country Risk Guide, and Business Environmental Risk Intelligence, incorporate governance in their indicators of creditworthiness.

These developments have led to global interest in measuring the performance of governments and of good governance using indicators of governance and institutional quality. Governance can be measured in terms of effort or in terms of results. Both have to happen simultaneously to give a true picture of the nature of governance. So measures of effort have to be complemented with measures of tangible progress towards objectives.

However, choosing the right combination of indicators raises many methodological issues. For example, laws and regulations enacted may not be enforced, or anti-corruption units may focus on harassing political opponents, expatriate advisors may be used to carry out policy formulation and coordination roles, thus sidelining competing counterparts. Staff given specialized training may be transferred to assignments, where the training is irrelevant, to clip their wings, thus perpetuating problems of low government effectiveness. Thus, the need for extraordinary care in selecting which aspect of governance we are seeking to measure and what inferences are drawn.

Most governance variables are proxies, whether qualitative (e.g. perception of corruption) or quantitative (e.g., percentage of total procurement, subject to open, competitive bidding). In the case of the former, subjective values are typically ranked on a continuous (ordinal) scale. The use of proxies runs a high risk of measurement errors and biased estimates. There is also a common problem that some explanatory variables used in studies are poor proxies to describe actual legal and political processes. A good proxy indicator should be relevant, and permit regular observation and reasonably objective interpretation to determine the change in its value or status.

Each aspect of governance can be measured in different ways, depending on the purpose. In the case of corruption, for instance, one can measure the perception of corruption, or experiences of corruption (e.g., frequency of bribery in procurement contracts, amount of bribes as a percentage of contract value, frequency of bribery in permit applications, percentage of public sector jobs bought, percentage of public servants income as bribes). Alternatively, one may use costs of public services, costs of construction, assets of leaders, etc. as more indirect, but 'harder', proxies of corruption, bad governance and wastage.

A good measurement of corruption can present hard data and generate public debate on the issue. But corruption is only a symptom of underlying weaknesses in governance. Thus, complementary diagnostic surveys on governance can potentially identify the underlying causes of corruption. For example, whether a public institution bases its hiring and promotion decisions on merit or favouritism and whether there are auditing and oversight mechanisms in budget allocation, may explain the extent of corruption in the institution. However, diagnostic surveys can depoliticize the discussion of corruption by shifting public attention from people to institutions.

Specific Indicators of good governance

These are indicators of 'good governance'. The reader will note that though these indicators claim to measure good governance, they are primarily meant to protect the interest of foreign investment.

Democracy and Governance Programme Indicators

Developed by the USAID to measure country and programme performance, they are organised under four governance objectives:

- Strengthened rule of law and respect for human rights- More genuine and competitive political processes
- Increased development of a politically active civil society
- More transparent and accountable governance institutions

Indicators are organised under programme 'objective' and 'intermediate results' being measured. Among the hundreds of indicators listed, one example is: 'Judicial salary and benefits as a percentage of what a comparable professional makes in private practice'. This comes under the first objective, and the intermediate result: 'Effective and fair legal institutions'. A suggested guideline is that compensation should ideally be

80-90 per cent of that in private practice. Compensation significantly less than that is a warning flag for rule of law.

Practical tools to enforce good governance

The emphasis of most of the methods is on the role of the poorest and most vulnerable groups of society in decision-making and their access to benefits. Because of the sustained effort of NGOs, activists, people's movements and citizen's groups, already a wide range of alternative methodologies and practical experiences exist at national and international levels.

These experiences should be used to determine future work to obtain a comprehensive coverage drawing on the strengths of each methodology and tool, and to exploit the potential for combining them to provide a more integrated analysis and practice of good governance.

Stakeholder Analysis

Stakeholder analysis is a vital tool for identifying those people, groups and organizations that have significant and legitimate interests in specific issues. Clear understanding of the potential roles and contributions of the many different stakeholders is a fundamental prerequisite for a successful participatory governance process, and stakeholder analysis is a basic tool for achieving this understanding.

Stakeholders are those :

- Affected by the issue or those whose activities strongly affect the issue;
- Possessing information, resources and expertise needed for strategy formulation and implementation, and
- Controlling relevant implementation instruments.

To ensure a balanced representation, the analysis should examine and identify stakeholders across a number of different dimensions. For example, the analysis should separately identify relevant groups and interests within the public sector, within the private sector, and within social and community sectors. In addition, the analysis can seek out potential stakeholders to ensure proper representation in relation to gender, ethnicity, poverty, or other locally relevant criterion.

A comprehensive long listing of stakeholders is the starting point for a stakeholder mapping and analysis. It can be seen in terms of five generally sequential stages of activity:

Specifying issuess to be addressed: Stakeholders are defined and identified in relation to a specific issue; people and groups only have a concrete 'stake' in a specific issue or topic.

- Long listing: With respect to the specified issue, a 'long list' of possible stakeholders, as comprehensive as feasible, should be prepared, guided by the general categories of stakeholder groups.

- Stakeholder mapping: The 'long list' of stakeholders can then be analyzed by different criteria or attributes. This will help determine clusters of stakeholders that may exhibit different levels of interest, capacities, and relevance for the issue.

- Identify gaps: Identify areas where capacity building is necessary for effective stakeholder participation, and highlight possible 'gaps' in the array of stakeholders. One of the several forms of stakeholder mapping is by degree of stake and degree of influence.

- Clarifying the rights: is an essential step in identifying those legitimate claims and entitlements that might be affected by decisions. It is also a pre-condition for effective identification of those stakeholder groups who should be entitled to a formal role in the process, and decisions on entitlements.

- Verify analysis and assess stakeholders' availability and commitment: With additional informants and information sources, review the initial analysis to ensure that no key and relevant stakeholders are omitted. Also, assess the identified stakeholders' availability and degree of commitment to meaningful participation in the process.

- Mobilise and sustain effective participation of stakeholders: Strategies should be tailored to the different groups of stakeholders as analyzed and classified above. For example, empowerment strategies could be applied to those stakeholders with high stake but little power or influence.

Social Audit

Social audit has attracted increasing attention in recent years. There are several reasons for this. Deregulation has left many government and non-government enterprises with much more independence and given managers significant control over decision-making. Yet many people are dissatisfied with the way such freedom is utilized. Another reason is that the techno-managerial approach gives primacy to the market and economic growth. Social audit stems from the conviction that human well being should be central to human endeavour and that human well being should be the basis of measurement. One major response has been a growth in concern about corporate governance and accountability. Another development has been a more intellectual one of communitarianism supported by Amitai Etzioni and a new discipline of socio-economics. This intellectual movement is based on the view that it is possible to develop more socially responsible organisations, ones that are more accountable to their stakeholders.

In particular there is a concern about social costs, an increasing disparity between income levels at the top and bottom of society, insecurity of employment and low quality of working life, neglect of effects of business decisions on communities, and that profit and costs are the overriding factors in determining product and service standards. An important part of addressing these issues is to develop the ability to monitor existing practices to determine their social costs and benefits both internally and externally. In this context, social audit and social reporting are becoming more established as important ways of doing this on a regular basis.

Social auditing is the process of verifying expenditures actually made as to their sufficiency with regard to stated social aims. Social auditing is a very important step in terms of accountability and making budgets responsive to the needs of those living in poverty. Social auditing and more generally participatory budgeting require transparent systems of information flow. Very often a significant obstacle to participatory budgeting is the lack of information made available to the public.

Seven Key Features of a Social Audit

Getting the evidence: Hard data from households, schools and communities, as well as from the service itself, are gathered systematically to guide planning and action.

Community participation: Communities not only co-produce the data, but, through focus groups and workshops involving community representatives, they help design local and national solutions.

Impartiality: Community-based audit by a neutral third party can help to build a culture of transparency and strengthen service credibility.

Stakeholder buy-in: All those who have a significant stake in service delivery are actively involved throughout the audit, from the initial stages of design to implementing community-led solutions.

No finger-pointing: A social audit is intended to focus on system flaws and programme content, rather than on individuals or organizations. Even negative endings can be framed as a starting point for improvement.

Repeat audits: Several audit cycles are usually needed to measure impact and progress over time, and to focus planning efforts where they can be most effective.

Dissemination of results: A communication strategy, including feedback to communities, mapping and media dissemination is part of every social audit design.

Financial audits aim to make businesses accountable to shareholders, governments to legislators and NGOs to donors for the finances they manage. Social audits make organizations more accountable for the social objectives they declare. Characterizing an audit as 'social' does not mean that costs and finance are not examined, the central concern of a social audit is how resources are used for social objectives including how resources can be better mobilized to meet those objectives.

Social accountability cannot be achieved by looking only at internal records of performance, however well and honestly an organisation, government or non-government may keep these. A social audit must include the experience of the people the organization intends to serve. Organizations that engage in the social audit process convey the message that they are serious about accountability, equity, effectiveness and value for money.

Social Audit: The MKSS Model

Mazdoor Kisan Shakti Sangathan (Movement for the Empowerment of Peasants and Workers, or MKSS) working in one of India's poorest areas in the State of Rajasthan, has given new meaning to the demand for the right to information in the context of development. MKSS has been active since the late 1980s, initially in trying to get the poor people of Rajasthan to pressurise the government to enforce minimum-wage regulations on its employment generation programmes in the drought prone area. Discrepancies had constantly arisen between the experience of the villagers hired in these projects or supposedly benefiting from them and the government's claims.

In 1990, MKSS began to take up the more general issue of government's transparency and accountability. It took four entire years for the MKSS to get the right to view bills, vouchers and employment rolls of development projects from the government at the panchayat (lowest unit of local self-government) level. It took another three years to get the right to copy documents, which was key since certified copies were needed to use as evidence when registering prima facie cases of corruption. Moreover, people needed time and assistance to interpret the technical details in these official documents. MKSS was active in cross-checking stories told in official documents with villagers.

They looked for discrepancies between the records and villagers' own experiences as labourers on public works projects, anti-poverty schemes, and as consumers in ration shops. In these shops, rationed amounts of basic staples and kerosene are made available to villagers at subsidized prices upon display of ration cards. Sales records were checked against ration cards to see whether the records had been falsified to conceal diversion to the open market where store owners could receive higher prices.

Starting in December of 1994, the MKSS began to hold public hearings (jan sunwayi). Elected representatives, local government officials and all local citizens are invited to attend these hearings, which have more resonance with popular notions of dispute settlement than with legal court hearings. Detailed accounts are read to villagers who then point out discrepancies between the accounts and their experiences. For example, some of these hearings revealed that sometimes villagers who had never received payment had been listed as beneficiaries of anti-poverty schemes and that local building con tractors received payment for works that were never performed.

In such cases of corruption, the MKSS attempted to use public pressure to force government and elected officials to return the amount that had they had embezzled to the same village and used public humiliation as a deterrent to future corruption. These hearings also served to reveal a connection between the wealth of the rich and the villagers' continuing poverty. Because of the testimonies and the analysis of public documents, people were able to trace how public moneys had actually been spent. Also the villagers gained confidence as a result of the mass mobilization around the hearings. They have gone on to raise other local issues of governance.

There is currently considerable interest in social audit and associated approaches internationally. However there is considerable diversity of approaches and methods and motives behind, social audit initiatives. One would not seek to homogenize and standardize practices since it carries a risk of undermining the richness and energy of a new set of initiatives. Clearly the social audit approach and process is emancipatory for those disempowered within the society.

Nature of Corporate Excellence

Excellence refers to a surpassing quality or merit and an outstanding performance in one's field of work. A vital force for growth, it sets an example or a standard of behaviour and galvanizes people into doing something remarkable. It can act as a bedrock of renaissance in developing countries. The varied forms of excellence can be seen in competitions, inventions and innovations. Breaking one's own past record, succeeding at an extremely difficult task, doing some unique or' pioneering work, developing one's multi-dimensional potentials, rendering selfless service and pursuing self-enlightenment, etc., are a few others. Excellence involves immense mental and physical effort and sacrifice and must be recognised and suitably rewarded. It flourishes in the nurseries of corporations, but is noticeably absent in corporations which are hotbeds of politics and cold-blooded commercialism. It requires both individual and team efforts.

Forms of Corporate Excellence

Khandwalla classifies corporate excellence into six categories:

- competitive
- rejuvenatory
- institutionalized
- creative
- missionary
- versatile.

Measurement of Corporate Excellence

Corporate excellence can be measured by using both objective and subjective criteria. Peters and Waterman offer six objective criteria for measuring corporate excellence: compound asset growth, compound equity growth, average ratio of market to book value, average return on total capital, average return on equity and average return on sales. While Fortune has adopted subjective criteria of ranking corporations: quality of management, quality. of products or services, innovation, value as a long-term investment, financial soundness, ability to attract, develop and keep talented people, community and environmental responsibility, and use of corporate assets. As Hickman and Silva observe, although both objective and subjective criteria help identify truly excellent corporations, they fail to do justice to every type of corporation at any given stage of development. For example, Dana Corporation, Boeing and Texas Instruments provided excellent models, yet subsequently each of these corporations suffered setbacks which according to the criteria suggested by Peters and Waterman might knock them off the list.

Blake and Mouton visualise that excellent corporations are able to accomplish and sustain outstandingly high returns on investments over long periods of time. Their managers have learnt to identify and create new opportunities while meeting present responsibilities. They plan for the future in the face of increasing competition and rising costs. Excellent corporations unceasingly strive to enhance the competence of their managers to manage and their employees to produce. The technological aspects of marketing, production, research and development are sophisticatedly upgraded. The concept of synergy is understood and exploited. Excellence involves the enrichment of the capacities of corporate members for using facts and data as the basis for thinking and analysing and for finding quick and valid solutions to the inevitable parade of problems of operational life. Excellent corporations aim to ensure that their members apply their unbounded energies in a self committed and enthusiastic way to contribute to reaching different but challenging corporate objectives. They suggest that grid OD can be used to accomplish corporate excellence. The six phases of grid OD provide a systematic yet aggressive approach to excellence. Phases 1, 2 and 3 provide for the development of individuals in their relationships with others, development of teams in which they work, and resolution of differences between work teams. Phases 4, 5 and 6 relate to planning. Likewise, Prasad shows that in the context of today's environment, OD can be used to accomplish corporate excellence.

Khandwalla found that the entrepreneurial (organic) mode of management (as opposed to the bureaucratic mode) was associated with excellent corporations in India. It is called pioneering innovative mode and is marked by commitment to novel, relatively sophisticated technologies, products or services, risk-taking, creativity and innovation, informality and adaptability. Maheshwari identified entrepreneurial orientation as an attribute of excellent corporations; good at anticipating problems, trends and emergencies and quick in responding to them; diagnosing weaknesses; making intuitively right decisions, commitment to getting results and getting people excited about big goals.

Jain and Ansari's brief compilation of the achievements of entrepreneurs revealed the competencies of risk-taking, self-confidence, persistence, persuasion/negotiation, innovativeness, coping with stress and hard work and commitment. Likewise, Manimala points out the following characteristics of pioneering and innovative entrepreneurs: search for new growth ideas and efforts to develop them in-house rather than just borrowing these ideas, preference for building up one's own resources, expertise and capabilities, efforts to enter new markets through new products, stress on quality and reliability of products, professionalism, reliance on expertise, teamwork, trust and paternalism, and network of industrial contacts.

Pitfalls to be Avoided

Labich identifies a set of six pitfalls which must be avoided to rule out possibilities of corporate failures. These are :

- identity crisis
- failure of vision
- the drastic squeeze
- the glue sticks, and slick notion
- anybody out there perspective
- enemies within.

First, corporate failure may result from an identity crisis. A lack of fundamental understanding of what the enterprise is all about, leads to capricious climate and corporation drifts.

4.10. THEORIES OF THE FIRM AND RELATED ISSUES

There are three definitions of the firm derived from economics, law and sociology. According to the Economics perspective, the firm is a production firm. The firm represents a technological transformation of inputs into outputs for which there is an economic value or market price. In the contemporary context when we refer to the firm, it is about the legal context or partnership or sole proprietorship. The legal view implies the duties and role of the directors of the board, while the economic definition emphasizes among the others, moral duty of directors.

So long as society regards the firm as the primary means of wealth creation, maximizing efficiency is the only legitimacy. Therefore, according to this view, the moral duty of the directors is to ensure the maximum efficient use of the firms' resources.

The second definition of the firm is that, it is a nexus of contracts. According to this view, the firm represents a series of contracts and the main corporate governance problems are to ensure the maintenance of integrity of contracts. It is uneconomical, and in fact, impossible to specify complete contracts because the informal ion required may not be available or may be costly to obtain, so good governance mechanism should provide space for reconstructing. In other words, the board of directors should protect the firms' implicit and explicit contracts.

The third definition of the firm is that of a social organization. In the transaction of the firm implicit the human dimensions are a number of non-economic considerations which play an important role.

The Theory of the Firm

The theories either belong to the principal agent or the incomplete contracting approach. The themes that are categorized in incomplete contracting model are the ones that stress the importance of employment relationship theories that stress the importance of ownership of assets for affecting incentives and some recent work on implicit contracts.

An obvious advantage of the corporation is its ability to raise capital for investment and the liquidity that share ownership provides. The corporation should give back to the society for the social costs of conducting business. The corporation gets certain privileges and, therefore, they should earn them by paying back to the society.

Limited liability was a feature of various companies acts of the nineteenth century. Britain and America, it had already emerged through free contracting.

Contractual approach essentially focuses on strategic decisions. Accordingly, it is essential to place strategic decision making in the centre of analysis.

Easterbrook and Fischel (1991) interpret law, as an attempt to write complete contracts in an optimal way supplying the rules that parties would agree upon; if they unite complete contracts.

Williamson (2002) describes human actors in more details, focuses on behavioural regularities; differences in alternative modes of governance and the importance of cooperative adaptation.

Most empirical evidence on corporate governance comes from USA. Governance structures and systems had not developed in Europe and Japan and most developing countries adopted the western model. The US has dominated the development of the theory of the firm.

Such aspects of board governance as board sub-committees and induction of independent directors, were addressed by way of regulative reforms by the governments through their stock exchanges and regulating mechanisms the securities regulators. However, the limitations of the unitary boards continue unabated. In a number of cases, independent directors did not have firm specific and industry specific knowledge to confront the management.

Financial Approaches align the behaviour of agents with the desires of the principals is the central theme (Hawley and Williams 1996). The theories outlines the rules and incentives as available in the US system. There are critical issues of principal and agency problem, as also incomplete contracts and efficacy of resources raised by the managers.

The Stewardship Model Concept

Ensures that the managers are capable of achieving high level of profits, and shareholder returns (Donaldson & Davis 1994). In other words, this refers to executive dominated board and it is advocated that the non-executive director is an ineffective control device.

Shareholder Versus Stakeholder Theory

The shareholder value maximization view of the firm's objective of the 1970s has come under increasing scrutiny of the stakeholders, government regulators and other voluntary organisations within civil society who are directly and indirectly impacted to the activities of the corporation. The scandals and corporate tragedies of the 1990s in the UK and more recently in the USA and other economies have reviewed focus of the debate on stakeholder view of corporate governance.

The corporation was meant to implement public policy and fulfill societal needs. The actions of the corporations were monitored by the state. After World War-I developments in America led to an emphasis on meeting shareholders views. Since the shareholders were property owners, managers should take into account their interest. The shareholders were the providers of capital.

The increasing adoption of Anglo-American model of governance led to a further renewal in the emphasis on shareholder primacy. But, at the same rime many states in the USA have passed strategies for meeting the interests of the stakeholders. Thus, the stakeholder view co-existed with the primacy of the shareholder view towards the end of the 20th century.

The corporation should integrate stakeholder interests in their business plan and strategy. Such an action should be reflected into their creation and distribution of health. But there is a problem of identifying and measuring the various stakeholder expectations and demands. There are several definitions of stakeholders which focus on legitimacy, power, dependence and urgency.

The stakeholder's value is also linked to firm performance. However, empirical work provides a mixed evidence of a connection between the stakeholder value and firm performance. One view is that the stakeholder management is complementary to shareholder value maximization. Also, it has been argued that the emphasis on shareholder value need not be construed as obstructing or neglecting stakeholder value. There can also be differences of values within the stakeholders, unless the case values of the key stakeholders are identified, it could be difficult for managers to incorporate the same in corporate plans.

Incorporating stakeholder concerns into corporate strategy is very essential. Resource Dependence Theory explains the important of primacy of stakeholders. An organization depends on external environments. Therefore managers have to make strategic decisions on managing the environments. Depending upon the stage of the organization life cycle, certain group of stakeholders will be more important than others. The moot question is, how much is the organisastion dependent upon those stakeholders who are likely to fulfill the resource values of the organization. Since the needs of the organization change overtime, the relative importance of stakeholder will also change.

Therefore, organizations will use different strategies for different group of stakeholders.

Government VS. Corporate Governance

The role of governments in favouring various constituencies, is critical in the political model. Corporate governance systems differ across countries because of variations in legal, regulatory institutional environments. Most researches on corporate governance systems and their efficacy have extensively analyzed the legal and regulatory aspects especially in USA and England. Most contemporary researches, have now focused on institutional environments including the political institutions. The political economy aspects include political parties, political institutions, coalitions and interest groups as important determinants of corporate governance systems. Political ideology is the underlying influence on public policy. The political system defines broadly the various constituents of market participants.

According to Mark J. Roe (2003) the competing claims arise from a dynamic social democracy and the claims on corporations manifest in terms of income security, social welfare and social stability etc.

The nature of product and market competition also matters as due to the type of available competition, firms may adopt specific rules and may even demand alterations of the existing rules, laws and regulations. Various interest groups will negotiate with the government for the change. Regulation is supplied in response to the demand for the same. Especially in large democracies like India, with weak enforcement and infant regulatory framework, the interest groups have exerted considerable influence over public policy.

Corporate governance systems differ significantly among nations. For instance, shareholder diffusion prevails in USA and UK; block holder concentration prevails in Germany and Japan and continental Europe.

According to public interest view; governments introduce reforms to correct market failures and to enhance economic efficiency. However market failures may persist over a period of time because of opposition to reform from the "entrenched political and economic elite."

Further in the emerging markets such as India, weak political executive comprising fragmented coalition often neither have the ability nor willingness to develop appropriate public policies for say public enterprise reforms and privatization. The role of political strategy in respect of trade policy, policy governing. Foreign Direct Investment (FDI), among others is important in that corporate governance practices are not developed.

Issues in Theories

Agency problems represent conflict of interests and influence expected cash flow accruing to the investors and the cost of capital. There are also issues of conflict of interest between majority and minority shareholders. Due to separation of ownership and control, there is tension between: decision-making and risk bearing functions. Dispersed shareholders have hardly any incentive to monitor the operating decisions of the managers. Therefore, there is the need for aligning interests of the managers with those of the shareholders by designing appropriate contracts. Further legal reforms should provide adequate protection for the enforcement of the shareholder rights. There are other indirect means of corporate control, such as managerial labour markets, capital markets and markets for corporate control.

One of the key assumptions of the agency theory is that agents are opportunistic and will engage in self-service behaviour. Therefore, the principals need to rely on control mechanisms available such as structures, procedures, information system, monitoring, performance evaluation, rewards and penalties, etc., the monitoring problem is very important because the principals cannot verify behaviours of the agent.

Because of asymmetric information there may be two problems: adverse selection and moral hazard. Adverse selection happens when the principals contact with an agent who is less capable. Moral hazard problem arises due to acts of commission or omission of actions in which situation the agents benefit but not the principal. Due to the problem of bounded rationality there is in reality incomplete contacts between principal and agent.

Agency Conflicts

There are mechanisms for control of agency problems between managers and shareholders: shareholding of insiders, institutions, and large blockholders, use of outside directors, debt policy, managerial and labour market, and market for corporate control.

Corporate governance is concerned with accountability and control over the firm. In the decentralized organization, there are problems of double and multiple agencies. Double agency problems occur when different groups in firms take strategic and operational decisions. Multiple agency situation arises when firm seeks competitive advantage through strategic alliances with other partner firms.

Recent codes of corporate governance and legislations like the Sarbanes-Oxley Act focus on the relationship between owner-shareholders and Corporate decision-makers. This is what the OECD defines as 'outsider' view of Corporate governance in which the owners rely on mechanisms external to the firm to ensure that their decision-making agents will act in accordance with their interests.

The board has a monitoring role in reducing agency conflict. The outside directors have an important role in monitoring management. Corporate governance literature examines performance measures such as accounting returns. Corporate boards with higher inside directors are less effective in monitoring management.

The efficiency of the board of directors depends upon the quality of the director. The quality of directors affects firms' stock performance and hence companies should look for superior quality of directors to monitor management, however, there are serious supply constraints. Board effectiveness is measured by the quality of debate and discourse at the board level.

According to Michael C. Jensen, events at the turn of the 20th century point to over evaluation of equity. It is necessary to understand the incentive and organizational effects of stock overvaluation. There has been an increasing number of cases of earnings management.

One implication of the agency theory is that agent's self serving behaviour can be monitored by information systems. Second; agency theory views control system aspects of compensation and incentive schemes as tools for aligning agents' motive with the organisations.

The agency theory brings into focus, uncertainty and risk considerations that need attention in devising control systems. The term agency means delegated decision rights, and such rights reconstitute control structures, process and procedures.

Transaction Costs Theory

The proponent of the transaction cost theory examined the economic explanation of the existence of firms (Case 1937). The nature of the firm is explained in terms of market imperfections and in terms of transaction costs of market exchange.

New institutional economics differs from agency theory, in that the Corporate governance problems of firms are perceived to proceed from a number of contractual hazards: self interest opportunism, information asymmetries, and the problems, bounded by rationality.

According to Kathleen Eisenhardt the agency theory offers unique insights into information systems, outcome uncertainty, incentives and risk, and therefore, these are novel contributions to organizational thinking and the empirical evidence is supportive of this theory, particularly when coupled with complementary theoretical perspectives.

Most frequently the agency theory is applied to organizational phenomenon such as compensation, acquisition and diversification strategies, board relationships, ownership and financing structures, vertical integration, and innovation.

Incomplete Contract Theory

Under incomplete contract theory, different governance systems have incentives structures that entail different trade-offs trade offs between ownership concentration and liquidity, between monitoring and management initiative, and between private rent seeking and activity benefiting shareholders as a group.

In the United States, proponents ask for deregulation of control on institutional investors, looking to encourage block holding and more effective monitoring. In Europe, proponents ask for stronger securities regulation, looking to encourage deeper trading markets.

A stakeholder's wealth perspective is unsatisfactory for the purpose of accurately answering two fundamental questions concerning the theory of the firm: that of economic value creation and the distribution of the economic value.

Law and Corporate Governance

A survey of literature analysing the relationship, between law, governance and economic outcomes, there are different scenarios.

Institutions have played a role in capitalist development. There are different results of studies on the degree to which legal regimes are more or less efficient than other regimes, scholars have debated why common and civil law institutions may have differential impacts on development.

Strong Corporate laws are necessary to protect stockholders. High quality Corporate law is insufficient to induce ownership to separate from control in the world's richest, most economically advanced nations.

Organization Theory

Organization theorists in the 1960s and 1970s combined research on inter- organizational relations and the political economy of organizations to develop a resource dependent view of the firm. Resources were described as tangible and intangible. The need for exchange of resources is highlighted.

It has been argued that a resource based view is more relevant for examining strategic partnerships because firms use partnerships to gain access to other firms' valuable resources. The resource-dependency view is of importance to technology strategy of the firm.

Why Do Firms Merge?

Usually mergers occur in a consensual (entering into by mutual consent) setting where executives from the target company help those from the purchaser in process to ensure that the deal is beneficial to both parties. Acquisitions can also happen by purchasing the majority of outstanding shares of a company in the open against the wishes of the target's board. The motives for the Merger and Acquisition activities differ across the industries and over time. However, certain motives stand out as the main drivers as are discussed below :

Motives Which Add to Shareholders Value

(a) Economies of Scale: This refers to the fact that the combined company can often reduce duplicate departments or operations, lowering the costs of the company relative to the same revenue stream, thus increasing profit.

(b) Increased Revenue Increased Market Share: This motive assumes that the company, will be absorbing a major competitor and thus increase its power (by capturing increased market share) to set prices.

(c) Synergy: This is the new financial maths that shows that $2 + 2 > 4$. That is, as the equation shows, the combination of two firms will yield a more valuable entity than the value of the sum of the two firms if they were to stay independent: Value (A + B) > Value (A) + Value (B).

(d) Taxes: A profitable company can buy a loss maker to use the target's tax write-off. In the United States and many other countries, rules are in place to limit the ability of profitable companies to "shop" for loss making companies, limiting the tax motive of an acquiring company.

(e) Geographical or other Diversification: This is designed to smoothen the earnings of a company, giving conservative investors more confidence in investing in the company. For instance, Tata Motors acquired Daewoo Commercial Vehicle Company (Republic of Korea) in 2003, and Infosys Technologies Ltd. acquired Expert Information Services Ltd. (Australia).

(f) Resource Transfer: Resources are unevenly distributed across firms and the interaction of target and acquiring firm resources can create value through either overcoming information asymmetry or by combining scarce resources. For example WIPRO acquired Nerve irelnc. (United States), and Reliance Infocomm acquired Flag Telecom (United Kingdom).

(g) Vertical Integration: Companies acquire part of a supply chain and benefit from the resources. For instance Hindalco acquired two copper mines in Australia and the ONGC acquired Sakhalin Oil and gas field in Russia.

(h) Increased Market Share to Increase Market Power: In an oligopoly market, increased market share generally allows companies to raise their market power.

Motives Which do not add to Shareholder Value

(a) Diversification: While this may hedge a company against a downturn in an individual industry it fails to deliver value, since it is possible for individual shareholders to achieve the same hedge by diversifying their portfolios at a much lower cost than those associated with a merger.

(b) Over extension: Tend to make the organisation unmanageable.

(c) Manager's Hubris: Manager's overconfidence about expected synergies from Merger and Acquisition results in overpayment for the target company.

(d) Manager's Compensation: In the past, certain executive management teams had their payout based on the total amount of profit of the company, instead of the profit per share, which would give them a bad incentive to buy companies to increase the total profit while decreasing the profit per share (which hurts the owners of the company, the shareholders); although some empirical studies show that compensation is rather linked to profitability and not mere profits of the company.

(e) Empire Building: Managers have larger companies to manage and hence more power. For example Tata Group, Infosys, Ranbaxy.

Synergies in Merger and Acquisitions

Synergy refers to the combination of two firms that yield a more valuable entity than the value of the sum of the two firms if they were to stay independent. The synergies, that arise from Mergers and Acquisitions are :

Financial Synergies

According to Ross, Westerfield, and Jaffe, there are many sources of financial synergy in Mergers and Acquisitions, which include :

Lower taxes: This can stem from the use of unused debt capacity, meaning when the target firm has a low debt/equity ratio. The acquiring firm, in this situation, can increase their debt, creating additional tax benefits, thus additional value to the shareholders. Another tax advantage comes from the use of tax losses from net operating losses. In this case, the losses from the target firm allow offsetting the taxable profits of the acquiring firm and vice-versa. The last tax advantage is the use of surplus funds and refers to a firm that has used all the positive net present value projects and still has excess capital. A firm in this situation has three options; payout the excess cash as dividends to the shareholders, buyback their own stock or purchase shares in another firm. From a tax point of view, the most favourable option is the last one.

Cost reduction: This builds on the assumption that the combined firm operates more efficiently due to economies of scale and economies of scope. In some cases, the acquisition can lead to better use of existing resources or to gain access to a critical success factors possessed by the target firm.

Get rid of inefficient and costly management: The reason could also be a need for new management ideas and competencies to make it possible to get rid of inefficient and costly management.

Revenue enhancement: One of the main reasons for a Merger and Acquisition is to enhance revenues or profits. By the Merger and Acquisition, competition is reduced and monopoly prices can be obtained.

Marketing gains can be won if one of the firms in the Merger and Acquisition possesses superior distribution networks, product mix, or media programming and advertising efforts.

Strategic benefits can be realized if the two firms' products can be combined in such a way that the end product is superior to those of the competitors.

The cost of capital: A larger company has greater chances of being granted loans at beneficial interest rates than the separate firms would have had.

Human Capital Synergies

In addition to the purely financial synergies that we have discussed up to this time, there could also be what we call human capital synergies. This is when the two firms can take advantage of the knowledge, the so-called intellectual capital, possessed by the employees in both firms. This intellectual capital could be expert knowledge in a particular area, as well as technological superiority and a range of other specific skills. The end goal of a Merger and Acquisition is almost always to become more successful in terms of earning money. The synergies gained through human capital should, in the long run, lead to increased efficiency and profitability.

There are, however, many problems associated with the human capital. Human beings are not like machines, their actions cannot be predicted and therefore it is very risky to estimate the human synergies in financial terms when first initiating the Merger and Acquisition. It has to be taken into account that people are creatures of habit and they are likely to oppose change if they do not feel that it is beneficial to them.

Impact of Merger and Acquisitions on Stakeholders

There are many parties who have interest, either direct or indirect, in the success of an organisation. The Merger and Acquisitions of organisations will have significant impact not only on the direct owners, i.e., shareholders, but also on various other groups like employees, consumers, government, general public, etc. The impact of Merger and Acquisitions on different stakeholders is discussed below:

Impact on Shareholders

Increasing the shareholders value is generally a prime objective of most of Mergers and Acquisitions today. The value to shareholders through Merger and Acquisitions could be increased either by cutting the costs by combining similar assets in the merging concerns or by enhancing the revenue by focusing on enhancing capabilities and revenues, and combining complementary competencies. However, most of the studies on the impact of Merger and Acquisitions on shareholders wealth reveal that on an average, Merger and Acquisitions consistently benefit the target company's shareholders, but not the acquirers' shareholders. Various consulting firms have also estimated that from one-half to two-thirds of Mergers and Acquisitions do not come up to the expectations of those transacting them, and many resulted in divestitures.

Impact on Employees

Human resource is another sensitive issue on the road to consolidation. Mergers and Acquisitions have profound impact on employment in all sectors of the economy. Mergers and Acquisitions invariably result in decline in the number of branches and thus leads to staff retrenchments. Consequently, mergers often lead to higher workloads being placed on remaining staff, with companies requiring flexibility in terms of working hours, mobility and skills, excellent and highly motivated employees of the merged entity may feel frustrated and may resign or they may not give their best to the organisation. So the retention of best talent and also motivation of the staff in the Merger and Acquisitions game has become a major challenge for the companies. Mergers also have brought about a change in the nature and quality of employment in the different sector.

Impact on Customers

In the changed economic scenario, the companies are providing a number of products to the customers at reduced prices. This is because the new market entrants are seeking to compete on the basis of the price. They are able to do this because new information and communication technology allows them to save cost by operating with fewer branches or without a traditional branch network (e.g. ICICI Bank and HDFC Bank). Customers are provided with new products and services with time flexibility. Customers need not stick on to bank working hours to conduct their business. Public Sector Banks have also initiated these steps in order to meet consumer needs.

Impact on Government

The Merger and Acquisitions to be successful should be created by market driven forces of synergy and other motives. However, some of the recent Mergers and Acquisitions are not market driven and created for mutual benefit of both acquirer and the target companies (mergers of Times Bank with HDFC Bank, 2000, Centurion Bank with Bank of Punjab, 2005, etc.). In this context, the Government and supervisory authorities have to provide an environment conducive for consolidation and convergence through appropriate fiscal and monetary policies supported by sound regulatory and supervisory framework. Otherwise, a few large institutions may create an oligopolistic structure of market. Thus, Government should remain alert in Merger and Acquisitions for a sound and balanced economy.

Impact on Organisation Culture

Every organisation has its own culture and some traditional activities. As organisation culture is the part of employee's identity, if the cultural issues are not effectively addressed, it may lead to loss of commitment among employees resulting in lost opportunities to retain qualified personnel and motivate individuals. A merger deal, which may appear to be perfectly sound from financial point of view, may fail miserably if cultural and human issues are not properly addressed in the newly created entity. Merger of New Bank of India with Punjab National Bank is a classic case of cultural differences on account of which the merged entity suffered a lot immediately following the merger.

Impact on Public

Local character of some of the organisations is another concern arising on account of Merger and Acquisitions. It is learnt that SBI never thought of merging its associates with itself because each of the associate banks has its own regional flavour, a clientele with which it is more comfortable after having nurtured it over many years, and therefore enjoys a niche presence. If the regional banks, for instance, State Bank of Travancore, or State Bank of Indore or, State Bank of Mysore, is merged into a single giant entity; it may affect the regional subsidiary and also the parent bank. Some of these banks have strong local or regional flavour to their operations that could get eroded if they were to merge with the parent.

CONCLUSION

Mergers and Acquisitions have become very popular over the years especially during the last two decades owing to rapid changes that have taken place in the business environment. Business firms are now facing increased competition not only from firms within the country but also from international business giants thanks to globalization, liberalization, technological changes, etc. Generally the objective of Merger and Acquisitions is wealth maximization of shareholders by seeking gains in terms of synergy, economies of scale, better financial and marketing advantages, diversification and reduced earnings volatility, improved inventory management, increase in domestic market share and also to capture fast growing international markets abroad. But astonishingly, though the number and value of Mergers and Acquisitions are growing rapidly, the results of the studies on the impact of mergers on the performance from the acquirers' shareholders perspective have been highly disappointing. This is because making the mergers work successfully is not that easy as here we are not only just putting the two organizations together but also integrating people of two organizations with different cultures, attitudes and mindsets. Meticulous pre-merger planning including conducting proper due diligence, effective communication during the integration, committed and competent leadership, speed with which the integration plan is integrated all this pave for the success of Merger and Acquisitions. While making the merger deals, it is necessary not only to make analysis of the financial aspects of the acquiring firm but also the cultural and people issues of both the concerns for proper post-acquisition integration.

5

CONTEXTUAL RELEVANCE IN CORPORATE GOVERNANCE

SCANNING THE RELEVANT CONTEXT

Whether an economy is already developed or emerging, owners and managers in that economy should scan the relevant context of their enterprise to identify the pressures it places on the enterprise, its employees, and its agents. These pressures take five forms: threats, opportunities, demands, constraints, and uncertainties.

Importance of Context

All enterprises-whether large or small-strive to meet enterprise goals and objectives in a context of legal, economic, political, environmental, socio-cultural, and technological elements. Known as the relevant context of the enterprise, these elements bring pressures to bear on the enterprise, its employees, and its agents.

i) Legal Element

Government creates the legal framework in which market processes operate. As a sovereign entity, a state regulates private activities to protect or promote the general well fare of its citizens and residents.

Government regulation also defines how, limited liability businesses, such as corporations, limited partnerships, and joint stock companies, are formed and what the limits of their liability are. Indeed, it may be that this tendency to define these roles and responsibilities in such detail is what discourages more discussion of corporate governance in terms of ethics and public policy. The very law that establishes minimum standards often comes to define the ceiling as well.

ii) Economic Element

An RBE must consider the nature of its market and the amount of trust that characterizes exchange transactions. In emerging market economies, the economic system is often unstable and characterized by frequent crises, causing businesses to seek short-term profits at the expense of long-term growth. Instability and lack of trust increase the cost of each transaction as the parties take expensive steps to protect themselves or avoid entering into a transaction at all. At the same time, the cost of regulation is often so high that shadow markets emerge, making it more difficult for the RBE to compete ethically.

Consumer expectations are a key economic factor. For example, consumers may be more or less discriminating. High quality may be more or less expected. Brand names may be more or less important.

Employee economic expectations are also important. Privacy expectations for e-mail usage on company time also differ, as do employee expectations about policies such as the hiring of relatives.

iii) Political Element

In scanning its relevant context, an RBE needs to understand the kind and degree of government influence in the market, how laws and regulations come about, who has practical access to influencing them, and the degree of government control over the economy. Government presence in the trade and investment sector often restricts trade: decreasing competition, creating opportunities for corruption, and increasing costs to business.

iv) Environmental Element

The physical world in which we live forms the widest relevant context for the business enterprise to consider and its most controversial element. The planet itself is a system of interdependent ecosystems that has evolved over millennia in a process of creation and destruction.

Human beings are an integral part of this system and the first to consciously influence   its evolution.

v) Socio-Cultural Element

In addition to organizational culture, each enterprise and its members operate within and are products of the broader cultures around them. Culture creates collective patterns of thinking, communicating, and acting that influence the decisions, processes, and activities of the enterprise itself. It influences the mental models that employees and agents bring to the job regarding fundamental issues of concern to managers, including the following five "dimensions of culture," developed by Geert Hofstede :

    i) Social inequality, including the relationship with authority.

    ii) Relationship between the individual and the group.

    iii) Concepts of masculinity and femininity.

    iv) Ways of dealing with uncertainty, relating to the control of aggression and the expression of emotions.

    v) Long-term versus short-term orientation.

    Culture influences the decisions of owners and managers regarding a range of specific issues, including policies regarding conflict of interest, privacy, and nepotism.

    vi) Technological Element

    An important consideration in planning and implementing a business ethics program is the technology available to the enterprise and its stakeholders. Particularly important are computer and telecommunications capabilities. Telecommunications capability will influence how owners and managers exercise control and relay communications.

The Leadership Transition

Rules ...

Rant

We fall back, in these crazy and chaotic times, on the command and control model of leadership – a model that no longer accords with how dynamic leaders actually operate.

    We seek shelter in the fantasy of a leader who has the Answers... who promises "change" or "success" or "Profit" in exchange for patient "followership" ("obedience"). But in an age when all value flows from creativity and initiative, we must imagine and embrace a model of leadership that is loose, open and perpetually innovation.

    We ask leaders to be "good stewards" of the assets they inherent. But in an age when permanence is a dangerous delusion, we must instead ask leaders to challenge the legacies that they have inherited, to create entirely new value propositions and then to get out before they get stale.

Vision

Leader envisions a Wondrous Opportunity to reinvent his/her company's chronically creaky customer-service operation. He/she tells everyone he/she meets about this exciting inkling, and everyone says, "Great idea, but good luck!" Even so, he/she works the problem and eventually cobbles together persons into project team. The team includes a Talent Developer and a Profit Mechanic; is Visionary and Head Cheerleader. Leading team on a Voyage of Mutual Discovery, he/she finds out that the original notion wasn't close to right, but the Unbounded Quest ultimately results in something ... far, far better strange.

Lead-Off Matter (A Muscular Definition)

Leadership is ... Joyous! It's matchless opportunity to Make a Difference by marshaling the talents of others to a ...Seriously Cool Cause.

Leadership is .... Horrible! It's an exercise in sorting through the mess of human relations, in all their gory detail, day after day. (After day.)

Leadership is ... Cool! It's a Glorious Adventure that enables us to magnify our impact on the world.

Leadership is ... Lonely! It's a battle against doubt and dread in which you have only your own judgment about human nature to fall back on.

Leadership is ... Different! It's a matter not of "doing" excellence but of "inspiring" excellence in others.

Leadership is ... The Ultimate Responsibility! It's an assumption of accountability... for people you cannot control, for actions that you do not perform, for institutions that may not share your sense of accountability.

Leadership is ... Not what you think! It's not about "command and control" or kingly charisma. It's about living in the depths (flourishing in the endless game of egos and institutions) and soaring to the heights (rallying others to invent and then pursue seemingly impossible dreams).

New Leadership is .., The Ultimate New Mandate! It's an apt prism through which to summarize this long journey that we have taken through our Disruptive Age. It's a never-ending project with a breathtakingly simple (and breathtakingly difficult) core objective: Re-imagine!

1. Leaders Are Rarely the Best Performers.
2. Leaders Are Talent Developers (Type I Leadership).
3. Leaders Are Visionaries (Type II Leadership),
4. Leaders Are "Profit Mechanics" (Type III Leadership).
5. Leaders Thrive on Paradox.

Forget what they taught you at the Harvard business school. The Illinois business school. The Stanford business school. The Wharton business school. Management is not science! It is. And 100 PERCENT OF THE TIME ... coping with complexity

MANAGEMENT IS AN ... ART. An art of paradox.

The Ultimate Paradox of ... LEADERSHIP: In order to be "excellent" you must be...CONSISTENT. (By most definitions: Excellence = Consistency of Superior Performance.) But the very moment you become excellently "consistent" ... you become... TOTALLY VULNERABLE ... to attack from the outside.

We must be constantly vigilant. Vigilant about ... OPPOSITES.

For example: Are we organized "enough"? If so.... don't WORRY. Are we disorganized "enough"? If so ... WORRY.

Worry, blends constantly... about the balance and the wobbleing the swing of the pendulum.

Well ... this idea is actually not about ... balance. It's about going one way... for a while ... & TOO FAR the other and then going back for a while ... & TOO FAR.

6. Leaders Do!

If you don't know what the hell is going on. If you don't know the shape or even the location of the playing field ... if you don't know the nature of the rule book or even if there is one ... then, in the immortal words of Old Man, "Thomas, don't just stand there. Do something."

7. Leaders Re-Do.

8. Leaders Know When to Wait.

9. Leaders Are Optimists.

10. Leaders Convey a Grand Design.

11. Leaders Attend to Logistical Details.

12. Leaders Side With the "Action Faction".

PROFESSION: THE "JOB" OF LEADING

13. Leaders Push Their Organizations into the Value-added Stratosphere.

In particular, they push their organizations to move up, up, up the Value-Added Ladder. Making "good stuff" is no longer enough. Not by a long shot. "Good stuff" has become but the Starting Point. In fact, even "Great Stuff" has assumed near-commodity status.

14. Leaders Create New Markets

NO ONE EVER MADE IT INTO THE ...BUSINESS HALL OF FAME ... ON A RECORD OF LINE EXTENSIONS.

Think Gates. (Microsoft.) McNealy. (Sun.) Ellison, (Oracle.) Dell. Jobs. (Apple.) Bezos. (Amazon.) Welch. (GE) Walton. (Wal Mart.) Blank & Marcus. (Home Depot.) Carnegie. Rockefeller. Sloan. Ford.

15. Leader Love New Technology.

The technology... is... changing ... everything. The leader, of whatever, also be the Chief Technology Officer. That's too much to ask, especially from 52-year-old Big Boss of Enormous Corp.

16. Leaders Are Salespeople Extraordinaire.

Leaders know It's ... ALL SALES, ALL THE TIME.

Leadership generates sales & creates competitive advantage.

Axiom: If you don't LOVE SALES ... find another life. And don't pretend to be a leader. (Harsh ... but true.)

17. Leaders Love "Politics"

LOVE POLITICS ... OR DON'T EXPECT TO GET ANYTHING DONE DURING YOUR TENURE.

Politics = Getting Things Done Through People,

Compromising. (True!)

Listening. (All the time!)

Standing your ground upon occasion. (Even if it costs you.)

People who govern in times of war ... do politics.

People who get scientific papers accepted at prestigious journals ... do politics. People who win Nobel prizes ... do politics.

The project manager of Boston's Big Dig does politics.

People who lead the effort to get a community center built ... do politics.

Sure it's sometimes downright dirty. (Sometimes you put your mirror away.) Yeti

politics ... is all about human beings coping and succeeding (or failing) ... in marriage I or in a business setting.

So: IF YOU DON'T LOVE POLITICS ... YOU'LL NEVER GET ANYTHING DONE, YOU ARE NOT A LEADER.

18. Leaders Are Great learners.

The best (and brightest) consultant I worked with in seven years at McKinsey had, I thought, one True Secret: He fearlessly and invariably asked ... "WHY?"

19. Leaders Are Great Performers.

If a leader attempts to induce risk taking ... she or he must embody risk taking, even if she or he is a naturally reticent person.

20. Leaders Focus.

The just retired chairman of CVS/pharmacy chairs the advisory board of an educational foundation. The very creative educational leader was about to embark on a program to expand his extraordinary school to a nationwide system of schools embodying his philosophy. My pal is at one of those "Inflection points" ... where the emphasis shifts from "making one great smooth" to "making a great system:" "Your number-one priority," this map said, "IS creating a 'to-don't' list:"

Nice. A "TO-DON'T" LIST!

21. Leaders Are the Brand.

Brand is a "character issue: Hence "branding" is ... personal. A Pure Leadership Issue. The leader Welch at GE, Goizueta at Coca-Cola, Gates at Microsoft, Jobs at Apple, reason at Virgin Group is the brand.

While a dozen programs may support the brand, it is the moment-to-moment actions of Nike's Knight or Oracle's Ellison or America's Bush that define the brand a host of publics.

Manage yourself! (Truly!) Watch yourself ! (Truly!) You will live ... or die ... as leader by the degree to which ... Your Calendar ... Precisely and Minutely... Reflects your Brand Priorities.

Summary:

1. You = Your Calendar.

2. You = The Brand.

3. The Brand = Your Calendar.

4. Q.E.D.

22. Leaders Know When to Leave.

There's a time to come. And ... a time to go.

People who are great at "roiling the waters and stirring up change" are, typically, woefully inept at "keeping the damn thing afloat" once it's been launched within "the system:"

None of us, it turns out, are "men [or women!] for all seasons:' We are men and women for a particular season... that is, at our best for a short period of time.

Think about it.

23. Leaders... (in concise or terse)

"Leadership is the process of engaging people in Creating a Legacy of Excellence."

"Leaders are living individuals whom employees can smell, feel, touch their presence."

"Leaders love their work. That passion is infectious."

"Leaders have a kid alive in them."

"Leaders ooze integrity:" [This one came up at least two dozen times amongst the 287 responses. And this was before Enron, World Com, Adelphia, et al.]

"Leaders are never afraid to walk away from [bad] business."

"Leaders communicate relentlessly."

"Leaders select their battles carefully."

"Real leaders don't always get their way."

"Leaders care."

"Leaders serve."

"Leaders Need to be the Rock of Gibraltar on Rollerblades:"

5.1. GOVERNANCE IN FAMILY-OWNED FIRMS AND DIFFERENT VIEWS

Introduction

The dominant form of business around the world is the family-owned business. In many instances, the family-owned business takes the form of a small family business whilst in other cases, it is a large business interest employing hundreds, or even thousands, of staff. The family-owned business can encompass sale traders, partnerships, private companies, and public companies. In fact, family ownership is prevalent not only amongst privately held firms but also in publicly traded firms in many countries across the globe. However, whatever the size of the business, it can benefit from having a good governance structure. Firms with effective governance structures will tend to have a more focused view of the business, be willing to take into account, and benefit from, the views of 'outsiders' (that is, non-family members), and be in a better position to evolve and grow into the future.

Ownership structures around the world

La Porta et al. (1999) analysed the ownership structure in a number of countries and found that the family-owned firm is quite common. Analysing a sample of large firms in 27 countries, La Porta et al. used as one of their criteria, a 10 per cent chain definition of control. This means that they analysed the share holdings to see if there were 'chains' of ownership: for example, if company B held shares in company C, who then held company B's shares. Of this 10 per cent chain definition of control, only 24 per cent of the large companies are widely held, compared to 30 per cent that are family-controlled, and 20 per cent state-controlled. Overall, they show that: 1) Controlling shareholders often have control rights in excess of their cash flow rights. 2) This is true of families, who are so often the controlling shareholders. 3) Controlling families participate in the management of the firms they own. 4) Banks do not exercise much control over firms as shareholders... 5) Other large shareholders are usually not there to monitor the controlling shareholders. Family control of firms appears to be common, significant, and typically unchallenged by other equity holders. La Porta et al.'s paper made an important contribution to our understanding of the prevalence of family-owned/controlled firms in many countries across the world.

A key influence on the type of ownership and control structure is the legal system. Traditionally, common law legal systems, such as in the UK and USA, have better protection of minority shareholders' rights than do civil law systems, such as those of France, Germany, and Russia. Often, if the legal environment does not have good protection of shareholders' rights, then this discourages a diverse shareholder base whilst being more conducive to family-owned firms where a relatively small group of individuals can retain ownership, power, and control. For example, in the UK and USA, where the rights of minority shareholders are well protected by the legal system, there are many more companies with diversified shareholder bases and family-controlled businesses are much less common.

However, recent research by Franks et al. (2004 and 2005) has highlighted that, in the UK in the first half of the twentieth century, there was an absence of minority investor protection as we know it today, and yet there still occurred a move from family ownership to a more dispersed share ownership. This was attributable to the issuance of shares through acquisitions and mergers, although families tried to retain control of the board by holding a majority of the seats. Franks et al. (2004) state the rise of hostile takeovers and institutional shareholders made it increasingly difficult for families to maintain control without challenge. Potential targets attempted to protect themselves through dual class shares and strategic share blocks but these were dismantled in response to opposition by institutional shareholders and the London Stock Exchange. The result was a regulated market in corporate control and a capital market that looked very different from its European counterparts. Thus, while acquisitions facilitated the growth of family controlled firms in the first half of the century, they also diluted their ownership and ultimately their control in the second half.

On the other hand, in many countries, including European countries such as France, many Asian countries, and South American countries, the legal protection of minority shareholders is today still either non-existent or ineffective, and so families often retain control in companies because non-family investors will not find the businesses an attractive investment when their rights are not protected.

However, many countries are recognizing that, as the business grows and needs external finance to pursue its expansion, then non-family investors will only be attracted to the business if there is protection of their rights, both in the context of the country's legal framework and also in the corporate governance of the individual companies in which they invest. This is leading to increasing pressure both for legal reforms to protect shareholders' rights and for corporate governance reforms within the individual companies. However, balanced against the pressures for reform are the, often very powerful, voices of family shareholders with controlling interests who may not wish to see reform to give better protection to minority interests because this would effectively dilute their control.

Family-owned firms and governance

Whilst a family-owned business is relatively small, the family members themselves will be able to manage and direct it. One advantage of a family-owned firm is that there should be less chance of the type of agency problems. This is because ownership and control, rather than being split. As a result of this overlap of ownership and control, one would hope for higher levels of trust and hence less monitoring of management activity should be necessary. However, problems may still occur and especially in terms of potential for minority shareholder oppression, which may be more acute in family-owned firms. Morek and Yeung (2003) point out some of the potential agency costs in family-owned firms:

In family business group firms, the concern is that managers may act for the controlling family, but not for shareholders in general. These agency issues are: the use of pyramidal groups to separate ownership from control, the entrenchment of controlling families, and non-arm's length transactions (aka 'tunneling') between related companies that are detrimental to public investors.

In some countries, remedies such as formal shareholder agreements, in which the nature of each shareholder's participation in the company is recorded, may be utilized as a mechanism to help resolve the potential oppression of minorities. Furthermore, Shareholder Rights Directive, is designed to help strengthen the role of shareholders by ensuring that they have relevant information in a timely manner and are able to vote in an informed way. Although the proposed directive is aimed at listed companies, and so likely primarily to affect institutional investors, Member States are free to extend to non-listed companies some or all of the provisions of the proposed Directive.

Another advantage of family-owned firms may be their ability to be less driven by the short-term demands of the market. Of course, they still ultimately need to be able to make a profit but they may have more flexibility as to when and how they do so.

However, even when a family business is still relatively small, there may be tensions and divisions within the family as different members may wish to take different courses of action that will affect the day-to-day way the business operates and its longer term development. In the same way as different generations of a family will have diverse views on various aspects of life, so they will in the business context as well. Similarly, as siblings may argue about various things, so they will most probably differ in their views of who should hold power within the business and how the business should develop. Even in the early stages of a family firm, it is wise to have some sort of forum where the views of family members regarding the business and its development can be expressed. One such mechanism is the family meeting or assembly, where family members can meet, often on a formal pre-arranged basis, to express their views. As time goes by and the family expands with family by marriage and new generations, then the establishment of a family council may be advisable. Neubauer and Lank (1998) suggest that a family council may be advisable once there are more than 30-40 family members.

When a business is at the stage where family relationships are impeding its efficient operation and development, or even if family members just realize that they are no longer managing the business as effectively as they might, then it is definitely time to develop a more formal governance structure. There may be an intermediate stage where the family are advised by an advisory board, although this would not provide the same benefits to the family firm as a defined board structure with independent non-executive directors. Figure illustrates the possible stages in a family firm's governance development.

Cadbury (2000) states that establishing a board of directors in a family firm 'is a means of progressing from an organisation based on family relationships to one that is based primarily on business relationships. The structure of a family firm in its formative years is likely to be informal and to owe more to past history than to present needs. Once the firm has moved beyond the stage where authority is vested in the founders, it becomes necessary to clarify responsibilities and the process for taking decisions'.

The advantages of a formal governance structure are several. First of all, there is a defined structure with defined channels for decision-making and clear lines of responsibility. Secondly, the board can tackle areas that may be sensitive from a family viewpoint but which nonetheless need to be dealt with succession planning is a case in point (deciding who would be best to fill key roles in the business should the existing incumbents move on, retire, or die). Succession planning is important too in the context of raising external equity because, once a family business starts to seek external equity investment, then shareholders will usually want to know that succession planning is in place. The third advantage of a formal governance structure is also one in which external shareholders would take a keen interest: the appointment of non-executive directors. It may be that the family firm, depending on its size, appoints just one, or maybe two, non-executive directors. The key point about the non-executive director appointments is that the persons appointed should be independent; it is this trait that will make their contribution to the family firm a significant one. Of course, the independent non-executive directors should be appointed on the basis of the knowledge and experience that they can bring to the family firm: their business experience, or a particular knowledge or functional specialism of relevance to the firm, which will enable them to 'add value' and contribute to the strategic development of the family firm.

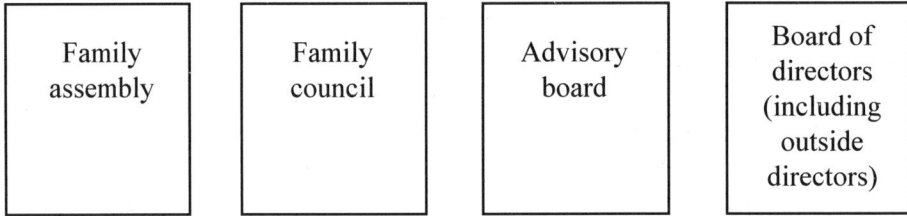

Fig. 5.1: Possible stages in a family firm's governance

Cadbury (2000) sums up the three requisites for family firms to manage successfully the impacts of growth: 'They need to be able to recruit and retain the very best people for the business, they need to be able to develop a culture of trust and transparency, and they need to define logical and efficient organisational structures. A good governance system will help family firms to achieve these requisites.

Smaller quoted companies

In the UK, many firms with family control will be smaller quoted companies, either on the main market or on the UK's Alternative Investment Market (AIM), which can be seen as a way for smaller firms to obtain market recognition and access to external sources of finance, often before moving on to the main market.

The Combined Code forms part of the UK Listing Authority's Rules and is applicable to all UK listed companies. This means that there should be no distinction between the governance standards expected of larger and smaller companies. The Combined Code states that it encourages smaller companies to adopt the Code's approach. However, smaller companies they should have at least two independent non-executive directors (rather than half the board being independent non-executive directors, which is the requirement for larger companies).

Another strand of the literature concentrates on firm-level characteristics that may serve to differentiate large and small firms. Larger firms tend to be more complex, whereas smaller firms adopt simpler systems and structures; smaller firms tend to have more concentrated leadership, whilst in a larger firm control may be more diffuse, or more subject to question by a larger board. In terms of the impact on corporate governance structures, it can be expected that in general, small and medium-sized firms will have simpler corporate governance structures than large firms this may include: combining various of the key committees (audit, remuneration, nomination); a smaller number of non-executive directors (NEDs); a combined chair/CEO; longer contractual terms for directors due to the more difficult labour market for director appointments into small and medium-sized companies.

The role and importance of NEDs was emphasized in the Cadbury Report (1992) and in the Code of Best Practice it is stated that NEDs 'should bring an independent judgement to bear on issues of strategy, performance, resources, including key appointments, and standards of conduct'. Similarly, the Hampel Report (1998) stated: 'Some smaller companies have claimed that they cannot find a sufficient number of independent non-executive directors of suitable calibre. This is a real difficulty, but the need for a robust independent voice on the board is as strong in smaller companies as in large ones'. The importance of the NED selection process is also emphasized: they 'should be selected through a formal process and both this process and their appointment should be a matter for the board as a whole'.

From Table, it can be seen that the areas where potential difficulties are most likely to arise tend to be those relating to the appointment of directors, particularly non-executive directors, which has implications for board structure. These differences arise partly because of the difficulties of attracting and retaining suitable non-executive directors in small companies.

In terms of the adoption of board committees, such as audit, remuneration, and nomination committees, small companies tend to have adopted audit and remuneration committees fairly widely but not nomination committees. In some smaller companies, the committees may carry out combined roles where, for example, the remuneration and nomination committees are combined into one committee; often the board as a whole will carry out the function of the nomination committee rather than trying to establish a separate committee from a small pool of non-executive directors.

Table Areas of the Combined Code that may prove difficult for smaller companies

Combined Code recommendations	Potential difficulty
Minimum two independent NEDs	Recruiting and remunerating independent NEDs
Split roles of chair/CEO	May not be enough directors to split the roles
Audit committee comprised of two NEDs	Audit committee may include executive directors
NEDs should be appointed for specific terms	NEDs often appointed for term

A word of caution should be sounded though in relation to quoted companies where there is still a large block of family ownership (or indeed any other form of controlling shareholder). Charkham and Simpson (1999) point out :

The controlling shareholders' role as guardians is potentially compromised by their interest as managers. Caution is needed. The boards may be superb and they may therefore be fortunate enough to participate in a wonderful success, but such businesses can decline at an alarming rate so that the option of escape through what is frequently an illiquid market anyway may be unattractive.

The points made are two fold: first, that despite a good governance structure on paper, in practice, controlling shareholders may effectively be able to disenfranchise the minority shareholders; second, that in a family-owned business, or other business with a controlling shareholder, the option to sell one's shares may not be either attractive or viable at a given point in time.

In many countries, family-owned firms are prevalent. Corporate governance is of relevance to family-owned firms, which can encompass a number of business firms including private and publicly quoted companies, for a number of reasons. Corporate governance structures can help the company to develop successfully they can provide the means for defined lines of decision-making and accountability, enable the family firm to benefit from the contribution of independent non-executive directors, and help to ensure a more transparent and fair approach to the way the business is organized and managed. Family-owned firms may face difficulties in initially finding appropriate independent non-executive directors but the benefits that such directors can bring is worth the time and financial investment that the family-owned firm will need to make.

- Family ownership of firms is the prevalent form of ownership in many countries around the globe.

- The legal system of a country tends to influence the type of ownership that develops. so that in common law countries with good protection for minority shareholders' rights, the shareholder base is more diverse, whereas in civil law countries with poor protection for minority shareholders' rights there tends to be more family ownership and control.

- The governance structure of a family firm may develop in various stages such as starting with a family assembly, then a family council, advisory board, and finally, a defined board structure with independent non-executive directors.

- The advantages to the family firm of a sound governance structure are that it can provide a mechanism for defined lines of decision-making and accountability, enable the family firm to benefit from the contribution of independent non-executive directors, and help to ensure a more transparent and fair approach to the way the business is organized and managed.

Example: Cadbury Schweppes pic, UK

This is an example of a family firm that grew over time, developed an appropriate governance structure, and became an international business.

Today, Cadbury is a household name in homes across the world. It was founded in the first part of the nineteenth century when John Cadbury decided to establish a business based on the manufacture and marketing of cocoa. His two sons joined the firm in 1861 and, over the years, more family members joined, and subsequently the firm became a private limited liability company, Cadbury Brothers Ltd. A board of directors was formed comprising of members of the family.

Example: Hutchison Whampoa Hong Kong

This is an example of an international company where there is controlling ownership by a family via a pyramid shareholding.

Hutchison Whampoa is a multinational conglomerate with five core businesses including ports and related services, property and hotels, telecommunications, retail and manufacturing, energy and infrastructure. It ranks as one of the most valuable companies in Hong Kong and over 40 per cent of it is controlled by Cheung Kong Holdings. The Li Ka Shing family owns some 35 per cent of Cheung Kong Holdings, which means that the Li Ka Shing family has significant influence over both companies, one through direct ownership, the other via an indirect, or pyramid, shareholding.

Fiat, Italy

This is a good example of a firm where the founding family still has significant influence through a complex shareholding structure. However, this may change in the future as there is considerable pressure in Italy to reform this type of control derived from complex shareholding structures.

Hierarchical Corporate Governance Framework

The figure shows that there are five levels applicable to the corporate governance hierarchy. Each one of them represents a stakeholder group. Level 1 represents the board of directors; Level 2 comprises the shareholders, directors and management, called the corporate governance tripod. Level 3 addresses other external stakeholders such as employees, customers and suppliers, Level 4 represents the society and Level 5 addresses the question whether a company's responsibility is to create shareholder value or prosperity for all stakeholders involved. Thus a broader societal role for the firm has emerged as a guide for food corporate governance. Its pyramid like view presupposes for dimensions to CG.

In the words of Dr Robert S. Kaplan, "It would be unfortunate if boards focused only on compliance and did not engage in functions that create value."

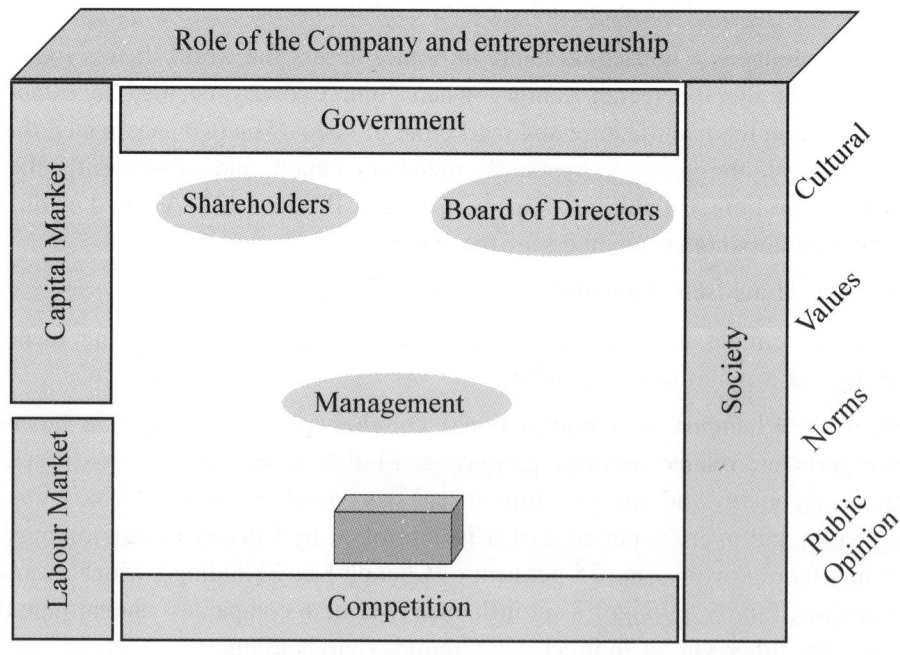

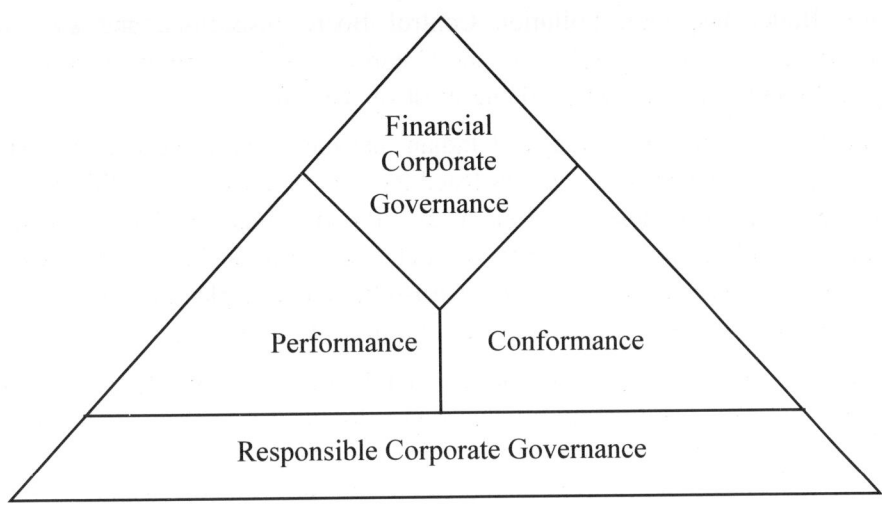

Corporate Governance Pyramid

5.2. ETHICAL APPROACHES TO CORPORATES IN INDIA

"Corporates in India Cannot Afford to be Ethical"

Indian companies face two types of corrupt practices: political corruption in which money is paid for favours done, and administrative corruption. A study on the ethical attitudes of Indian managers conducted by Arun Monappa (1977) reported that business executives listed three major obstacles to ethical behaviour, namely, company policies, unethical industry climate and corruption in government.

When questioned about unethical practices, many companies claim that the conditions in India are not conducive to allow them the luxury of being completely ethical. Thousands of underhand deals are struck everyday and go unreported. There is hardly a company which has not at sometime or the other been either involved or suspected of some foul play. Even companies that started off with intentions to do business in an ethical manner have had to compromise their principles due to the highly politicized and bureaucratic business environment in the country. Growing corruption, increasing disparity between people and rapidly reducing profit margins add to the woes of organisations that want to be ethical.

Indian companies face two types of corrupt practices: (i) political corruption in which money is paid for favours done, and (ii) administrative corruption. In the early days of Independence, companies had to grease the palms of bureaucrats to make them do things they were not supposed to do, but now corruption has graduated to such an extent that companies have to bribe bureaucrats to make them do things they are supposed to do. Examples of this sort of corruption include "gifts" to the Factory

Inspector, Boiler Inspector, Pollution Control Board Inspectors, and assessors for customs, excise, income tax, sales tax and Control. It is the administrative corruption, which most companies find unavoidable most of the times.

A study on the ethical attitudes of Indian managers conducted by Arun Monappa (1977) reported that business executives listed three major obstacles to ethical behaviour, namely: (i) company policies, (ii) unethical industry climate and (iii) corruption in government. Company policies tend to be unethical due to socio-cultural environment, and get reinforced because of the sense of frustration and helplessness that comes from the prevalent and all pervading unethical environment.

With regard to the socio-cultural reasons underlying the tendency of Indian corporates to be unethical are, the low priority accorded to business ethics in newly formed democracies as it seems there are more urgent demands that have to be dealt with first (Rossouw, 1998): The imperatives of the day-to-day survival for businessmen and the law-makers are to be unduly concerned about the ethical and moral implications of their actions. This situation has been sharpened by the opening up of the economy wherein Indian corporates find it increasingly difficult to compete in a dog-eats-dog kind of global markets. Another factor that has contributed to the lack of ethical ethos and behaviour is the county's aspiration to build a strong and economically powerful nation in a short time.

The other factors affecting ethical dilemmas of corporates are: (i) socio-cultural factors such as the sense of hospitality (not inviting a business associate could be construed as impolite, and once invited, showering him with gifts is an accepted custom) and reciprocity (You gave me a license with which I make money, and there is nothing wrong in sharing a part of it with you); (ii) the psychological fear of losing jobs; (iii) lax government structures and regulations; (iv) sanctions and discriminations in society that can be offset with accumulation of wealth by fair or foul means; (v) uncertainties and fears about the future; (vi) strong family traditions and laws of inheritance in which parents want to leave substantial assets to their progeny; (vii) overall scarcity of resources and the difficulty of amassing wealth through normal and legitimate means; (viii) an inequitable and scorching tax system (almost an unbelievable 97.75 per cent in terms of both direct and indirect taxes at the highest bracket in the 60s and 70s of last century) which discourage hardworking and honest tax-payers and lead them to bribe tax-collectors; (ix) a belief that business and ethics are irreconcilable; and (x) a tendency to adopt an easy option when confronted with difficult ethical choices "Well, if I can't beat them, I may as well join them" becomes a natural choice.

Lea (1999) gives another explanation to the deviant ethical behaviour found among corproates in developing societies. Transition from subsistence culture to the commercial enterprise of capitalistic culture can result in a moral chaos in which behaviour falls short of ethical expectations. In traditional sub-cultures rituals govern life, these rituals are insufficient behavioural guides in capitalism, which increases individual autonomy and responsibility and generates surpluses and wealth. Rapid economic growth leads to the development of a distorted understanding of capitalism and growth, in which money power, survival and profitability at any cost are considered as the primary goals of any business. The manifestation of this idea is very apparent in India and, especially in the case of some famous "rags to riches" stories.

The need to adapt to the unethical environment is so strong that even large multinationals setting up facilities in India have been unable to avoid cutting corners. In their eagerness to capture the Indian market and beat the competition, many companies have grossly broken their stringent codes of conduct, which in the West would be unthinkable. This was apparent when a major portion of the top management of a leading FMCG multinational in India was removed on grounds of violation of the ethical code of conduct. However, there was no visible effort on the part of the company to own up or reverse some of the unethical actions performed by its erstwhile employees.

Roots of Unethical Behaviour

Some of such influencing factors that make the employees think and act in unethical ways are: "pressure to balance work and family, poor communications, poor leadership, long work hours, heavy work load, lack of management support, pressure to meet sales or profit goals, little or no recognition of achievements, company politics, personal financial worries, and insufficient resources." It is evident that conflicting interests lead to most of the unethical practices.

People often wonder why employees indulge in unethical practices such as lying, accepting bribery, coercion, conflicting interest, etc. There are certain factors that make the employees to think and act in unethical ways. Some of such influencing factors are: "pressure to balance work and family, poor communications, poor leadership, long working hours, heavy workload, lack of management support, pressure to meet sales or profit goals, little or no recognition of achievements, company politics, personal financial worries, and insufficient resources." The statistical data given by Ethical Officers Association in 1997 shows how certain practices or factors contribute to unethical behaviour.

Balancing work and family	52 percent
Poor leadership	51 percent
Poor internal communication	51 percent

Lack of management support   48 percent

Need to meet goals   46 percent

From the above statistics it is very much evident that conflicting interests lead to most of the unethical practices.

Why Does Business Have Such a Negative Image?

Competitive pressures, individual greed, and differing cultural contexts generate ethical issues for organizational managers. Further, in almost every organization some people will have the inclination to behave unethically (the ethical egoist) necessitating systems to ensure that such behaviour is either stopped or detected (after unethical behaviour occurs), and remedied. Ethics (also called moral philosophy) involves systematizing, defending, and recommending concepts of right and wrong behaviour.

Why Should Businesses Act Ethically?

Businesses should act ethically to protect their own interest (prudence) and the interests of the business community, keep their commitment to society to act ethically, meet stakeholder expectations, prevent harm to the general public, build trust with key stakeholder groups, protect themselves from abuse from unethical employees and competitors, protect their own reputations, protect their own employees, and create an environment in which workers can act in ways consistent with their values.

There are a number of reasons given below as to why businesses should act ethically:

- Protect its own interest (prudence).
- Protect the interests of the business community.
- Keep its commitment to society to act ethically.
- Meet stakeholder expectations.
- Prevent harm to the general public.
- Build trust with key stakeholder groups.
- Protect themselves from abuse from unethical employees and competitors.
- Protect their own reputations.
- Protect their own employees.
- Create an environment in which workers can act in ways consistent with their values.

Moreover, if a corporation is not there to on its agreement and expects others to keep theirs, it will be unfair. It will also be inconsistent on its part, if business agrees to a set of rules to govern behaviour and then to unilaterally violate those rules. Moreover, to agree to a condition where business and businessmen tend to break the rules and also they can get away with it is to undermine the environment necessary for running the business.

Additionally, an organisation has to be ethical in its behaviour because it has to exist in the competitive world. We can find a number of reasons for being ethical in behaviour, few of them are given below. Most people want to be ethical in their issues.

- Values give management credibility with its employees. Only perceived moral uprighteousness and social concern brings employee respect.

- Values help better decision-making.

- Hard decisions which have been studied from both an ethical and an economic angle are more difficult to make, but they will stand up against all odds, because the good of the employees, the public interest, and the company's own long term interest and those of all stakeholders have all been taken into account.

Ethics within organizations is a must, as only then that can be conveyed through the activities they perform. Ethics should be initiated from the top management to the bottom of the hierarchy. "Ethical behaviour starts at the top. Before a company can expect to be viewed as ethical in the business community, ethical behaviour within its own walls to and by employees is a must, and top management dictates the mood. Ethical behaviour by the leaders of an organization will inevitably set the tone for the rest of the company values will remain consistent. Further, a well-communicated commitment to ethics sends a powerful message that ethical behaviour is considered to be a business imperative. If the company needs to make profit and to have a good reputation, it must act within the confines of ethics. The ethical communication within the organisation would be a healthy sign that the company is marching towards the right path. Internalization of ethics by the employees is of very much importance. If employee has properly internalized ethics, then the activities that he or his organisation caries out will have ethics in it.

Ethical Decision Making

Ethical decision making is a very tough prospect in world. However, in the long run, all will have to fall in and play fair. The clock is already ticking for the unscrupulous corporates. In this age of liberalisation and globalisation, the old dirty games and unethical conduct will no longer be accepted and tolerated.

Norman Vincent Peale and Kenneth Blanchard in their book, "The Power of Ethical Management", have prescribed some suggestions to conduct ethical business.

- Is the decision you are taking legal? If not legal, it is not ethical.

- Is the decision you are taking fair? In other words, it should be a win-win-equitable risk and reward.

- The Eleventh Commandment – "Thou shall not be ashamed when found", meaning when you are hauled up over some seemingly unethical behaviour, if one's conscience is clear, then there is nothing to be ashamed of.

How Corporates Observe Ethics in their Organisations?

Organizations have started to implement ethical behaviour by publishing in-house codes of ethics which are to be strictly followed by all their associates. They have started to employ people with a reputation for high standards of ethical behaviour at the top levels. They have started to incorporate consideration of ethics into performance reviews. Corporations that wish to popularize good ethical conduct have started to reward ethical behaviour. Codes promulgated by corporations and regulatory bodies continue to multiply. Some MNCs like Nike, GM and IBM and Indian Companies like Infosys, ICICI, TISCO, ONGC, Indian Oil and several others want to be seen "socially responsible" and have issued codes governing all types of activities by their employees. SEBI, the capital market regulator, CII, Assocham, and such organizations representing corporates have issued codes of best practices and enjoin their members to observe them. These normative statements make it clear that corporate leaders anxious for business growth should not make plans without looking at the faces and lives of those oppressed by poverty and injustice. In fact, today managers and would-be entrepreneurs are groomed to be ethical and socially responsible even while being educated. The Indian Schools of Management (IIMs) and highly rated B-schools like the Xavier Labour Relations Institute (XLRI) and the Loyola Institute of Business Administration (LIBA) have core courses in their curriculum and give extensive and intensive instruction in business ethics, social responsibility and corporate governance. Many corporations conduct an ethics audit and, at the same time, they are continuously looking for more ways to be more ethical.

Some Unethical Issues

As we discuss business ethics, it is necessary to address the ethical issues that are involved in business. Right from the Harshad Mehta scam till the recent insider trading of L&T versus Reliance, we see unethical practices taking place even in reputed organisations. Researches and studies show that several ethical issues are faced by an organisation, they are: bribery, coercion, deception, theft, unfair discrimination, insider trading, conflicts of interest, etc. Some of these are dealt in detail below.

Bribery

Bribery is a manipulative method where one buys the power or the influence of other person in order to satisfy his selfish need. Bribes create a conflict of interest between the person receiving bribe and his/her organization. This conflict would result in unethical practices. When somebody is bribed for something his thinking and actions are oriented towards his personal goals. This direction towards personal goals always results in a mismatch between the interest of the organization and of the individual. When there is a mismatch between the goals, naturally he cannot be loyal to the organization, and in turn, he will indulge in unethical practices. Bribery undermines market efficiency and predictability, thus ultimately denying people their right to the minimal standard of living. "Bribery does more than destroy predictability; it undermines essential social and economic system."

For example, companies like Boeing and GE (General Electric) have well formed policies to deal with this issue. These policies of the company protect the employees from indulging in such practices. The statement of GE is worth mentioning, "No matter how high the stakes are, no matter how great the 'stretch is', GE will do business only by lawful and ethical means. When working with customers and suppliers in every aspect of our business, we will not compromise our commitment to integrity".

Likewise, Boeing is categorical with regard to this issue: "It is the policy of The Boeing Company to deal with its suppliers and customers in a fair and impartial manner, business should be won or lost on the merits of the Boeing products and services. A business courtesy may never be offered under circumstances that might create the appearance of impropriety or cause embarrassment to Boeing or the recipient. An employee may never use personal funds or resources to do something that cannot be done with Boeing resources. Accounting for business courtesies must be in accordance with approved company procedures and practices.The Indian engineering giant, L & T makes its business policy clear thus: "All marketing personnel will adhere to the highest standards of personal and corporate integrity and thereby maintain and promote our reputation as an outstanding company with which to do business."

Bribes create a conflict of interest between the person receiving bribe and his/her organization, which result in unethical practices. When somebody is bribed for something his thinking and actions are oriented towards his personal goals. This always results in a mismatch between the interest of the organisation and of the individual.

Coercion

"Coercion is forcing a person to act in a manner that is against the person's personal beliefs." It is an external force or a man-made constraint created in circumstances asking the other to act against his free will. Authority of the person who demands certain activity plays an important role, that is, blackmailing or arm-twisting an individual in an organization. This may be in the form of threat of blocking a promotion or the loss of a job. This sort of unethical practice in the organization will lead to further unethical behaviour of an individual. For example, the Tylenol tampering case of Johnson and Johnson was done with an intention of damaging the image of the company and forcing it to incur heavy financial expenses in correcting the problem.

Insider Trading

This is one form of misuse of official position by an individual in the organisation. Here, the employee leaks out certain confidential data to outsiders or to other insiders, which in turn ruins the reputation of the company. Insider trading may lead to the bad performance of the company.

This is how it is done: If the employees trade the confidential matters, the competitor may intervene and make use of the opportunity. Inside traders often defend their actions by claiming that they don't injure anyone. It may be true with non public information but certain moral concerns arise because of this Act. For example, the report of L&T versus Reliance (The Hindu, 23 November 2003) issue which was reported in the media shows that such practices are taking place in reputed companies and at the top management level.

Tax Evasion

There are major unethical practices towards tax evasion. Many large corporations hire the services of professional tax consultants to take advantage of loopholes in the law and evade taxes to the extent possible. The reason they attribute for such behaviour is the prevalent rate of corporate taxation, which is very high. In fact, this has generated a parallel economy in spite of government's continuous endeavours to channellise this money towards legitimate purposes.

The well-known tax consultant, Dinesh Vyas, says that J. R. D. Tata never entered into a debate over tax avoidance, which was permissible, and "tax evasion," which was illegal; his sole motto was "tax compliance." On one occasion a senior executive of a Tata company tried to save on taxes. Before putting up the case, the chairman of the company took him to JRD and Vyas explained to JRD: "But sir, it is not illegal." JRD asked, softly: "Not illegal, yes. But is it right?" Vyas says that during his decades of professional work, no one had ever asked him that question. Vyas later wrote in an article: JRD would have been the most ardent supporter of the view expressed by Lord Denning. "The avoidance of tax may be lawful, but it is not yet a virtue."

Conflicts of Interest

The CEO and the senior leadership of the finance department bear a special responsibility for prompting integrity throughout the organisation, with responsibilities to stakeholders both inside and outside of companies. They have to act with the honesty and integrity, avoiding actual or apparent conflict of interest in personal and professional relationships.

Even the most loyal employees can find that their interests collide with that of the organisation. Sometimes, this clash of goals and desires can take the serious form of conflicts of interest. In an organisation, conflict of interest arises when employees at any level behave with the private interests that are substantial enough to interfere with their job or duties. This conflicting interests in the individual and the decisions taken may act against the desire of the employer. Conflicts of interests are morally worrisome not only when an employee acts to the detriment of the organisation but also when the employee's private interests are significant enough that they could easily tempt the employee to do so. Great men like J. R. D. Tata had been trying all their lives to reduce such conflicts of interest in the work place. JRDTs' strong point was his intense interest in people and his desire to make them happy. Towards the end of his life, he often said: 'We don't smile enough". Once he told a friend about his dealings with his colleagues: "With each man I have my own way. I am one who will make full allowance for a man's character and idiosyncrasies. You have to adapt yourself to their ways and deal accordingly to draw out the best in each man. At times it involves suppressing yourself. It is painful, but necessary. To be a leader you have got to lead human beings with affection." It is a measure of his affection that even after some of them retired, he would write to them. He was always grateful and loyal. To him, ethics included gratitude, loyalty and affection. It came about because he thought not only of business, but also of people.

In dealing with his workers he was particularly influenced by Jamshetji Tata, who at the height of capitalist exploitation in the 1980s and the 1990s gave his workers accident insurance and a pension fund, adequate ventilation at the workplace and other benefits. He wanted workers to have a say in their own welfare and safety, and he wanted their suggestions on the running of the company. A note that he wrote on personnel policy resulted in the founding of a personnel department. As a further consequence of that note came about two pioneering strokes by Tata Steel: a profit-sharing bonus and a joint consultative council. Tata Steel has enjoyed peace between management and labour for 70 years.

Pollution

The unethical practice towards pollution affects society and population to a major extent. The high levels of pollution due to the indiscriminate and improper disposal of effluents by industries has rendered the world a highly unsafe place for progeny. In his last years, J. R. D. Tata was very conscious of the environment and industry's part in spoiling it. He wrote in his Foreword to The Creation of Wealth in 1992: "I believe that the social responsibilities of our industrial enterprises should now extend even beyond serving people to the environment." The J. R. D. Tata Centre for Ecotechonology at the M.S. Swaminathan Research Foundation was created in furtherance of his desire.

Corporate Governance Ethics

Though the concept of Corporate Governance may sound a novelty in the Indian business context and may be linked to the era of liberalization, it should not be ignored that the ancient Indian texts are the true originators of good business governance as one important sloka from the Rigveda says: "A businessman should benefit from business like a honey-bee which suckles honey from the flower without affecting its charm and beauty."

As a public company, it is of critical importance that companies' information reporting with the regulators be accurate and timely. The chief executive officer and the senior leadership of the finance department bear a special responsibility for prompting integrity throughout the organisation, with responsibilities to stakeholders both inside and outside of companies.

1. Act with honesty and integrity, avoiding actual or apparent conflict of interest in personal and professional relationships.

2. Provide information that is accurate, complete, objective, relevant, timely and understandable to ensure full, fair, accurate, timely, understandable disclosure in reports and documents that companies file with, or submit to, the regulators.

3. Comply with applicable laws, rules and regulations of federal, state, and local governments, and other appropriate public and private regulatory agencies in all material respects.

4. Act in good faith, responsibility, with due care, competence and diligence without misrepresenting material facts or allowing one's independents judgment to be subordinated.

5. Respect the confidentiality of information acquired in the place of one's work except when authorised or otherwise legally obligated to disclose. Confidential information required in the course of one's work will not be used for personal advantage.

6. Share knowledge and maintain skills important and relevant to stake-holders needs. Proactively promote and be an example of ethical behaviour as a responsible partner among peers, in the work environment and the community.

7. Achieve responsible use of and control over all assets and resources employed or entrusted with.

Benefits from Managing Ethics in Workplace

(a) Attention to business ethics has substantially improved society: Establishment of anti-trust laws, unions, and other regulatory bodies has contributed to the development of the society. There was a time when discriminations and exploitation of employees were high, the fight for equality and fairness at workplace ended up in establishing certain laws which benefited the society.

(b) Ethical practice has contributed towards high productivity and strong teamwork: Organisations being a collection of individuals, the values reflected will be different from that of the organisation. Constant check and dialogue will ensure that the employee matches to the values of the organisation which will in turn result in better corporation and increased productivity.

(c) Changing situations require ethical education: During turbulent times, where chaos becomes the order of the day, one must have clear ethical guidelines to take right decisions. Ethical training will be of great help in those situations.

(d) Ethical practices create strong public image: Organisations with strong ethical practices will possess a strong image among the public. This image would lead to strong and continued loyalty. Conscious implementation of ethics in organisations becomes the cornerstone for the success and image of the organisation. It is because of this ethical perception, that the employees of TISCO and the general public protested in 1977 when the then Minister for Industries in the Janata Government, attempted to nationalize the company.

(e) Strong ethical practices act as insurance: Strong ethical practices of the organisation are an added advantage for the future function of the business. In the long run, it would benefit if the organisation is equipped to withstand the competition.

Characteristics of an Ethical Organization

Mark Pastin in his work, The Hard Problems of Management: Gaining the Ethical Edge provides the following characteristics of ethical organizations:

(a) They are at ease interacting with diverse internal and external stakeholder groups. The ground rules of these firms make the good of these stakeholder groups part of the organization's own good.

(b) They are obsessed with fairness. Their ground rules emphasize that the other persons' interests count as much as their own.

(c) Responsibility is individual rather than collective, with individuals assuming personal responsibility for actions of the organization. These organizations' ground rules mandate that individuals are responsible.

(d) They see their activities in terms of purpose. This purpose is a way of operating that members of the organization highly value. And purpose ties the organization to its environment.

(e) There will be clear communications in ethical organizations. Minimized bureaucracy and control paves way for sound ethical practices.

Recognizing Ethical Organizations

There are certain principles by which we will be able to identify the ethical organization.

(a) On the basis of corporate excellence: Corporate excellence mainly centres on the corporate culture. Values and practice of such values constitute the corporate culture. Values of the organization give a clear direction to the employees. Values are found in the mission statement of the organizations. Often they remain as a principle and never put into practice. Only the practised value creates the organization culture. When values act in tune with the goals of the organization, we call it as the corporate culture of that organization. Often we see conflicting interests between the value and the organizations' goal. An organization must eradicate such impediments to be identified as an ethical organization.

(b) In reference to the stakeholders: Meeting the needs of the stakeholders by the activities of the managers determines whether the organization is ethical or not. The top management is the representation of the stakeholders and every decision taken must satisfy the needs of the stakeholder. The management in taking decisions must see that the stakeholders enjoy the maximum benefit of that decision. For example, Marico, the makers of Parachute Oil discovered a harmless tint in the oil from one of its production lines. The company withdrew the batch from the market, shut down the production line, but kept the workers on payroll and involved them in the investigation of the cause. Shortly, the workers located the cause, rectified it and resumed production.

(c) In relation to corporate governance: Managers are only stewards of the owners of the assets. Thus they are accountable for the use of the assets to the owners. If they perform well in the prescribed manner, then there would not be much question of corporate governance. Such behaviour of the top managers would generate ethical practices or at least would encourage ethical practices in the organisation. If only the top management is paid as per their performance, this approach would work.

How Ethics Can Make Corporate Governance More Meaningful?

(i) Corporate governance is meant to run companies ethically in a manner such that all stakeholders creditors, distributors, customers, employees, the society at large and governments are dealt in a fair manner.

(ii) Good corporate governance should look at all stakeholders and not just shareholders alone. Otherwise, a chemical company, for example, can maximize the profit of shareholders, but completely violate all environment laws and make it impossible for the people around the area to lead a normal life. Ship-breaking in Valinokkam, near Arantangi in Tamil Nadu, leather tanneries in South, Arcot and hosiery units in Tirupur, have brought about too much of environmental degradation that has unleashed untold miseries to people in and around their locations.

(iii) Corporate governance is not something which regulators have to impose on a management, it should come from within. There is no point in making statutory provisions for enforcing ethical conduct.

(iv) There is a lot of provisions in the Companies Act, for example, (i) disclosing the interest of directors in contracts in which they are interested; (ii) abstaining from exercising voting rights in matters they are interested; and (iii) statutory protection to auditors who are supposed to go into the details of the financial management of the company and report the same to the shareholders of the company. But most of these may be observed in letter, but not in spirit. Members of the board and top management should ensure that these are followed both in letter and spirit.

(v) There are a number of grey areas where the law is silent or where regulatory framework is weak, which are manipulated by unscrupulous persons like Ketan Parikh and Harshad Mehta. In the US, for instance, the courts recognize that new forms of fraud may arise, which may not be covered technically under any existing law and cannot be interpreted as violating any of the existing laws. For example, a clever conman can try to sell a piece of the blue sky. In order to check such crooks, there is the concept of the "blue sky" law. However, such wide-ranging processes are not available to courts in developing countries.

(vi) The Securities and Exchange Board of India (SEBI) has jurisdiction only in cases of limited and listed companies and are concerned only with their protection. What about the shareholders of other unlisted Limited companies?

(vii) The Serious Fraud Investigation Office (SFIO) in the Department of Company Affairs (DCA) has been investigating several "Vanishing Companies". By 2003, SEBI has identified 229 as "vanishing companies" which tapped the capital market, collected more than Rs. 800 crores from the public and subsequently became untraceable. However, thousands of investors have lost their hard-earned money and no agency has come to their rescue so far.

With the globalization of business, monopolistic market condition or state patronage for any business organization has become a thing of the past. A business organization has to compete for a share in the global market on its own internal strength, in particular on the strength of its human resource, and on the goodwill of its other stakeholders. While its state-of-the-art technologies and high level managerial competencies could be of help in meeting the quality, cost, volume, speed and breakeven requirements of the highly competitive global market, it is the value-based management and ethics that the organization has to use in its governance that would enable it to establish productive relationship with its internal customers and lasting business relationship with its external customers. It is for these reasons that in present day's environment the value based management and practice of ethics have become imperatives in corporate governance, and also in the foreseeable future. "If values are the bedrock of any corporate culture, ethics are the foundation of authentic business relationships." The glaring examples are TISCO, ITC & J&J.

Mission of TISCO

The fundamental mission of TISCO is to strengthen India's industrial base through increased productivity, effective utilization of manpower and material resources, and continued application of modern scientific managerial methods as well as through systematic growth in keeping with national aspirations. The company recognizes that while honesty and integrity are the essential ingredients of a strong and stable enterprise, profitability provides the main spark for economic activity. It affirms its faith in democratic values and in the importance of success of individuals, collective, and corporate enterprise for the economic emancipation and prosperity of the country. Guided by its basic philosophy, the company believes in discharging its responsibilities towards share holders employees, customers and community.

Management Philosophy of ITC Limited

The Management philosophy of ITC Limited consists of the following concerns:

i. Concern for their ultimate customers millions of customers.

ii. Concern for their intermediate customers the trade.,

iii. Concern for their suppliers their source of raw materials and ancillaries.

iv. Concern for their employees their most valued asset.

v. Concern for their competitors whom they wish well for healthy competition ultimately benefits the customer.

vi. Concern for their shareholders the investing public.

vii. Concern for national aspiration India's future.

The company undertakes various programmes to fulfill its obligations to various constituents of its business.

(Source: 'Principles and Practice of Management'; L.M. Prasad; S. Chand & Sons; 1992) The second instrument for corporate excellence is the Code of Ethics.

The Ethics Code of Johnson and Johnson, 'Our Credo'

We believe our first responsibility is to the doctors, nurses and patients, to mothers and all others who use our products and services. In meeting their needs everything we do must be of high quality.

We must constantly strive to reduce our costs in order to maintain reasonable prices.

Customers' orders must be serviced promptly and accurately.

Our suppliers and distributors must have an opportunity to make a fair profit

We are responsible to our employees, the men and women who work with us throughout the world. Everyone must be considered as an individual.

We must respect their dignity and recognize their merit.

They must have a sense of security in their jobs.

Compensation must be fair and adequate, and working clean orderly and safe.

Employees must feel free to make suggestions and complaints.

There must be equal opportunity for employment and promotion in the organisation.

We must provide competent management and their actions must be just and ethical.

5.3. CORPORATE GOVERNANCE AND ETHICS

Corporations Must Act Ethically

This article is drawn from a speech William S. Kanaga. the former Chairman of CIPE's Board of Directors, delivered in Bucharest in 1998 at a conference on the "Role of the Corporation in Today's Society."

What is a corporation's duty to the society in which it functions, and how can its responsibilities be reconciled with the ever-increasing demands to be competitive in the global economy? The importance of corporate citizenship in the modern world should not to be taken lightly because it represents a business strategy for survival in global markets.

The more appropriate term for "corporate citizenship" is "corporate leadership." This concept suggests much more than the amount of money a company can donate to a worthy cause. Corporate leadership represents the recognition that actively participating in its community's life enhances its operating environment. The world's most successful corporations have learned that leadership is a necessary tool to overcome the challenges that face management in the global and local markets. Corporate leadership is a strategy that needs to be implemented at many levels, both within and outside the company.

For their employees

First and foremost, the corporate leader pays attention to its most valuable asset, its people. An organization that cannot treat employees in a fair and ethical manner weakens its own ability to respond and innovate when it faces competitive challenges.

For their communities

Corporate leaders recognize the need to strengthen their own community base. A vibrant local economy can become as important a factor in business growth as export development. Corporations can take leading roles in developing local economic conditions through the following activities :

- improving the local labour base by investing in education and training;
- working with local governments to enhance the business environment by improving infrastructure, regulations and enforcement; and
- working with local suppliers to encourage regional economic development and the growth of an integrated and competitive economy.

For democracy

Finally, the good corporate leader is also a corporate citizen in the democratic process. If businesses do not take an active role in the policy development process, their needs and concerns may go unheeded or even damaged by others who participate in the political process. By joining an association or chamber of commerce, or working with other interested groups to effect change, businesses can encourage democracy and the growth of market economies.

Policy measures

As with all business activities, the notion of corporate leadership should undergo a cost-benefit analysis. From a strategic perspective, by building a broad-based public/private commitment to economic growth and development, the benefits of corporate leadership affect the community as well as the corporation. The US-India Business Council has outlined several ways in which an active corporate citizenship strategy can bring about needed change :

- Improve the policy environment for business. Active participation in the policy development process will ensure that private sector interests are taken into consideration.

- Develop a local identity and strengthen community linkages. Corporate partnerships with local non-profit organizations and governments help to build trust and communication.

- Enhance the ability to respond to grassroots sentiment and concerns. By building trust and working with established community leaders, corporations can develop sensitive lines of communication that allow a problem-solving approach to emerging challenges.

- Improve public understanding of business-centered approaches. Corporate involvement with community groups can directly and effectively broaden the social understanding of entrepreneurship and sound business management practices.

- Establish a forum for public discourse. Partnerships can create a friendly community environment in which the benefits and costs of industrial development can be rationally discussed.

- Strengthen the human resource base of a community. Training, outreach to local education groups and small capital investments can enable a community to develop vocational employment programs. These benefit the local company as well as the entire region.

- Leverage financial and other investments in the community. Because non-profit organizations mobilize vast reserves of goodwill, corporate investment in the community can have tremendous reach in building a better corporate profile and in strengthening public support of the private sector.

The recent Asian economic crisis has highlighted the urgent need to adopt these measures. In that region, the world has witnessed the costs of corruption, self-interest, and poor corporate governance. Many people have interpreted this crisis of bad corporate leadership as a crisis of capitalism. Throughout Asia, people have perpetuated a system doomed to failure because they avoided transparency, ethics and democratic principles, and they now seek to discredit the very policies that represent the remedy for their Asian flu.

The crisis could have been avoided if the business community had been willing to take a leading role in society to call for higher standards in government and commercial activity, push for better economic development strategies, demand more of all participants in society and lead by example.

The policy community and business associations should ask themselves additional questions. They may already be aware of the costs and benefits of strong corporate leadership in society, but how can they promote these ideals both within their organizations and membership? How can they utilize these values as a recruitment tool and a marketing strategy?

As we enter the twenty-first century, we are witnessing the rise of the global economy and global corporation. As a matter of strategic survival, corporations must develop smaller markets. Never has the term "think global, act local" been so apt. By applying responsible concepts of corporate leadership in all communities in which it acts, business not only improves its competitive edge, but works to secure new markets, and provides social, political and economic stability for its market place.

Good Corporate Citizenship

Good corporate citizenship implies that the business unit complies with the rules of the land, pays taxes to the government regularly, discharges its obligations to society and cares for its employees and customers.

Corporate Citizenship

Indian industry, for many years, has been doing community-related work such as medical centres, rural development, educational activities and several other areas. But, this is more an exception than the rule.

More recently, efforts have been made by Indian industry to extend the frontiers of its involvement in such key issues as environment protection, population management,

AIDS prevention etc. Gone are the days when industry could merely concern itself with production and profit and industry has realized that Corporate Citizenship is an important character and responsibility, to be fulfilled whole heartedly, as a current and future responsibility.

In shaping India's future, the integration of the corporate sector into community development in all its aspects is a reality and the managerial expertise of industry will help to build a stronger community at the micro level in a rural area and at the macro level of nation-building.

Bending rules of the land, evading tax payments by under-invoicing exports and dubious tax-planning; cornering licenses at the cost of others; adulterating quality of products; and indulging in other unethical practices may earn money. But such practices hardly speak highly of corporate citizenship. The Tatas are a contrast to the general trend. Unethical practices are anathema to the Tatas. The best way to substantiate this claim is to quote J.R.D. Tata. "This factor has also worked against our growth. What would have happened if our philosophy was like that of some other companies which do not stop at any means to attain their ends. I have often thought of that and I have come to the conclusion that if we were like these other groups, we would be twice as big as we are today. What we have sacrificed is a 100 per cent growth."

5.4. CORPORATE PHILOSOPHY

Philosophy and Religion

The word 'religion' is derived from the two Latin words: 'Re' and 'Ligare', 'Re' means 'again' and 'Ligare' means 'to join', therefore religion means to join and go back to that from which a person has separated, There are three main aspects which characterize all religions. These are mythology, rituals and philosophy. Ritual and mythology differ from religion to religion even though philosophy remains the same. The tenets of philosophy include the following :

1. All religions accept the fact that there is an ultimate reality.

2. This reality is called by various names in various religions. According to Hindus it is 'Narayana', according to Christianity it is 'Father in Heaven', and according to Islam it is 'Allah'.

3. The mission of a human being is to attain this reality and to strive towards that goal by prayer and worship.

The necessity and motive for seeking this reality is to obtain absolute happiness, or bliss (ananda). This absolute happiness is obtained when the individual self merges with the "Universal Self".

In the earthly world, desires dominate the individual and give rise to miseries, because the individual's life becomes dominated by maternal objects and goals. He therefore, becomes dejected, depressed and is always tense, nervous and sad, since his goals themselves are shallow, transient and give only temporary satisfaction. All religions, therefore, suggest the same philosophy that desires should be overcome through prayer and love. Each religion lays down different principles for salvation. According to Hinduism and Bhagwad Gita, the recommended spiritual courses consist of Gnana Yoga for the intellectuals, Bhakti Yoga for the emotional, Karma Yoga for all those involved in performing various actions and are active. Hatha Yoga is prescribed for the indolent. The practice of Yogas brings about a proper control on desires and in this respect different religions preach in various forms and in various proportions, the substance contained in these Yogas.

The Centre of all the religions is God; all religions exhort their adherents to worship God, all the paths of all religions converge towards this centre. All religions admit that the body perishes, but the eternal light continues in the form of the soul. All religions profess that the realization of God is through rite soul and love. Hence religious methods, creeds, rituals, ideas may differ but no religion is above the universal, eternal truth to infinity.

India a secular country which does not mean that it is against religion. The following are the interpretations of secularism :

1. The original western sense, as separation of state and religion;

2. The impartiality of state arbitration between competing religions;

3. A belief system opposed to religion, and

4. Equal recognition to all religions.

The second and fourth definitions are unique to India's socio-historical context. According to Mahatma Gandhi, it is necessary that we keep our windows open winds from other climes and cultures but not to be swept off our feet. Hence, secular India is open to both tradition and modernity.

Morality and Spirituality

Morality is a great human characteristic that can be deemed as an ability and willingness to stick to upright, good conduct and behaviour and perform action in a righteous manner. Spirituality is the ability of an individual to understand the innate divine nature of the human self that can realize its potential to reach the "Higher Self". This is also known as the spiritual quest. Morality and spirituality comprise the greatest of all human treasures. It is through these guiding measures that a human being can understand the process of transformation from humanity to divinity. Great sages, seers, saints, mahatmas and holy men have emphasized the practice of these two venues in every human being as a means of attaining divinity.

Human society can adopt principles of morality and adopt spiritual needs towards the attainment of moral goals of life. Human society needs proper governance to avoid chaos and misery.

For the promotion of morality human conduct needs the exercise of discretion and understanding of wisdom, the philosophy of Hinduism in India spiritualizes human existence through spiritual awakening and wisdom. The great saints and sages through their spiritual and divine powers experienced the eternal truth and recorded their experiences and wisdom for the benefit of posterity. These records constitute the four Vedas. The Hindu scriptures consist of (i) Srutis, (ii) Smritis, (iii) Itihasas, (iv) Purands (v) Agamds and (vi) Darshans. To practise morality and spirituality and to guide oneself towards the goal of divinity, it is necessary to overcome the enemies such as lust, anger, greed, attachment, ego and hatred, and embrace virtues such as truth, righteousness, peace, love and non-violence. Harmonious conditions of human existence can alone promote ethical functioning for the good and welfare of oneself and all living creatures.

Value Education

Value education is true education. The end of education is character. Education should be for building up good character. 'True education is that which gives the knowledge of the self'. It originates from the spiritual heart. Physical hearts may lead to worry and anxiety whereas spiritual heart is the fountain of true knowledge. Consider 'character as your very life-breath' and spiritual heart as your conscience. True spirituality lies in knowing the 'self'. Secular knowledge should lead to unity. From unity to purity and from purity to divinity, secularism should promote unity, harmony, love and peace. Attachment and hatred lead to ego and all these three lead to a person's doom. Mahatma Gandhi stated: "By education I mean an all-round drawing out of the best in child and man's body, mind and spirit. Education must be of a new type for the sake of the creation of a new world."

Proper education broadens a businessman's vision towards a benevolent, customer-oriented mission. Business thrives when people's development and national integration usher prosperity and plenty into the lives of millions of haves and have-nots. Much progress in India depends on the percolation of the ideals of secularism and communal harmony throughout the length and breadth of the country. It is necessary to emphasize proper educational values that would promote unity and mutual tolerance while excluding those that breed and promote ill-will or hatred between groups and communities. Narrow communal feelings or improper feelings based on caste, creed or religion should be eliminated. It is necessary to imbue the people with a sense of national purpose and fraternal feeling. Enlightened education alone can bring about a proper understanding between different regional linguistic groups.

Some of the values and behavioural codes of conduct towards human betterment and happiness can be classified as :

(a) Human Values

(b) Social Values

(c) Business Values

(d) Community Values

(e) Family Values

(f) Professional Values

(g) National Values

Unethical Practices Vis-a-Vis Cheating

Are the following not unethical practices? How did they arise? Why could they not be prevented? These are the questions highlighted below.

(a) As many as 11 brokers of a certain stock exchange were debarred from participating in trading by the stock exchange authorities because these brokers did not fulfill the capital adequacy norms laid by the Securities and Exchange Board of India some months ago.

(b) The kingpin of the incredible money doubling scheme, belonging to a certain consultancy firm vanished without a trace after collecting huge sums of money in a certain city. Many employees of well-known companies were victims, undone by their own greed.

(c) Video shops trading in blue films which are replete with sadomasochist acts, bestiality and violence, are polluting the minds of college students. A certain TV network that shows numerous amorous scenes is another. Pornographic films which come from some foreign countries are sold in India. Drugs, mainly brown sugar, sold secretly to young students are causing a serious problem.

(d) Large speculative investments by public sector undertakings, irregularities in mutual funds, misuse of Portfolio Management Scheme (PMS) and serious breach of rules and norms by some banks and financial institutions are some of the malpractices besmirching the image of the financial sector.

(e) Some brokers misused the resources of financial institutions, mutual funds and banks to artificially boost the values of shares, many middle class investors purchased shares in the hope that when the values of the shares rose further, they would sell off the shares and earn a fortune. Many were ruined when the prices of such shares crashed.

Cultural Ethos

Successful countries like America, Japan, Germany, France and Korea have significantly developed their own management styles in consonance with their own cultural ethos, while India is still in the process of evolving Indian management processes in accordance with Indian ethos and culture. This is due to India's fascination for western management systems and styles. Though a fervent admirer of Japanese, Taiwanese and American systems, India has made significant efforts to develop its own management styles based on its rich heritage and cultural ethos.

India's Vedantic philosophy based on Upanishads considers the entire human race as "Vasudeva Kutubakam" which means "a single family of God". Indians recognized each individual as a form of divinity, hence the greeting 'Namaskar' which means 'salute to the divinity'. Further. 'Tatwamsi'. and 'Aham Brahmasmi' are the guiding tenets of Indian culture. Lord Srikrishna in The Bhagwad Gita states "I am the self, O Gudakesha, seated in the hearts of all creatures; I am the beginning, the middle and also the end of all beings".

In Indian philosophy, the 'Individual' is, therefore, all important. If his conduct is good, then everything is good. Hence Swami Vivekananda exhorted individuals; "Be good, do good". The Bhagwad Gita contains many lessons on enlightened and appropriate management.

According to Arnold Toynbee, "It is already becoming clear that a chapter which had a western beginning in business management, will have to have an Indian ending, when the world adopts rich thoughts of Indian ethos and wisdom, if it is not to end in the self-destruction of the human race".

5.5. THE FOUR STRATEGIC DECISIONS IN TRANSPARENCY

In the opening chapters of this book, we have focused on the ways in which perfect information and interdependent decision making will undermine our most cherished economic assumptions, sanctify extraordinary profits, and usher in a new era of power law dynamics that will produce dramatic changes in behaviour in many marketplaces. In this section, we will translate these macro trends into strategy prescriptions for businesses and investors. Specifically, we will strive to understand how transparency influences four critical decisions each company must make concerning its focus, business model, competition, and marketplace strategy.

The four critical decisions are:

1. What part of the business do we want to own?
2. What is the most profitable way to organize businesses?
3. What competitive threats do we face and how do we win?
4. Which behaviours do we encounter in the marketplace for our products and which strategies will yield the highest returns?

Regarding the first decision, consider that the trend toward perfect information will spur a continuing shift away from vertically integrated companies and toward focused companies. As a number of writers have already pointed out, corporate leaders and investors will enjoy a vast increase in their options for pursuing narrowly focused activities. For these business leaders and investors, the most important issue is strategic: which of these focused companies should I own? Our answer: to maximize profits, own the businesses with a power node.

The second decision is about which business model is most conducive to increased returns. In transparency, the ease of monitoring and coordination made possible by perfect information will facilitate many new kinds of arrangements between companies, without the requirement of shared ownership. These new relationships will enable companies to capture many of the same benefits that vertical integration once offered including protection from such problems as motley crews and hazardous markets without the accompanying drain on performance and profits. Particularly when they are organized around power nodes, these distributed business arrangements will allow to varying degrees the long-term creation of positive sums for all concerned. In transparency, such a nexus of relationships will replace vertical integration as the prevailing archetype for "the firm."

This idea is an extension of ongoing thinking about business models. Several authors have said, in effect: Perfect information is here. All companies will be stand-alone companies. In this view, companies have only two options: being focused and going solo in the marketplace or remaining vertically integrated. But in transparency, companies have much greater freedom to compose or arrange themselves. The two options of being focused and stand-alone or being vertical will emerge as extremes. Perfect information will permit so much in the way of oversight and enforcement that the vast majority of companies will be able to structure a wide variety of high performing and profitable relationships that fall somewhere along the continuum between stand-alone and vertically integrated.

With such a wide assortment of choices, transparencies managers and entrepreneurs will have a great deal to think about as they determine the shape of their companies. It is important to note that the decision about business model is a matter of trade-offs among a long list of factors, not just a few. Indeed, the identified list of 14 distinct factors that must be considered when deciding how to organize or shape a business.

A step-by-step review of the factors is something that you can build into your routine each time, you look at buying and turning around or reorganizing a company.

These are based on a great deal of pre-existing theory empirical research, as well as on my own findings about the justifications for distributed power relationships. For the most part, the factors included in are neither new nor controversial. What is different is our categorization in and our discussion of how the considerations are influenced by perfect information and therefore how the choice to organize is altered by transparency.

How does transparency alter the evaluation of these trade-offs? With perfect information, the arguments that were in favour of vertical integration tend to fade away, and second considerations in favour of focus gain more weight. Therefore, in transparency I and II consideration tends to be replaced in favour of focused companies. But III issues still remain: assembly of an entire value chain under one roof offers benefits and protections that focused companies operating alone in the marketplace simply cannot reproduce. But thanks to perfect information, focused companies will for the first time enjoy a wide range of options for recouping those benefits including the development of power relationships that can be used to manage distributed business arrangements.

Regarding the third decision, how does transparency affect competition?

Before transparency, competition was discussed primarily as a fight for market share among the vertically integrated gladiators in industries whose boundaries were clearly defined. Now, with surging numbers of free standing focused companies at every level of the value chain, we are seeing exponential increases in the number of points of competition among companies providing intermediate goods and services. Once, competition was thought of primarily as a one-dimensional struggle for market share in markets for final goods and service. Now it is becoming a three-dimensional fight against horizontal rivals in the same industry, against horizontal rivals outside of the industry and around the globe, and against vertical rivals up and down the value chain. In transparency, the key determinant of profitability will be the ability to defend or extract returns from vertical competitors. Those companies that win vertical battles will be the owners of power nodes. The emphasis of competitive strategy needs to, be on profitability rather that on market share.

Finally, we turn to the fourth decision about how to successfully navigate the new dynamics in marketplaces. How will transparency change our approach to the marketplaces with a large number of participants, which will include most mass consumer markets for goods and services? Profoundly. As we now know, under conditions of perfect information and interdependent decision making, the behaviours of market participants are likely to display the same dynamics found in scale-free networks. This means two things. First, we will see huge concentrations of choices around the small proportion of points (products, services, people, etc.) that become hubsand this phenomenon will guarantee that a few select companies will be in the fortunate position of being able to create profits that may be more extraordinary than ever before. Second, in transparency, people will reach out for information (or products or, services) rather than waiting for it to reach them. Companies in transparency will thrive by anticipating and responding to consumer choices rather than forcing them into being a new approach that, I call "information aikido."

In the following, we will take a closer look at the impact of transparency on each of the four essential strategic decisions in turn. By the conclusion of this section, readers will have a clear sense of how to frame their answers to the key business questions in transparency.

Unless you plan, you cannot succeed. CEO of a company was given world-wide recognition for promoting his products. However, he always had a dream to start his own business. But what business, he did not know. He collected facts, figures and information about the product that he was working for. He envisioned a market, the profit, margin and a life cycle for his product and finally established his business. He sold his products on-line and through retail outlets making them easily accessible to customers; He spent lot of time on planning his venture and, therefore, it picked up fast. As the business expanded, he wanted to raise money from investors which was not difficult as investors had faith in that business. Effective planning, thus, helped the CEO in successfully launching a business and managing its growth.

Planning is the first function of management and lays the foundation for successful business.

Strategic intuition

There's something missing between strategic analysis and planning. Once you analyse your situation, you need a creative idea and then you go into strategic planning. That missing step is the key. And that key lies in understanding how the flashes of insights work in the human brain, flashes which we call intuition.

We all need to innovate. But how? The recent financial "innovations" that led to the US recession show us that innovation without constraints is a recipe for disaster. Yet, putting in place creative constraints can be just as bad: A company-wide "innovation process," with rules and procedures and forms to fill out, will surely kill the golden goose before it takes a breath.

So what can we do?

The usual methods of strategic analysis offer no help. They tell you how to analyse your industry, your competitors and your own company. But they're silent about how to come up with a creative idea for what to do next. They tell you how to analyse a strategy, not how to make one. There are many popular methods of analysis Five Forces, SWOT, Blue Ocean and Core Competence. They are all very useful for understanding your situation. None of them gives you the next step. Now what do we actually do?

The usual methods of strategic planning are no help either. They tell you how to get in a room and come to an agreement on vision, goals, objectives, and so forth. They're great for coming to a consensus on strategy. But they offer no way to tell if the strategy is a good one or not. These are really methods of strategic agreement, to get everyone rowing in the same direction. But is it the right direction? Strategic planning has no answer. It does not tell you how to come up with a good strategic idea.

There's something missing between strategic analysis and strategic planning. Once you analyse your situation, you need a creative idea for what to do. Then you go into strategic planning to figure out how to do it. That missing step is the key. In the past two decades, an array of innovation methods has grown to fill that gap. You get six thinking hats, brain-storming sessions, blue sky thinking, bean-bag chairs, and lego blocks in the lobby. These all aim to make you creative instead of analytical which, in other words, means to turn off your left brain and turn on your right one.

There are many other complementary techniques that combine analysis and creativity in the same moment of insight. You can make them a regular part of your basic problem-solving and planning methods at all levels of your company, for any kind of problem, large and small. Innovation then becomes not a special event or a rare exception, but everyday business for one and all.

Strategic Innovation and Entrepreneurship

As world economy gets more and more entrepreneurial, the same has been the movement of education systems. But there are still some flaws in them and that can only be removed by increasing activities, which allow them to face real-world situations. Can we use technology on which Gen Y is so much more dependent as the best medium for involving more up-to-date means of education?

Entrepreneurship education has become an increasingly important aspect of the entrepreneurial endeavour, especially in science and engineering. While the US graduate education system is the envy of other developed countries around the world, with independent minded graduate students and post-doctoral fellows advancing the frontiers of engineering, physical and life sciences, that system has done little to help researchers evaluate, which ideas and discoveries have commercial potential. Providing scientists and engineers with useful entrepreneurial skills will accelerate the translation of ideas from lab to product to enterprise. The authors are developing an innovative method for teaching scientists and engineers how to develop and harness the value of their discoveries by using digital media based collaborative support to overcome the various obstacles that arise in launching technological startup companies.

5.6. THE INNOVATION X FRAMEWORK

Armed now with an understanding of the type of problem you are facing, you can explore constructive ways of solving it. This is the Innovation X framework, because it is specifically about how to explore the landscape of an X-problem and then create valuable innovation based on it. As illustrated in Figure, four methods make up the framework: immersion, convergence, divergence, and adaption.

The four methods can be thought of as bundles of specific research and analytical tools, coupled with principles for how to use them. The Innovation X methods work together to help you understand the X-problem, identify and prioritize innovations, and guide development of marketable solutions based on the innovations. They do this in several ways:

- By creating clarity about the X-problem by dimensionalizing it and giving it as much structure as possible at any given time

- By enabling deep insight into the surrounding context customers, competitors,

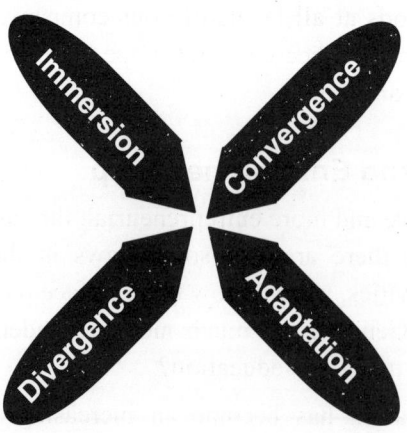

brand, retail, organization, and so on that gives a thorough understanding of possible innovation options

- By showing how your products and their surrounding experiences are meeting or not meeting customer needs, and where there are opportunities to meet new needs, expand the ways you can engage customers, and reach new customers that you have never addressed before

- By describing the boundaries of your business domain, and where it can be strengthened and stretched in new directions

- By creating adaptability as solutions are put out into the world, the X-problem becomes further understood, and new solutions are created and modified.

- By providing ongoing tracking of emerging opportunities and threats.

Immersion

Because opportunities are getting harder to identify, customers more complex in their expectations, and competitors more sophisticated in their offerings, it takes more effort to gain true insights into what are their offerings, it takes more effort to gain true insights into what the X-problem is and ultimately how to focus innovation efforts. I call this method immersion because it truly requires a good soaking in a broad variety of investigations and inputs that go beyond conventional market research (though that remains an essential tool). Traditional market research often treats customers as data, while real immersion in people's lives uncovers the nuances and unmet needs that inform innovation.

Convergence

Convergence is crucial to addressing the challenge of customers seeking integrated solutions and systems rather than isolated products. The term convergence has been in use for some years now and in fact at frog we made it the backbone of our business back in the 1990s by building up software and Web design capabilities to complement our traditional industrial design and engineering offering. Companies have recognized its importance and have put more effort into creating offerings that converge products, online experiences, and software. Yet confusion is still widespread over what convergence really involves and how to make it work successfully. Do you find yourself puzzling over vague and conflicting definitions of ecosystems, customer touch points, customer journeys, and customer experience? If so, you are not alone.

Convergence, as I use it here in the context of X-problems, means the integration of multiple components (hardware, software, and services), customer interaction points, and enabling technologies to deliver functionality, benefits, and experiences that would be impossible from stand alone products.

Adaptation

Adaptation is a term from biology that refers to the process by which organisms gradually adjust as their environment changes. Here, the changing environment is the emerging X-problem. Adaptation is vital for ensuring that innovations match up to opportunities and to business goals.

While the other three Innovation X methods help you build your understanding of your environment and identify opportunities and threats, the adaptation method focuses on flexible development of new innovations and feedback loops to course-correct over successive iterations of prototypes and launches.

To sum up, you can think of the four methods like this: immersion develops an understanding of how the world is; convergence and divergence conceptualize how the would could be; and adaptation looks at what the world is becoming (that is to say, the ever-changing gap between is and our desire could be).

5.7. ROLE OF INFORMATION SYSTEMS IN BUSINESS

The Role of Information Systems in Business Today

It's not business as usual in India anymore, or the rest of the global economy. In 2009, Asia-Pacific Small and Medium Businesses spent US$153 billion on IT and telecom. Chinese, Korean, and Indian SMBs make up more than 50% of Asia-Pacific spending, according to the latest study by New York-based Access Markets International (AMI) Partners Inc. In addition, they will spend another $900 billion on business and management consulting and services much of which involves redesigning firms' business operations to take advantage of these new technologies.

As managers, most of you will work for firms that are intensively using information systems and making large investments in information technology. You will certainly want to know how to invest this money wisely. If you make wise choices, your firm can outperform competitors. If you make poor choices, you will be wasting valuable capital.

How Information Systems are Transforming Business

You can see the results of this massive spending around you every day by observing how people conduct business. More wireless cell phone accounts were opened in 2008 than telephone landlines installed. Cell phones, BlackBerrys, iPhones, e-mail, and online conferencing over the Internet have all become essential tools of business. Fifty-eight percent of adult Americans have used a cell phone or mobile handheld device for activities other than voice communication, such as texting, emailing, taking a picture, looking for maps or directions, or recording video (Horrigan, 2008).

By June, 2008, more than 80 million businesses worldwide had dot-com Internet sites registered (60 million in the U.S. alone) (Versign, 2008). According to the report, about 85% of users online are engaging in e-commerce. In Ireland, 91% of shoppers are undertaking transactions online, 86% in Turkey are shopping online, and India and the UAE take up the joint third place where about 84% of online shoppers are buying things online. That is, 84% of roughly 3.2% of India's population (the number of people in India on the Internet).

In 2007, FedEx moved over 100 million packages in the United States, mostly overnight, and the United Parcel Service (UPS) moved 3.7 billion packages worldwide. Businesses sought to sense and respond to rapidly changing customer demand, reduce inventories to the lowest possible levels, and achieve higher levels of operational efficiency. Supply chains have become more fast paced, with companies of all sizes depending on just-in-time inventory to reduce their overhead costs and get to market faster.

As newspaper readership continues to decline, more than 64 million people receive their news online.

About 67 million Americans now read blogs, and 21 million write blogs, creating an explosion of new writers and new forms of customer feedback that did not exist five years ago (Pew, 2008). Social networking sites like MySpace and Facebook attract over 70 and 30 million visitors a month, respectively, and businesses are starting to use social networking tools to connect their employees, customers, and managers worldwide.

E-commerce and Internet advertising are booming: Google's online ad revenues, surpassed $16.5 billion in 2007, and Internet advertising continues to grow at more than 25 percent a year, reaching more than $28 billion in revenues in 2008.

New federal security and accounting laws, requiring many businesses to keep e-mail messages for five years, coupled with existing occupational and health laws requiring firms to store employee chemical exposure data for up to 60 years, are spurring the growth of digital information now estimated to be 5 exabytes annually, equivalent to 37,000 new Libraries of Congress.

What's New in Management Information Systems?

Lots! What makes management information systems the most exciting topic in business is the continual change in technology, management use of the technology, and the impact on business success. Old systems are being creatively destroyed, and entirely new systems are taking their place. New industries appear, old ones decline, and successful firms are those who learn how to use the new technologies. Table summarizes the major new themes in business uses of information systems. These themes will appear throughout the book in all the chapters, so it might be a good idea to take sometime now and discuss these with your professor and other students. You may want to even add to the list.

In the technology area there are three interrelated changes: (1) the emerging mobile digital platform (think iPhones, BlackBerrys, and tiny Web-surfing net-books), (2) the growth of online software as a service, and (3) the growth in "cloud computing" where more and more business software runs over the Internet. Of course these changes depend on other building-block technologies described in Table, such as faster processor chips that use much less power.

YouTube, iPhones and Blackberrys, and Facebook are not just gadgets or entertainment outlets. They represent new emerging computing platforms based on an array of new hardware and software technologies and business investments. Besides being successful products in their own right, these emerging technologies are being adopted by corporations as business tools to improve management and achieve competitive advantages. We call these developments the "emerging mobile platform."

Managers routinely use so-called "Web 2.0" technologies like social networking, collaboration tools, and wilds in order to make better, faster decisions. Millions of managers rely heavily on the mobile digital platform to coordinate vendors, satisfy customers, and manage their employees. For many if not most Indian managers, business day without their cell phones or Internet access is unthinkable.

Technology	Business Impact
Cloud computing platform emerges as a major business area of innovation	A flexible collection of computers on the Internet begins to perform tasks traditionally performed on corporate computers.
More powerful, energy efficient computer processing and storage devices	Intel's new PC processor chips consume 50% less power, generate 30% less heat, and are 20% faster than the previous models, packing over 400 million transistors on a dual-core chip.
Growth in software as a service (SaaS)	Major business applications are now delivered online as an Internet service rather than as boxed software or custom systems.

Technology	Business Impact
Netbooks emerge as a growing presence in the PC marketplace, often using open source software	Apple opens its iPhone software to developers, and then opens an Applications Store on Tunes where business users can download hundreds of applications to support collaboration, location-based services, and communication with colleagues.
A mobile digital platform emerges to compete with the PC as a business system	Small, lightweight, low-cost, energy-efficient, net-centric notebooks use Linux, Google Docs, open source tools, flash memory, and the Internet for their applications, storage, and communications.

MANAGEMENT

Managers adopt online collaboration and social networking software to improve coordination, collaboration, and knowledge sharing	Google Apps, Google Sites, Microsoft's windows Share point Services and IBM's Lotus Connections are used by over 100 million business decision makers worldwide to support blogs, project management, online meetings, personal profiles, social bookmarks, and online communities.
Business intelligence applications accelerate	More powerful data analytics and interactive dashboards provide real-time performance information to managers to enhance management control and decision making.
Managers adopt millions of mobile tools such as smartphones and mobile Internet devices to accelerate decision making and improve performance	The emerging mobile platform greatly enhances the accuracy, speed, richness of decision making as well as responsiveness to customers.
Virtual meetings proliferate	Managers adopt telepresence video conferencing and Web conferencing technologies to reduce travel time, and cost, while improving collaboration and decision making.

ORGANIZATIONS	
Web 2.0 applications are widely adopted by firms	Web-based services enable employees to interact as online communities using blogs, wikis, e-mail and instant messaging services. Facebook and MySpace create new opportunities for business to collaborate with customers and vendors.
Telework gains momentum in the workplace	The Internet, wireless laptops, iPhones, and BlackBerrys make it possible for growing numbers of people to work away from the traditional office. Between 2007 and 2011, the worldwide corporate teleworking population of individuals that spend at least one day a week teleworking from home is expected to show a CAGR of 4.4 percent. This population is likely to reach 46.6 million by the end of 2011.
Outsourcing production	Firms learn to use the new technologies to outsource production work to low wage countries.
Cocreation of business value	Sources of business value shift from products to solutions and experiences and from internal sources to networks of suppliers and collaboration with customers. Supply chains and product development become more global and collaborative; customer interactions help firms define new products and services.

Intranets and Extranets

Enterprise applications create deep-seated changes in the way the firm conducts its business, and they are often costly to implement. To coordinate their activities, many different departments and employees must change the way they work and the way they use information. Companies that do not have the resources to invest in enterprise applications can still achieve some measure of information integration by using intranets and extranets.

Intranets and extranets are really mere technology platforms than specific applications, but they deserve mention here as one of the tools firms use to increase integration and expedite the flow of information within the firm, and with customers and suppliers, Intranets are internal networks built with the same tools and communication standards as the Internet and are used for the internal distribution of information to employees, and as repositories of corporate policies, programs, and data. Extranets are intranets extended to authorized users outside the company.

An intranet typically centers on a portal that provides a single point of access to information from several different systems and to documents using a Web interface. Such portals can be customized to suit the information needs of specific business groups and individual users if required. They may also feature e-mail, collaboration tools, and tools for searching for internal corporate systems and documents.

For example, Indian Airlines' corporate intranet for sales provides its sales-people with sales leads, fares, statistics, libraries of best practices, access to incentive programs, discussion groups, and collaborative workspaces. The intranet includes a Sales Ticket capability that displays bulletins about unfilled airplane seats around the world to help the sales staff work with colleagues and with travel agents who can help them fill those seats.

Companies can connect their intranets to internal company transaction systems, enabling employees to take actions central to a company's operations, such as checking the status of an order or granting a customer credit. Indian Airlines' intranet connects to its reservation system.

Extranets expedite the flow of information between the firm and its suppliers and customers. Indian Airlines uses an extranet to provide travel agents with fare data from its intranet electronically. GUESS Jeans allows store buyers to order merchandise electronically from ApparelBuy.com. The buyers can use this extranet to track their orders through fulfillment or delivery.

How Information Systems Impact Organizations and Business Firms

Information systems have become integral, online, interactive tools deeply involved in the minute-to-minute operations and decision making of large organizations. Over the last decade, information systems have fundamentally altered the economics of organizations and greatly increased the possibilities for organizing work. Theories and concepts from economics and sociology help us understand the changes brought about by IT.

Economic Impacts

From the point of view of economics, IT changes both the relative costs of capital and the costs of information. Information systems technology can be viewed as a factor of production that can be substituted for traditional capital and labour. As the cost of information technology decreases, it is substituted for labour, which historically has been a rising cost. Hence, information technology should result in a decline in the number of middle managers and clerical workers as information technology substitutes for their labour.

As the cost of information technology decreases, it also substitutes for other forms of capital such as buildings and machinery, which remain relatively expensive. Hence, over time we should expect managers to increase their investments in IT because of its declining cost relative to other capital investments.

IT also obviously affects the cost and quality of information and changes the economics of information. Information technology helps firms contract in size because it can reduce transaction costs the costs incurred when a firm buys on the marketplace what it cannot make itself. According to transaction cost theory, firms and individuals seek to economize on transaction costs, much as they do on production costs. Using markets is expensive because of costs such as locating and communicating with distant suppliers, monitoring contract compliance, buying insurance, obtaining information on products, and so forth. Traditionally, firms have tried to reduce transaction costs through vertical integration, by getting bigger, hiring more employees, and buying their own suppliers and distributors, as both General Motors and Ford used to do.

Information technology, especially the use of networks, can help firms lower the cost of market participation (transaction costs), making it worthwhile for firms to contract with external suppliers instead of using internal sources. As a result, firms can shrink in size (numbers of employees) because it is far less expensive to outsource work to a competitive marketplace rather than hire employees.

For instance, by using computer links to external suppliers, the Chrysler Corporation can achieve economics by obtaining more than 70 percent of its parts from outside. Information systems make it possible for companies such as Cisco Systems and Dell Inc. to outsource their production to contract manufacturers such as Flextronics instead of making their products themselves.

Information technology also can reduce internal management costs. According to agency theory, the firm is viewed as a "nexus of contracts" among self-interested individuals rather than as a unified, profit-maximizing entity. A principal (owner) employs "agents" (employees) to perform work on his or her behalf. However, agents need constant supervision and management; otherwise, they will tend to pursue their own interests rather than those of the owners. As firms grow in size and scope, agency costs or coordination costs rise because owners must expend more and more effort supervising and managing employees.

Information technology, by reducing the costs of acquiring and analyzing information, permits organizations to reduce agency costs because it becomes easier for managers to oversee a greater number of employees. Information technology enables firms to increase revenues while shrinking the number of middle managers and clerical workers. IT expands the scope of SME by enabling them to perform coordinating activities such as processing orders or keeping track of inventory with very few clerks and managers.

Because IT reduces both agency and transaction costs for firms, we should expect firm size to shrink over time as more capital is invested in IT. Firms should have fewer managers, and we expect to see revenue per employee increase over time.

Organizational and Behavioral Impacts

Theories based on the sociology of complex organizations also provide some understanding about how and why firms change with the implementation of new IT applications.

IT Flattens Organizations

Large, bureaucratic organizations, which primarily developed before the computer age, are often inefficient, slow to change, and less competitive than newly created organizations. Some of these large organizations have down-sized, reducing the number of employees and the number of levels in their organizational hierarchies.

Flattening organisations

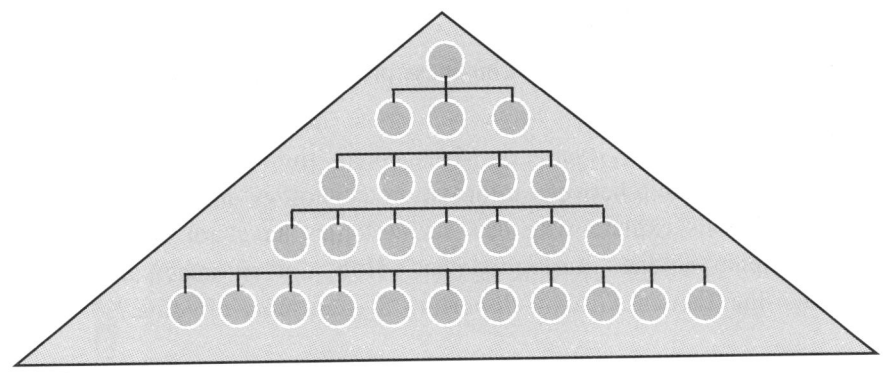

A traditional hiearchical organisation with many levels of management

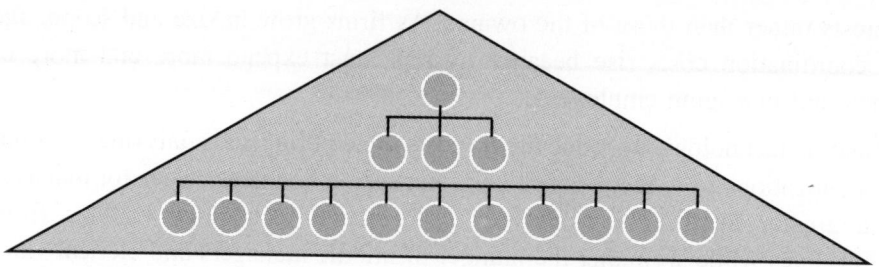

An organisation that has been "flattened" by removing layers of management

These changes mean that the management span of control has also been broadened, enabling high-level managers to manage and control more workers spread over greater distances. Many companies have eliminated thousands of middle managers as a result of these changes.

Implications for the design and understanding of information systems

To deliver genuine benefits, information systems must be built with a clear understanding of the organizational factors to consider when planning a new system which are given below :

- the environment in which the organization operates
- the structure of the organization: hierarchy, specialization, routines, and business processes.
- The organization's culture and politics
- the type of organization and its style of leadership
- The principal interest groups affected by the system and the attitudes or workers who will be using the system
- The kinds of tasks, decisions, and business processes that the information system is designed to assist.

As you read the Interactive Session on Management, think about what you have just learned about the relationship between information technology and organizations. What features of organizations explain why new technologies have not been as useful as envisioned in, helping soldiers during combat? How did an understanding of this relationship make the Tactical Ground Reporting System more effective?

Using information systems to achieve competitive advantage

In almost every industry you examine, you will find that some firms do better than others. There's almost always a stand-out firm. In the automotive industry, Toyota is considered a superior performer. In pure online retail, Amazon is the leader, in off-line retail Walmart, the largest retailer on earth, is the leader. In online music, Apple's iTunes is considered the leader with more than 75 percent of the downloaded music market, and in the related industry of digital music players, the iPod is the leader. In web search, Google is considered the leader.

Firms that "do better" than others are said to have a competitive advantage over others: They either have access to special resources that others do not, or they are able to use commonly available resources more efficiently usually because of superior knowledge and information assets. In any event, they do better in terms of revenue growth, profitability, or productivity growth (efficiency), all of which ultimately in the long run translate into higher stock market valuations than their competitors.

But why do some firms do better than others and how do they achieve competitive advantage? How can you analyze a business and identify its strategic advantages? How can you develop a strategic advantage for your own business? And how do information systems contribute to strategic advantages? One answer to that question is Michael Porter's competitive forces model.

Porter's Competitive Forces Model

Arguably, the most widely used model for understanding competitive advantage is Michael Porter's competitive forces model. This model provides a general view of the firm, its competitors, and the firm's environment. Earlier in this chapter, we described the importance of a firm's environment and the dependence of firms on environments. Porter's model is all about the firm's general business environment. In this model, five competitive forces shape the fate of the firm.

Managing Strategic Transitions

Adopting the kinds of strategic systems described in this chapter generally requires changes in business goals, relationships with customers and suppliers, and business processes. These socio-technical changes, affecting both social and technical elements of the organization, can be considered strategic transitions a movement between levels of socio-technical systems.

Such changes often entail blurring of organizational boundaries, both external and internal. Suppliers and customers must become intimately linked and may share each other's responsibilities. Managers will need to devise new business processes for coordinating their firms' activities with those of customers, suppliers, and other organizations. The organizational change requirements surrounding new information systems are so important that they merit attention throughout this text.

Improving Decision Making : Using a Database to Clarify Business Strategy

Software skills : Database querying and reporting; database design

Business skills : Reservation systems; customer analysis

In this exercise, you'll use database software to analyse the reservation transations for a hotel and use that information to fine-tune the hotel's business strategy and marketing activities.

Business has grown steadily during the past 10 years. Now totally renovated, the inn uses a romantic weekend package to attract couples, a vacation package to attract young families, and a weekday discount package to attract business travelers. The owners currently use a manual reservation and book-keeping system, which has caused many problems. Sometimes two families have been booked in the same room at the same time. Management does not have immediate data about the hotel's daily operations and income.

At the Laudon Website you will find a database for hotel reservation transactions developed in Microsoft access.

Develop some reports that provide information to help management make the business more competitive and profitable. Your reports should answer the following questions :

- What is the average length of stay per room type?
- What is the average number of visitors per room type?
- What is the base income per room (i.e. length of visit multiplied by the daily rate) during a specified period of time?
- What is the strongest customer base?

Group dimensions of Organizational Structure and Culture

When discussing corporate culture, we tend to focus on the organization as a whole. But corporate values, beliefs, patterns, and rules are often expressed through smaller groups within the organization. Moreover, individual groups within organizations often adopt their own rules and values. We will therefore look next at several types of groups as well as the conflicts that can occur between the norms of these groups and individuals.

Types of groups

Two main categories of groups affect ethical behaviour in business. A formal group is defined as an assembly of individuals that has an organized structure accepted explicitly by the group. An informal group is defined as two or more individuals with a common interest but without an explicit organizational structure.

Formal groups can be divided into committees and work groups and teams.

5.8. CORPORATE GOVERNANCE IS TO BUILD FINANCIAL VALUE

What is a business for? Why was it formed in the first place? What is the difference between success and failure? We suggest that a business exists to generate long-term financial value for its investors.

If that makes the reader yawn, it is understandable. It has become commonplace. However, yawns may be premature.

In this chapter we first give 'value' a precise and practical meaning, and set out the criteria that should be applied. The chapter then describes and discusses the main objection to value as the objective, made by those who prefer the stakeholder view of what business is for. We clarify that in rejecting the stakeholder view we do not advocate that business managers should act unethically or even uncaringly.

Although the stakeholder view has found favour with a number of senior managers, few of its originators and principal advocates have experienced the decisions and dilemmas that face practising managers. Because the stakeholder approach still causes so much discussion in the early years of the twenty-first century, we give it a thorough airing.

The chapter then discusses two alternative objectives which are popular with managers rather than with outsiders: diversification and size. Neither of these is necessarily believed by its champions to be incompatible with the means of achieving value. We state our has the following shortcomings for the purpose of evaluating new projects :

1. Profit (or loss) relates to one period (e.g. a year) at a time, and cannot be related to all the cash flows out and in over the lifetime of the project.

2. Profit does not measure the time value of money, as does NPV. The time value of money recognizes that cash returns in year 5 are not as valuable as those in year 4.

3. Income and expenditure are not cash concepts, whereas investment is a cash event. Returns to an investment can only be measured in cash flows, not in accrual terms including depreciation and other provisions, which serve the purpose of allocating income and expenditure to appropriate accounting periods.

4. The accounting measurement of profit often requires subjective judgements at

the end of each period, such as inventory valuations, doubtful debt provisions and the like. Cash receipts and payments do not.

5. Profit tends to be contrasted with loss, implying a performance yardstick of zero profit. Financial value's yardstick or hurdle is the expected rate of return that is the cost of capital. Its threshold level is therefore at a point above zero profit. A positive rate of return, which falls short of the cost of capital, represents a loss of financial value.

Ratios like return on capital employed (ROCE) or return on sales (ROS) are indispensable as control ratios for overall results in a given accounting period. Such controls ratios need to be accounting ratios, as only accounting data are available. However, they are inappropriate for assessing a new project, like a new offering, where the critical hurdle is its zero NPV.

Financial Value : Valuation Criteria

Our valuation of a project should aim to apply the criteria which the stock market would apply to it, if it had our internal information about its prospects. The cost of capital should be taken as the rate the market applies now to any project with a similar risk profile and with the same time-frame. The time-frame depends on the payback period of the project. A new automobile model might have a payback period of 10 years from the decision to invest; a decision by a shop to invest in a fax service for customers might pay back in a matter of months. Merry's option 2 might have a payback period of 3 years.

The market's discount rate, like the price earnings ratio, is often volatile. The peaks and troughs of the market tend to overshoot, as for example at the end of the 1990s with the 'dotcom' boom. This is due to bandwagon effects caused by market participants watching each other rather than the underlying supply and demand. At times like that it may be advisable to make some normalizing adjustment to the current discount rate by consulting economists, or by reading experienced economic commentators in the press. During such exaggerated peaks and troughs economists tend to be surprisingly unanimous. The adjustment will necessarily be a guesstimate, but better than no adjustment. The discount rate has a powerful effect on the evaluation of any project, and its selection needs all the care we can give it.

Some important features of this NPV test are sometimes overlooked :

• The cash flows consist of expectations.

• The cash flows are incremental: they are the difference between the expected future cash flows of the company (a) with, and (b) without the offering. In the Merry example, option 1 is the reference point.

- The comparison is between cash flows, which exclude, for example, allocated costs or depreciation charged in the accounts.

- The cash flows must take into account any favourable or unfavourable effects on other parts of the company. For example, a new Starbuck outlet may not only take business away from other local coffee outlets, but also from the nearest Starbuck only a few streets away. Microsoft Windows 2020 may take business away from Windows 2018. In our alcotrop example the most important item is the effect of each option on sales of Merry lager.

- The discount rate varies with the risk of each individual new offering. An important element of that risk is that customers may prefer competing substitutes, some of them not yet on the market and thus unknown at the time of evaluation. What is the risk for Merry of liquor stores importing drinks from Australia?

- The test will have different results at different stages in the life of an offering. Expectations of future cash flows will change with fresh news, and so may the discount rate with reassessments of the risk factor.

- At the original evaluation of any option such as option 2 in the alcotrop case, any negative NPV would mean that the company would be better off without that option, quite apart from the comparison with the alternatives.

- As the new offering goes through its lifespan, a time will inevitably come when the test will show the cumulative total value generated to date no longer growing, but declining. The offering has completed its useful life, and should at that point be discontinued or divested, as it has now begun to destroy value.

What conflicts in fact face real managers?

All these conflicts have two features in common. First, they are all between a manager's personal interests and the value objective of the company. Secondly, some ambivalence is here caused by the company's two very different roles, which each serve the interests of a defined group of people :

1. as servant of its investors; and

2. as the employer, to whom the manager owes a contractual and professional duty.

In other words, managers' real conflicts are not between loyalties to different groups of people, such as stakeholders, but between their own and their colleagues' personal interests and those of the company they serve.

Managers are of course ethical human agents, and have their own value systems and consciences, but these are a part of each manager's personal interests, which may also point to conflicting courses of action. More about managers' personal value systems and interests later.

Yet when discussing managers' real conflicts, we need to separate out two sub-categories.

1. facing CEOs or all managers?

2. at times of a survival crisis or in calmer times?

The CEOs in relatively calm times face a conflict of interest mainly between investing in risky projects of high financial value, or safer ones of lower value. Riskier projects meet CEOs' duty to their employing companies; safer ones appear to be better for their own job security. Again, CEOs may be tempted to abuse their position of power to obtain pay, employment conditions and fringe benefits in excess of their market value to the company.

In a survival crisis CEOs might be tempted to take damaging steps designed either to bring about short-term improvements in the company's share price or to frustrate hostile bidders, for example by poison pills. Their own consciences usually prevail against these temptations. In any case financial market regulations and legal requirements have in many countries been tightened to prevent such actions and to safeguard the interests of investors.

All managers (including CEOs) may in calmer times be tempted to act to the detriment of the company's financial value by claiming reimbursement of unnecessary expenses, by personally consuming the company's property, or by obtaining other personal benefits at the company's expense.

Priorities tend to change in a survival crisis. In such a crisis all managers have a natural tendency to become preoccupied with their own and their closer colleagues' and their dependants' future, and to give more priority to these personal interests than to those of the company and its investors. This temptation is reinforced by the fact that at such times investors are apt to become less committed to the retention of the present management team.

If these reflections are correct, then the conflicts of interest are between manager Ann's own contractual and professional duty and her personal interests, which include her own and her closer colleagues' jobs and livelihood. This dilemma is not of course clear-cut between good and bad, between selfishness and conscience. In a survival crisis, if Ann chooses the path of duty, that will improve her chances of finding managerial employment elsewhere. If she chooses the other path, she may be motivated by compassion for close colleagues and their families.

The survival crisis is a very common experience in modern times. It forces a person into some agonizing decisions. In an extreme case her duty to the company and its investors is to recommend a hostile takeover bid, if it represents better value than whatever NPV the present management team's realistic plans provide. However, that recommendation may well lead to her own and her colleagues' redundancy, and to severe hardship for their families and dependants. The existing management team may well be close friends and deeply committed to each other's vital interests. The dilemma is moral, and very real. Much more real than any assumed choice between loyalty to investors and loyalty to all stakeholders.

Which of the two views is more desirable?

Investors are interested in financial value, the other stakeholder groups have different and often mutually conflicting priorities. For example, the interests of trade unions are often opposed to those of customers. Hence the stakeholder view requires managers to hold some balance between those conflicting views. Investors want profits and dividends, employees want good pay, good working conditions and job security. Customers want low prices and efficient performance. All these conflicts can raise costs and reduce profit. Similarly, the interests of customers and suppliers and of the environment often reduce profit by raising costs or reducing selling prices.

Of course, looking after the interests of these other stakeholders also has some benign effects on profits. It avoids the costs of conflict, of pressures and of friction. British Airways' profits, for example, have been slashed more than once by damaging strikes. A company which cultivates the goodwill of its non-financial stakeholders and of their champions can operate more efficiently and smoothly. This is good for value creation. There is thus some built-in overlap between the interests of all the various stakeholders. The conflict is not head-on.

We believe the stakeholder view has one fatal drawback. Competitive business is a hard challenge. It needs to be directed by people who understand the parts played by all links in the value chain, to weld all parts of the effort together for single-minded thrust in the tough arena of competition. Management needs to be single-minded, and should not have to serve conflicting interests. Financial value provides a single criterion of success. If managers are not focused on it, they are bound to fail. The Economist in a survey by Clive Crook puts it as follows: 'Managers. . . ought not to concern themselves with the public good: they are not competent to do it, they lack the democratic credentials for it, and their day jobs should leave them no time even to think about it.'

The task of restricting the freedom of businesses to act against the public good is detrimental to that of governments. Managers are the agents of the owners and must discharge their duty to them by serving the company's goal of financial value within the constraints imposed by governments and the law.

We asked which view is more desirable. Both have desirable features, and managers should not neglect those interests of non-financial stakeholders which promote the company's financial aims, such as good relations with employees. However, where the two views clash, the financial interest must in our view prevail.

Is the stakeholder view valid?

The stakeholder view may not be desirable as a statement of ultimate priority. However, is it valid? Does it fit the facts of competitive business?

The stakeholder view often sees the business or company as a collection of human beings, consisting of groups with various potentially conflicting interests. This is a social model. The financial view on the other hand sees the company as an inanimate object, as one or more investment projects. The business is 'born' as a vehicle to generate financial value returns in excess of the cost of capital. It 'dies' when investors believe it is no longer capable of earning those returns. At that point the investors will either sell it, perhaps by accepting a takeover bid, or break it up to turn the assets into cash. In either case the managers will lose their power to look after the interests of employees or of any other stakeholders. This inanimate model of the company is a financial one.

Short-term vs. long-term objectives

The survival crisis raises another big issue. Are investors interested in short- or long-term returns? It is not a matter of short or long, but of how long. Investors apply a discount factor to different future periods. Their concern is the NPV of expected future cash flows, discounted at the appropriate rate. Cash flows in distant periods are less valuable than imminent ones. What happens in the survival crisis is that the more distant periods attract a much higher risk factor and therefore lose even more of their attraction. This is equally true for investors and managers, both become preoccupied with the present.

There is nevertheless in the survival crisis also a conflict between the interests of managers and investors. Managers are bound to be concerned to preserve their and their colleagues' jobs; investors are only concerned with financial value.

Are we then left with the law of the jungle?

Our conclusion that the stakeholder view is neither tenable nor desirable is not widely shared. It seems to argue for an uncaring world of business in which the weak are left unprotected against brutal exploitation by the strong, in which managers are encouraged to be ruthless in satisfying greedy investors, and in which business is free to destroy the planet. We saw earlier that this is why so many object to the creation of value as the overriding purpose of business.

The Objective of Risk Diversification

We now turn to the two internal objectives at conflict with that of financial value. In the second half of the twentieth century two false gods were in vogue among managers, risk diversification and size. The less influential of these was risk diversification. The idea was to make the company safer by diluting its risk profile. A caricature of this would be a civil contractor acquiring a dealer in bankrupt stocks. Contracting is a highly cyclical business, vulnerable to recessions, whereas dealing in bankrupt stocks is countercyclical; it thrives in a recession. The same logic might prompt a silicon chip maker to add potato chips to its collection. A classic example was BAT Industries in the mid-twentieth century. To diversify its exposure to the risks of anti smoking regulation or consumers shifting away from cigarettes, BAT acquired a number of companies, among them a cosmetics producer and Eagle Star, an insurance company. The acquisitions were unsuccessful for many reasons.

The theoretical case for risk diversification was in any case undermined by the recognition that the risks could be more cheaply diversified by investors in their portfolios than physically by the companies in which they invested. In short, if the intention of risk diversification was to make the companies more valuable to investors, that purpose fell flat on its face. It made them less attractive to investors.

The Objective of Size

The more influential false god, however, was size. The unquestioned belief was that companies should aim to become bigger. Bigger in what sense? Size was thought of in a number of ways, including number of management units, physical size of real estate or plants, volume of sales, number of employees, financial size or market capitalization (number of shares issued multiplied by share price), economies of scale, market power or dominance in either buying or selling customer markets.

The last two objectives, scale economies and market power in either buying or selling, can of course be genuine sources of financial value. They can give a company better margins in its markets. Whether they will in any given case have that effect depends, however, on other factors like the reactions of customers and competitors. By themselves they do not guarantee success. Will customers, for example, resent a supplier's takeover of a competitor and show their resentment by defecting to a smaller independent firm or lobbying for anti monopoly intervention by the authorities? Customers do not like having their choice restricted.

These two apart, the listed items have no inherent power to boost long-term financial performance. Raising turnover will not raise earnings if the extra sales are loss making. New shares issued for a bad acquisition will reduce earnings per share. Extra physical capacity may raise or reduce unit costs. So may the occupation of extra space. The track record suggests that on balance growth in sheer size reduces rather than adds value.

5.9. SIGNIFICANCE OF CORPORATE RESTRUCTURING

Corporate restructuring is a strategic decision leading to the maximization of company's growth by enhancing its production and marketing operations, reduced competition, free flow of capital, globalization of business, etc. Section 391 and 394 of the Companies Act constitute a legislative tool that facilitates corporate restructuring in a variety of ways. Corporate enterprise must be armed with the ability to be efficient and to meet the requirements of a rapidly evolving business reality. Corporate Restructuring is one of the means that can be employed to meet the challenges and problems which confront business. The law should be slow to retard or impede the discretion of corporate enterprise to adapt itself to the needs of changing times and to meet the demands of increasing competition.

Some of the most common reasons of corporate restructuring are discussed briefly hereunder :

(i) **Economies of Scale:** Economies of scale arise when increase in volume of production leads to a reduction in cost of production per unit. For instance, overhead costs can be substantially reduced on account of sharing central services such as; accounting and finance, personnel and legal, sales promotion and advertisement, top level management and so on. So when two or more companies combine, certain economies are realized due to the larger volume of operations of the combined entity. These economies arise due to increased production capacity, strong distribution networks, effective engineering services, research and development facilities, data processing system and others.

(ii) **Operating Economies:** Apart from economies of scale, a combination of two or more firms may result in reduction of costs due to operating economies. A merged firm may avoid overlapping. Various functions may be consolidated and duplicate channels may be eliminated by implementing proper planning and control system.

(iii) **Synergy:** Synergy refers to a situation where the combined firm is more valuable than the sum of the individual combining firms. It refers to benefits other than those related to economies of scale. Apart from operating economies, synergy may also arise from enhanced managerial capabilities, creativity, innovativeness, R&D, productivity improvements improved procurement and the elimination of duplication.

(iv) **Reduction in Tax Liability:** In India, a profitable company is allowed to merge with a sick company to set-off against its profits, the accumulated losses and unabsorbed depreciation. A large number of companies in India have merged to take advantage of this provision. The conditions to claim this tax benefit have further been relaxed by amendment made in Section 72A of the Income Tax Act.

(v) **Managerial Effectiveness:** One of the potential benefits of merger is an increase in managerial effectiveness. This may occur if the existing management team, which is performing poorly, is replaced by a more efficient one.

(vi) Other reasons of corporate restructuring are appended below :

(a) To return to the shareholders of the surplus cash, which is not required in the near foreseeable future.

(b) To enhance the earnings per share of the company.

(c) To provide to shareholders/investors that the market is presently under valuing the share of the company in relation to its intrinsic value and the proposed buyback will facilitate recognition of the true value.

(d) To increase the promoters' voting power.

(e) To maintain shareholders' value in a situation of poor state of secondary market by a return of surplus cash to the shareholders.

(f) Eliminating the takeover threats.

(g) An opportunity to grow faster, with a ready-made market share.

(h) To eliminate a competition by buying it out.

(i) Diversification with minimum cost and immediate profit.

(j) To forestall the company's own takeover by a third party.

Kinds/Forms of Corporate Restructuring

It is argued that managers should restructure companies to improve value; otherwise, external raiders will get an opportunity to takeover the company. Therefore, they claim that it is in the best interest of both managers and shareholders to keep the gap between potential and actual value as close as possible. Management can improve operations by increasing revenue or reducing cost, acquiring or disposing of assets and improving the financial structure of the company. Moreover, executives often restructure their companies for enhancing productivity, reducing costs or increasing shareholder's wealth. They classify restructuring activities into four categories: portfolio restructuring, financial restructuring, organizational restructuring and technological restructuring. These are briefly discussed below :

(a) **Portfolio Restructuring:** It includes significant changes in the mix of assets owned by a firm or the lines of business in which a firm operates, including liquidation, divestitures, asset sales and spin-offs. Company management may restructure its business in order to sharpen focus by disposing of a unit that is peripheral to their core business and in order to raise capital or rid itself of a languishing operation by selling off a division. Moreover, a company can involve on an aggressive combination of acquisitions and divestitures to restructure its portfolio.

(b) **Financial Restructuring:** Financial structure refers to the allocation of the corporate flow of funds cash or credit and to the strategic or contractual decision rules, that direct the flow and determine the value-added and its distribution among the various corporate constituencies. It includes significant changes in the capital structure of a firm, including leveraged buyouts, leveraged recapitalizations and debt for equity swaps, mergers, acquisitions, joint ventures, strategic alliances, etc. The elements of the corporate financial structure include the scale of the investment base, the mix between active investment and defensive reserves, the focus of investment (choice of revenue source), the rate at which earnings are reinvested, the mix of debt and equity contracts, the nature, degree and cost of corporate oversight (overhead), the distribution of expenditures between current and future revenue potential, and the nature and duration of wage and benefit contracts. Financial restructuring generates economic value.

(c) **Organizational Restructuring:** Organizational restructuring includes significant changes in the organizational structure of a firm, including redrawing of division boundaries, flattening of hierarchic levels, spreading of the span of control, reducing product diversification, revising compensation, streamlining processes, reforming governance and downsizing employment. It is observed that lay-offs unaccompanied by other organizational changes tend to have a negative impact on performance. Downsizing announcements combined with organizational, restructuring are likely to have a positive, though small effect on performance.

Studies reveal that generally, there is a statistically significant improvement in the organizational performance after a restructuring event. However, it may not be the case always. It is cited that the average percentage change in performance is positive for finance and portfolio restructuring, while it is negligible or sometimes negative in case of organizational restructuring.

The portfolio and financial restructuring display higher percentage of positive returns than organizational restructuring. In five out of six cases (i.e. 86%) of financial and portfolio restructuring, the impact on performance was positive, while for organizational restructuring the impact was positive for 50% of the cases.

(d) Technological Restructuring: An alliance with other companies to exploit technological expertise is termed as technological restructuring.

Choice of Corporate Restructuring

The term "corporate restructuring" is quite wide covering various aspects. It may be chosen from among four broad groups, viz.:

(a) Exposition

(b) Contraction

(c) Corporate Control and

(d) Changes in Ownership Structure

Exposition

Expansion basically implies expanding or increasing the size and volume of business of the firm. It generally includes Mergers and Acquisitions (Merger and Acquisitions), Tender offers, Assets Acquisition and Joint Ventures.

(a) **Tender offers:** In the case of a tender offer, a public offer is made for acquiring the shares of the target company. Here, the acquisition of shares of the target company indicates the acquisition of management control in that company. For instance, India Cements gives an open market offer for the shares of Raasi Cements.

(b) **Asset Acquisitions:** Asset acquisitions imply buying the assets of another company. Such assets may be tangible assets like; a manufacturing unit of the firm or intangible assets like brand, trade mark, etc. In the case of asset acquisitions, the acquirer company may for instance limit its acquisitions to those parts of the firm which match with the needs of the acquirer company. For instance, Laffarge of France acquired only the cement division of Tata group. Laffarge actually acquired only the 1.70 million tonne cement plant and the assets related such division from Tata group. Such assets may also be intangible in nature. For example Coca-Cola acquired some popular brands like Thumbs-Up, Limca, Gold Spot, etc., related to soft drinks from Parle and paid a total consideration of Rs1.70 crore. Ranbaxy Laboratory's brand acquisition from Gufic Laboratories must be one of the few cases, where the revenues from the brands have matched projections in the first year after the acquisition. The four brands Max, Exel, Zole and Roxythro acquired from Gufic helped notch up sales of Rs 72 crore in first year, a 20% improvement over their sales figure under Gufic.

(c) **Joint Venture:** In the case of joint venture two companies enter into an agreement and accumulate certain resources with a view to achieve a particular common business goal. It generally involves fusion of only a small part of the activities of the companies involved in the agreement and usually for limited period of time duration. The returns arising out of such venture are shared by the partners according to their prearranged agreement. While entering into any foreign market, multinational companies pursue this strategy of joint venture. For example, in order to manufacture automobiles in India, Daewoo Motors and DCM group entered into a joint venture programme.

Contraction

Contraction is the second form of restructuring. In the case of contraction, generally the size of the firm gets reduced. Contraction may take place in the form of spin-off, split-off, divestitures, split-ups and equity-carved out.

(a) **Spin-offs:** A spin-off is a type of transaction in which a company distributes all the shares owned by it or its subsidiary to its own shareholders. Such distribution of shares among the shareholders is made on a pro-rata basis. As result, the proportional ownership of shares of the shareholders becomes the same in the new legal subsidiary as well as the parent company. The new entity has its own management and is operated independently without the interventions of the parent company. A spin-off generally does not result in an infusion of cash to the parent firm. For example, by spinning off its investment division, Kotak Mahindra Capital Finance Ltd. formed a subsidiary known as Kotak Mahindra Capital Corporation.

(b) **Split-offs:** In the case of split-off, a new company is created in order to takeover the operations of an existing division or unit of a company. A portion of the existing shareholders of the company obtains stocks in the subsidiary (i.e. the new company) in exchange for stocks of the parent company. As a result, the equity base of the parent company is reduced representing the downsizing of the firm. Thus, shareholding of the new entity (i.e., subsidiary) does not imply the shareholding of the parent company. In the case of a split-off, there is no question of cash inflow to the parent company. For example, the Board of Directors of Dabur India Ltd. decided to split-off the Pharma segment (including assets) and transfer it to a new company for the Financial Year 2002-03. The demerger proposal was a significant strategic decision reflecting corporate restructuring initiative and was expected to provide greater focus on independence to the company's two main segments. The FMCG business, which would remain within Dabur India Ltd, would concentrate on its core competencies in personal care, Health care and Ayurvedic Specialities. The new pharmaceutical company Dabur Pharma Ltd. will focus on its expertise in Allopathic. Oncology, Formulations and Bulk Drugs.

(c) **Divestitures:** A divestiture involves the sale of a portion or segment of the company to an external party. Such sale may cover assets, product lines, subsidiaries or divisions of the undertaking. A divestiture generally results in an infusion of cash to the parent company. A company may choose to sell an undervalued operation which according to the company is unrelated or non-strategic to its core business activities. The sale proceed arising out of such sale may be utilized for investing in profitable investment opportunities that are expected to offer potentially higher returns. Divestiture is considered to be a form of expansion on the part of the buying company and a form of contraction on the part of the selling company.

(d) **Equity Carved-out:** An equity carved-out implies the sale of a segment or portion of the firm through an equity offering to the external parties. Here, new shares of equity are sold to outsiders who, in turn, give them ownership of a portion of the previously existing firm. In that case, a new legal entity is created. The equity holders in the newly generated entity need not be the same as the equity holders in the original seller.

(e) **Split-ups:** In the case of a split-up, the entire company is broken up in series of spin-offs. As a result, the parent company no longer exists and only the new offsprings continue to survive. A split-up basically involves the creation of a new class of stock for each of the parent's operating subsidiaries, paying current shareholders a dividend of each new class of stock, and then dissolving the parents company. Stockholders in the new companies may be different as shareholders in the parent company may exchange their stock for stock in one or more of the spin-offs. Restructuring of the Andhra Pradesh State Electricity Board (APSES) is a good example of split-up. APSEB was split-up in 1999 as a part of the power sector reforms. The power generation division and the transmission and distribution division of APSES was transferred to two different companies namely APGENCo and APTRANACo respectively. As a result of such split-up, the APSES ceased to exist.

Corporate Control

Corporate restructuring may be done without necessarily acquiring new firms or divesting existing organizations. Corporate control is another type of restructuring which involves obtaining control over the management of the firm. Controlling here, is basically defined as a process through which top managers influence other related members of an entity to implement the predetermined organizational strategies. The top managers and promoters group who stand to lose from competition in the market for corporate control may use the democratic rules to benefit themselves. Ownership and control are not always separated. A large block of shares may give effective control even when there is no majority owners. Corporate control generally includes anti-takeover defence, share repurchases, exchange offers and proxy contests.

(a) **Anti-Takeover Defences:** It is a technique followed by a company to prevent forcefully acquiring of its management. With the high level of hostile takeover activity in recent years, various companies are resorting to the strategy of takeover defences. Such takeover defences may be pre-bid or preventive defences and post-bid or active defences. Pre-bid or preventive defences are generally employed with a view to prevent a sudden, unexpected hostile bid from obtaining control of the company. When preventive takeover defences become unsuccessful in preventing an unwanted bid, the post-bid or active defences are implemented. Such takeover defences attempt at changing the corporate control position of the promoters.

(b) **Share Repurchases:** It involves repurchasing its own shares by a company from the market. Shares may be repurchased by following either the tender offer method or through open market method. Shares repurchased are also called Buyback of shares, leading to reductions in the equity capital of the company. Shares buyback facilitates in strengthening the promoters' controlling position in the company by increasing their stake in the equity of the company. It is also used as a takeover defence to reduce the number of shares that could be purchased by the potential acquirer.

(c) **Exchange Offers:** Exchange offer generally provides one or more classes of securities, the right or option to exchange a portion or all of their holding for a different class of securities of the firm. The terms of exchange offered necessarily involve new securities of greater market value than the pre-exchange offer announcement market value. Exchange offer includes exchanging debt for common stock, which enhances the degree of leverage or conversely, exchanging common stock for debt, which reduces leverage. Exchange offers help a company to change its capital structure while holding the investment policy unaltered.

(d) **Proxy Contests:** The proxy contest is a way to take control of a company without owning a majority of its voting right. So it is an attempt made by a single shareholder or a group of shareholders to undertake control or bring about other changes in a company by the use of the proxy mechanized of corporate voting. In a proxy might, a bidder may attempt to use his voting rights and garner the support from other shareholders with a view to expel the incumbent board or management. Proxy contests are less frequently used than tender offer for effecting transfer of control. It provides an alternative means of corporate control but cost of proxy challenges is high. Inefficiency in proxy context raises the question of adverse selections which is the main disadvantage of corporate restructuring through proxy contest.

Present Scenario of Corporate Restructuring

Today, a restructuring wave is sweeping the corporate sector all over the world, taking within its fold both big and small entities, comprising old economy businesses, conglomerates and new economy companies and even in the infrastructure and service sector. Mergers, acquisitions, and takeovers have become an integral part of new economic paradigm. Conglomerates are being formed to combine businesses and where synergies are not achieved, demergers have become the order of the day. With the increasing competition and the economy heading towards globalization, the corporate restructuring activities are expected to occur at a much larger scale than at any time in the past, and are stated to play a major role in achieving the competitive edge for India in international market place.

The process of restructuring through mergers and acquisitions has been a regular feature in the developed and free economy nations like Japan, the US and European countries with special reference to the UK where hundreds of mergers take place every year. The mergers and takeovers of multinational corporate houses across the borders have become a normal phenomenon.

Corporate restructuring being a matter of business convenience, the role of legislation, executive and judiciary is that of a facilitator for restructuring on healthy lines. In this era of hyper competitive capitalism and technological change, industrialists have realized that corporate restructurings are perhaps the best route to reach a size comparable to global companies so as to effectively compete with them. The harsh reality of globalization has dawned that companies which cannot compete globally must sell out as an inevitable alternative.

Indian Scenario

In the last few years, India has followed the worldwide trends in consolidation amongst companies through mergers and acquisitions. Companies are being taken over, units are being hived off, joint ventures tantamount to acquisitions are being made and so on. It may be reasonably be stated that the quantum of corporate restructurings in the last few years must be more than the corresponding quantum in the four and a half decades post independence.

Today Indian economy is passing through recession. In such a situation, corporates which are capable of restructuring can contribute towards economic revival and growth. Despite the sluggish economic scenario in India, merger and amalgamation deals have been on the increase. The obvious reason is as the size of the market shrinks, it becomes extremely difficult for all the companies to survive, unless they cut costs and maintain prices. In such a situation, merger eliminates duplication of administrative and marketing expenses. The other important reason is that it prevents price war in a shrinking market. Companies, by merging, reduce the number of competitors and increase their market share.

Supreme Court of India in the landmark judgment of HLL-TOMCO merger has said that "in this era of hypercompetitive capitalism and technological change, industrialists have realised that corporate restructurings are perhaps the best route to reach a size comparable to global companies so as to effectively compete with them."

Corporate India is facing new flexibility requirements due to restructuring, some of which are discussed below :

   (i) **Employment:** Companies are moving away from relying on workers on open-ended, full-time contracts and, increasingly, use part-time, temporary, 'contigent' and contract workers;

  (ii) **Jobs:** The content of jobs is being expanded to encompass a greater variety of tasks;

 (iii) **Skills:** New work practices are raising skill levels and requirements, and workers thus are required to continuously upgrade their skills so as to be able to cope;

 (iv) **Workplace:** Home and tele-working is growing, thanks to new information and communication technologies.

(v) **Working time:** Increases in demand are met by overtime work or by "a more flexible approach as to when and how to work", so as to extend operating hours without having to pay overtime rates;

(vi) **Remuneration:** Profit-sharing, ESOPs and various types of bonuses are becoming more and more common;

(vii) **Management:** Indians are beginning to understand that a more flexible world is demanding a different kind of management philosophy; and

(viii) **Organisation:** There is a trend towards flatter organisational structures. It is clear that corporate restructuring is a deep and pervasive phenomenon across India. The increasing trend of mergers and acquisitions is one of the clearest and most readily measurable manifestations of restructuring.

Limitations of Corporate Restructuring

Corporate restructuring is the buzzword for those companies which are in dire straits. The best thing that may be done by these companies is to restructure operations so that things may improve. But how practicable are the Corporate Restructuring activities? Is it the panacea for all corporate ills? A number of limitations can be associated with Corporate Restructuring, they are :

(a) **Work Assurance:** Before the announcement of Corporate Restructuring, especially during integration the employees of the ailing firm feel relief, but in most cases the reality becomes better than the employees used to experience before integration. The management of the acquiring firm takes policy for performance improvement resulting closure of certain divisions or departments affecting a number of jobs.

(b) **Retention of Best Management:** Retaining the best management combination is always an uphill task, with differing pay scales, responsibility levels, product and service portfolios, and organizational vision. Companies pay more attention to the financial figures and benefit is weighed only numerically but they remain very unprofessional in handling the transition process, since corporate restructuring is still a new phenomenon. Joint committees to determine new management provisions are rare, and often ineffective.

(c) **Delay in Deal Finalization:** New structures are announced after long delays, and communication is woefully lacking. A gag order is a knee-jerk reaction when corporate senses that things are going wrong. After the initial announcement of Corporate Restructuring it takes to finalize the deal as it involves so many issues like boardroom tussles, labour trouble, and queries from shareholders.

(d) **Executive Stress:** In most cases the divisions and products get closure after Corporate Restructuring. After restructuring the control of the companies goes to separate set of management creating a stress on the executives. While the restructuring takes place through contraction by way of separation the distribution of management brings a large number of denials from the management.

(e) **Workers' Woes:** At union levels, there is outright opposition to restructuring activities. Number of mergers awaits legal clearance months after announcement. This is because unions protest pay changes, proposed lay-offs, outsourcing and asset liquidation.

(f) **Cultural Mismatch:** The situation mostly arises in mergers and takeovers when the organizational culture of one firm gets mismatched with other firms. This results in the destroying efficiency of the worker as well as management of both the firms.

(g) **Inability to Create Value:** Corporate restructuring is aimed at generating value for the firm and finally for the shareholders. But very frequently it is observed that the organizations could not create value, instead they have destroyed it. Sudden change in the management and organizational vision creates a gap leaving certain capacity idle and promoting inefficiency in utilization of resources.

5.10. CHALLENGES OF CHANGE

We have come a long way since the 70s and 80s. The challenges that organizations face today are very different from the yesteryears. Let us look into some of the concerns.

Pace of Change

Toffler in his classic book: "Future Shock" has said, "As interdependency grows, smaller and smaller groups within society achieve greater and greater power for disruption. However, as the rate of change speeds up the length of time in which they can be ignored shrinks to near nothingness.

Change has affected all aspects of our lives, social, cultural, political and economic. Never before in the history of mankind has change been so rapid, intense and widespread. In the last five years alone the Indian market has been flooded with consumer goods and new services resulting in wider choice for consumers. Suddenly organizations in India find themselves in an environment of stiff competition. Organizations are vying with each other to increase their market share by providing value for money and variety of promotional packages. Organizations have come to realize that in the current scenario either they excel or they perish, thereby leaving no room for complacency.

The fast pace and complexity of change has resulted in increased uncertainty in the world of business. It is becoming more and more difficult to predict the developments in various spheres of human endeavour. Due to the increased interdependencies amongst societies and nations in the world change in any one field or any one corner has far reaching impact on the global economy. One might recall the event of September 11, 2001 in which the World Trade Centre in New York was destroyed by a handful of people and subsequent events in Afghanistan and Iraq. The impact of these episodes on global economic order was so devastating that economies of many nations are yet to recover from the repercussions.

Organizations thus have to prepare themselves to survive in a turbulent, orderless, chaotic and complex environment.

Liberalization, Privatization and Globalization (LPG)

Faced by the serious financial crisis as also the then emerging trends in developed and developing countries all over the world, India chose to introduce economic reforms with a view to opening up its economy. The economic reforms included three important processes: Liberalization, Privatization and Globalization.

Liberalization was aimed at easing barriers to entry and exit of businesses and other sectors of economy through deregulation of market and doing away with the "license raj" in the early 90s. Before economic reforms it was the government that decided what the people would eat, drink or drive. Gone are the days when the consumer was left with little or no choice. De-regulation now lets the consumer decide on what will sell. A favourable climate for foreign direct investment minus the license hassles has helped create a better business environment. Domestic firms also have made entry in variety of economic sectors and in sizeable numbers. Deregulated markets have contributed to lowering of prices and quality improvement particularly in consumer products and services. For example, the market is flooded with a variety of cars ranging from Maruti to Hyundai to Mercedes to a recently launched Arnage R by the Volkswagen group.

Privatization meant transfer of assets of public sector undertakings by the government to private hands through the process of outright sale or disinvestment of equity by the government. In keeping with the spirit of socialism, public ownership of the means of production and distribution became an important objective of the national policy in India. According to the latest estimates available, a total investment of over rupees two lakh thirty thousand one rundred forty crores (Rs. 2, 30,140 crores) at the end of March 1999 in the public sector, the overall rate of return on capital employed (net profit to capital employed was only 4.8% in 1998-1999). This poor profitability has thus not contributed to developing a vibrant economy but has also imposed severe budgetary strains on the government.

It is in this context that privatization of public sector undertakings by different means including disinvestment acquires legitimacy as one of the options for revitalizing the Indian economy. Management of these enterprises in private hands will not only usher in changes in the business perspective but will also generate surplus essential for growth of business and contribute to the development of the economy. The government has already initiated the process of transferring ownership through disinvestment in the joint sector of Maruti Udyog Ltd. and in the public sector in ITDC, BPCL and many are in the offing. With the change in the patterns of ownership, the organizations will have to reorient themselves to function in a competitive environment, reset their priorities and work towards benchmarking themselves with the best in the industry globally.

The process of globalization entailed removing trade and tariff barriers so as to facilitate foreign direct investment, entry of multi-national corporations and opportunities for Indian organizations to conduct their businesses and operate in global market. "It's a small world after all" is a line, which has become a reality owing to the rapid advances in technology and communication. Companies can now take advantage of this situation and promote their products and services in a global domain of market, which needs them the most leading to increased profitability.

No doubt globalization brings with it its share of disadvantages. The environment is becoming competitive, complex and is rapidly changing. There is a pressure on companies to meet the international standards of performance in parameters like productivity, cost, quality, delivery and service and continuous innovation. For example, high labour cost in the US has inspired them to set up call center operations in India where labour is cheaper and technically qualified.

Downsizing

Organizations in developed as well as developing economies are under pressure to reduce cost and improve quality of their products and services. Doing more with less and less, i.e. maximum output with minimum input is a trend towards that end. It is imperative for organizations to downsize all their resources including manpower in order to remain competitive. A wide variety of methods are being used by organizations such as reducing the cost of input material, process improvement, low inventory, controlling wastages, eliminating wasteful practices and cutting down expenses.

The downsizing of manpower has led to massive lay-offs, a trend that is likely to continue with the introduction of automation and high technology for increased efficiency. While in the U.S. thousands of employees have been laid off in various sectors like Airlines and Information Technology, Indian organizations have also followed suit in order to bring productivity level to global standards so that they remain competitive.

This has been achieved through a variety of methods like Voluntary Retirement Scheme, attrition, relocation in subsidiaries and units and redeployment, reduction in manpower has resulted in remarkable gains for the organizations.

Outsourcing

Trade barriers are being dismantled, entry restrictions have been done away with, the world is moving towards creating a boundaryless economic order. It is natural for organizations therefore to look for green pastures for accessing cheaper and high quality products and services. Organizations therefore are outsourcing a number of business processes, products or subparts and services to organizations which are able to provide low cost and high quality service with speed.

For developed countries India is an attractive destination for outsourcing Information Technology related services. This trend is increasingly being adopted by practically all public and private sector organizations in India also. For example automobile majors like Maruti and Hyundai outsource a wide number of automobile parts like wind-shield, wipers etc. to other organizations. In public sector undertakings, usually the canteen, transport, maintenance, security jobs are being outsourced. The trend is going to grow at a faster pace as the pressure of global competition becomes more intense.

Information Technology

Information Technology tears down the boundaries of space and time. The Internet boom has enabled organizations to tap new markets and reach markets they never could have before. A direct link has been established between the company and the consumer. This change has come about due to increased speed, high accessibility and networking. The focus is now on information and knowledge. This would necessitate seeing the businesses differently.

- ◆ "as generators of resources, that is, as organizations that can convert business costs into yields. It is yield per customer both the volume of services a customer uses and the mix of those services that determines costs and profitability;

- ◆ as links in an economic chain, which managers need to understand as a whole in order to manage their costs;

- ◆ as society's organs for the creation of wealth; and

- ◆ as both creatures and creators of a material environment, the area outside the organization in which opportunities and results lie but in which the threats to the success and the survival of every business also originate";

There is a need to constantly acquire new skills and expertise without which one would be left behind in the Information Technology Revolution, The emphasis is on information sharing and knowledge management.

Knowledge-Capital

Amongst all the resources available to the organization, knowledge capital is unique to the organization which can neither be copied nor borrowed. Human resources as custodian of knowledge thus are being considered as assets capable of growth with no known limits. Individually acquired knowledge and intellect will need to be converted into collective knowledge and collective intelligence. It is this collective knowledge that provides distinct competitive advantage to the organization. Organizations therefore, need to continually invest in utilizing, nurturing and developing their intellectual capital.

Knowledge Workers

Linked to advances in information technology, is the emergence of knowledge workers who are replacing the manual workers. In recent years, a number of old jobs have been eliminated rendering the skills associated with those jobs redundant. For example shorthand and typing with typewriters, clerical jobs and the like no longer are in existence. They have been replaced by computer related jobs. Their aspirations of knowledge workers stem from the belief that knowledge is the basis for accomplishment. They work as professionals with a highly intellectual orientation. What motivates them is similar to what motivates volunteers. They require challenges and a continuous upgradation of job responsibility. They look for a means of growth which is performance driven and not for merely a means of earning a livelihood via manual skills. It is these workers that process present information to create new information, which could be used to identify and solve future problems.

Decision Response Time

In a highly competitive scenario characterized by constant change, organizations have to respond quickly to the internal pressures within the system and demands of the external environment. With the advances in Information Technology, time has shrunk to nothingness. At the click of a button one can have access to information on any subject from any part of the world. Mass opinions can be mobilized instantly. Business transactions can be carried out without any time lag. Speed by which organizations respond to these demands will have a decisive role to play in their survival vis-a-vis their competitors. This will call for continuous and progressive reduction in decision response time in organizations.

Price Led Strategy

A monopoly usually thrives on a cost-led strategy. The consumer buys at a price quoted by the enterprise as he has no choices available in the market. In such a situation there is no pressure on organization to reduce cost and improve quality. Cost reduction and quality improvement were conceived as contradictory. In public sector enterprises, prices of many products and services were administered by the government of India. The needs of consumers and the price they can afford to pay were not considered.

However with growing competition providing wider choices of products and services in the markets there is pressure on organizations to produce or provide service at a price which is competitive and affordable by the consumers. Organizations therefore have to develop price led strategy in place of cost led strategy.

Example: Imagine buying a bottle of coke for Rs. 100. Seems unimaginable, doesn't it? Keeping the market's sensitivity to price in mind, coke has launched its 200 ml bottle for an amount of Rs. 5. Likewise the incoming calls on Hutch and Airtel have been made free owing to pressure from competitors such as Reliance and Tata.

Thus it is clear that the consumer's role in determining the price has become a reality.

Life Cycle of Organization

The organization's life cycle has undergone a complete transition. Organizations are made up of their products and procedures, people and their relationships, knowledge and technology. Innovation has gained importance. Manufacturers are thus producing products with new variations in features at a phenomenal rate, making the product life cycle shorter. Expertise and technology go hand in hand with innovation. Rapid technological advancements and the need to constantly upgrade one's skills have led to shortening of the technological life cycle.

Thus there is a constant need for retraining and redeployment of manpower. The life span of organization is increasingly becoming shorter. The growing trend towards acquisition, merger even bankruptcy is indicative of this trend.

Customer Driven

"The customer is the King" and "The customer is always right" are phrases that have become today's business mantras. Organization processes need to be developed in response to the customers needs, aspirations and expectations. There are a variety of products and services to choose from and customization of products and services is what will attract consumers today. The emphasis lies on services more than products. The organization providing a feel good factor using etiquette and effective communication will scale the heights of success.

Diversity

There is a shift in the profile of the workforce. The number of women entering the corporate world is on the rise. There is a jump in the number of young achievers. With the entry of MNCs, a culture of diversity has come into being. The so called MNC culture is performance based. The old school of thought where seniority and loyalty were rewarded is no longer valid. People belonging to different ethnic, cultural and professional groups have to work together. Diversity arising out of cultural differences results in clash of priorities, conflicting values, varied work habits and different expectations. The need for effective cross-cultural manpower thus has become imperative. This has led to a very competitive atmosphere where there is no place for complacency.

Changing Expectations of Stakeholders

As the organization owes its existence to the support it is able to mobilize from various components of the environment; socio-cultural; politico-legal; economic and ecological, various constituents of the environment whose contribution is required by the organization and inducements that the constituents expect are called stakeholders. Stakeholders can be internal to the organization such as organization members, the unions and the top management which draws its authority from the Board of Directors. Examples of external stakeholders would include investors, legal and financial institutions, suppliers of resources/ contractors, customers, government and non-government regulating agencies and community at large. Since the community, the government and various interest groups of the society are important stakeholders, their expectations must be met. The Figure (Stakeholders) illustrates the major internal and external stakeholders.

Corporate Governance and Business Ethics

Xerox, Enron and Arthur Anderson are some of the biggest names in the business, which have supposedly failed in adhering to Corporate Governance and norms of business ethics. Corporate Governance is attracting the concern of investors and the public in general. There is a need for transparency, social accountability and social responsibility. Organizations are required to comply with legal requirements, ensure respect for people, communities and the environment. An example of such an initiative is of an NGO called Parivartan which has decided to promote openness in government. It empowers the citizens to access government files through incredibly simple mechanism of filling a form and submitting it to the department concerned.

Corporate governance philosophy of Infosys International Inc. is based on the following principles :

1. Satisfy the spirit of the law and not just the letter of the law. Corporate governance standards should go beyond the law.

2. Be transparent and maintain high degree of disclosure levels. When in doubt, disclose.

3. Make a clear distinction between personal conveniences and corporate resources.

4. Communicate externally, in a truthful manner, about how we run our company internally.

5. Comply with the laws in all the countries in which we operate.

6. Have a simple and transparent corporate structure driven solely by the business needs.

7. Management is the trustee of the shareholders' capital and not the owner.

Source: http://www.infosys.com/investor/governance.asp

Emergent Paradigm

Organizations of today need to respond proactively to the challenges of change mentioned above. In order to do so, organizations will have to understand the changing realities of business. The old ways of doing business are no longer relevant in the current and emerging global scenario. Adopting new approaches to looking at business will require change in their systems of beliefs, values and assumptions about the organization and its relationship with environment. It would further necessitate a shift in paradigm often called as mental models or mindsets by which we define, relate and respond to environment. The focus of organizations in pre-liberalization era was on maintaining internal efficiency and control aimed at achieving higher volumes of production. Most organizations therefore were inward looking responding to changes as and when it became a threat to their survival.

In post liberalization era however, with increased competition, choices available to the consumer and high magnitude of change; organizations have to be market and customer driven. The focus has shifted towards providing value added products and services to suit the consumers' tastes rather than on volume of production. Organizations are being compelled to continuously align themselves with the environment through change in their internal processes and continuous adjustment. They have become outward looking.

Organization Design

Organizations are designed to ensure effective utilization of resources, to facilitate attainment of goals and to seek alignment with their relevant environment. Organization design is the process of integrating and coordinating various resources particularly people, task, technology and information to achieve the purpose for which the organization has been established. It involves developing an organizational architecture to find the optimum fit amongst its components and facilitate synchronization of people's efforts towards a set of common goals.

Typically the design of organization involves developing structures, systems, processes and culture appropriate to the objectives of the organization and expectations of its stakeholders.

Structure includes

- ♦ activity or task analysis
- ♦ grouping of activities
- ♦ allocation of roles, responsibilities and accountability
- ♦ relationship amongst roles

Systems include

- ♦ policies
- ♦ procedures
- ♦ guidelines
- ♦ decision making and control systems

Processes include

- ♦ Business processes
- ♦ Integration of technology, information and manpower
- ♦ Behavioural processes

Culture includes

- ♦ value system
- ♦ philosophy and ideology
- ♦ work ethics, belongingness

The design of structure, systems, processes and culture developed in traditional organizations will undergo change with the emergence of new paradigms of business. Flexibility rather than rigidity, changeability rather than stability, effectiveness rather than efficiency will be the hallmark of the design of organizations in the new economic order.

Organization Development

Organization development aims at developing such internal capabilities and competencies as are desirable for the organization to adapt to changes within and outside the organization. It provides a body of concepts, tools and techniques of behavioural sciences for revitalization and renewal of organization, enhancing their adaptive capacity and developing appropriate designs of organization. Organizations will have to develop their internal systems, procedures and mindset to be able to cope with increased pace of change and uncertainties arising thereof.

CASE STUDY

CASE - 1: SATYAM COMPUTERS

Satyam means truth but Rs. 7.000 crore scam is the largest corporate scam confessed by Satyam Computers. Satyam is the country's fourth largest IT Company after TCS, Infosys and Wipro. It was cooking its books for last several years by inflating its revenues and profits. It boosted its cash and bank balance and showed interest income which did not actually exist and overstated its debtors to promote its sales figures.

Ramalinga Raju was the company's promoter Chairman based at Hyderabad. He kept lying the company's financial position to its shareholders, employees and public around the globe. On 7th January, 2008, Wednesday, the scam was unfolded when confession was made to SEEI which knocked down the stock market and created a black shadow over the industrial sector, particularly the IT sector. Questions were raised about corporate governance, role of auditors and independent directors.

Most of the shareholders made on exit from the share market and by the end of the day, massive sale by FIIS drove the stock market down by 7800.

Doubts over financial health and corporate governance of sensex constituents resulted in panic selling and experienced the 10th worst fall in the stock market index (down by 749 points) since the 30-share index was launched in 1986.

In order to cover the misdeeds, Raju wanted to sellout $1.6 billion to acquire his sons' companies, Maytas properties and Maytas Infra. He wanted to make fictitious cash transfers to come out of the financial scam.

Merrill Lynch, the company's financial advisor terminated its agreement with Satyam to advise it on strategic options because of material accounting irregularities. By the end of Wednesday, there was huge hue and cry by investors to probe against Raju, Satyam and its auditors with SEBI, the Exchanges, ICAI and department of company affairs.

In his letter to the Satyam Board, Ramalinga Raju said "neither myself nor the management director, (including our spouses) sold any shares in the last eight years excepting for a small proportion declared and sold for philanthropic purposes." The promoters wanted to suggest that they have not sold the shares of the company for several years and, therefore, have not benefited from inflating the share price by overstating its performance. Despite the claim, however, the fact remained that promoters' holding fell to 8.2% of the total equity by June 2006 from 25,600 of the total shares of the company in March 2001. It fell from 7.2 crore shares to 3 crore shares during the same period. The promoters realized more than Rs. 1,000 crore from sale of shares at the time when average share price of the company was about Rs. 700. The share price was about Rs. 225 in mid-December, 2007 and on Wednesday, the 7th January, 2008 the share price fell to Rs. 40. It becomes difficult to accept from these figures that Raju's family members have not gained from inflated results.

The Chairman, Ramalinga Raju has admitted that the company had been cooking figures in its financial statements and showing higher revenues and lower liabilities. It has been overstating its profits and reserves. From September 2007 to December 2008, the company's bank balance was overstated by Rs. 5,040 crore and accrued interest by Rs. 376 crore. Liabilities were understated by Rs. 1,720 crore. Thus, the overall financial statement was overstated by Rs. 7,136 crore. The purpose of the scam was to keep the financial position of the company look good. As the promoters had a small percentage of equity, the poor performance of the company could result in a takeover. It was, therefore, important that the company was presented as a growing company.

Acquisition of Maytas, a real estate firm owned by Raju's family was seen as the last attempt to fill company's fictitious assets with real ones. Investors of Maytas viewed it as a good divestment opportunity. Maytas was proposed to be bought for about Rs. 6,400 crore and turned into a wholly-owned subsidiary of Satyam. It was hoped that payment to Maytas could be delayed. Since payment had to be made to Maytas, which was owned by Raju family, for its shares only on paper, Rs. 6,400 crore would move out from Satyam's books into Raju family's hands. Satyam would have acquired Maytas shares, in turn. The payment was just a formality where Raju family technically received Rs. 6,400 crore. This would result in Satyam's inflated reserves and surplus to come down to actual level without showing that the books were overstated. The Raju family would have actually got much less than Rs. 6,400 crore but then, they would not complain and nobody but Raju family would know about this. It would only result in passing of non-existent reserve of Satyam to Maytas to deflate the inflated figures in Satyam's books.

However, investors created a furore over the proposed merger as they felt that Maytas was being purchased at an over-valued figure. This closed the opportunity for Satyam to engage in bungling with Satyam's paper transfers.

The initial investigations by the Registrar of Companies into the Satyam scam has revealed large-scale selling of the company's shares by institutional investors just days ahead of Ramalinga Raju's confession of manipulating company accounts.

Highly placed official sources said, The Serious Fraud Investigation Office (SFIO) would closely scrutinize some of these transactions since they have raised the suspicion of possible insider trading.

The sales took place after the Satyam-Maytas merger fiasco which, too, might have encouraged these big players to exit the IT major.

The sources said five sales between December 23 and January 5, totaling 2.45 crore shares are particularly under the scanner. These sales were all involving shares pledged by the Raju family with various entities to raise loans.

Questions

1. Analyze the case in terms of requirements of corporate disclosure.

2. What are the principles of corporate governance and how have they been violated by Satyam Computers.

CASE - 2: ITC: VISION, VALUES AND VITALITY

The potential of an enterprise for wealth creation is set apart by the distinctive amalgam of its Vision, Values and Vitality. It represents a mix of constancy and change; of a timeless core and constantly evolving strategies and processes built around the core. The effectiveness of interplay between these complementary elements determines the extent to which latent potential is realized. It is the role of leadership to nurture a unique combination of the 3 Vs towards ensuring that the enterprise sustains superior wealth generating capacity in an environment of competitive pressure. Such leadership in a multi-business context needs to extend beyond the corporate level to the strategic business units and their constituents. In line with this thought, the following illustrates the unique blend of Vision, Values and Vitality that has powered the transformation of the Company (ITC).

Vision

A compelling Vision creates and forges corporate identity. It imparts a larger purpose and meaning to individual endeavour. It is aspirational, unifying and motivational. Envisioning a larger societal purpose has always been a hallmark of ITC, described in the past as "a commitment beyond the market". This compelling Vision of enlarging its contribution to the Indian society has powered the Company over the past decade. Such a Vision is manifest in multiple forms, significantly reshaping ITC's profile. The Vision requires each of ITC's business to attain leadership on the strength of international competitiveness. Simultaneously, it has driven the Company to also consciously contribute to enhancing the competitiveness of the larger value chains beyond its own operations. This broader commitment has led to the creation of unique business models that synergise long term shareholder value enhancement with fulfilment of the larger societal purpose. It has expanded corporate consciousness in the practice of trusteeship to ensure sustainable wealth creation. Above all, this superordinate purpose of creating growing value for the Indian society has inspired the company's human resource and aligned their collective endeavour to provide unity of purpose across the organisation.

Values

Values refer to the institutional standards of behaviour that strengthen commitment to the Vision, and guide strategy formulation and purposive action. The core values of the company are shaped around the belief that enterprises exist to serve society. In terms of this belief, profit is a means rather than an end in itself, a compensation to owners of capital linked to the effectiveness of contribution to society and the essential ingredient to sustain such enlarged societal contribution. Thus, the company has embraced an extended role of trusteeship that reaches beyond the assets reflected in the balance sheet to encompass societal assets. An unwavering commitment to integrity, ethical conduct, teamwork and abiding concern for stakeholders are at the heart of the company's value system. The defining trait of ITC however, is its deeply 'Indian' character that aligns corporate strategy to national priorities. Such a character flows from the Indianness of its soul rather than the origin of its capital. As a premier 'Indian' enterprise, ITC consciously engages across the value chains towards maximizing benefit for the Indian society. Such a combination of values determines, choice of corporate strategy, orients such strategy in favour of Indian value chains wherever feasible, and engages the organization willingly in confronting the larger societal challenges of inclusive and sustainable growth. The company's abiding commitment to society provides depth of moral content and infuses energy across the enterprise, thus elevating collective corporate effort to the mission for, the ultimate benefit of all stakeholders, including the shareholders.

Vitality

Compelling vision and strong values by themselves could not have radically transformed the company without the vitality that enables robust strategy formulation and world-class strategy execution. Vitality in ITC is manifest in many ways including the strengthening competitive capability, the deepening consumer insight, the breakthrough innovations in products; and processes the ability to rapidly absorb knowledge and harness technology, manage change and addictiveness to continuously leverage market opportunities. ITC's robust strategy of organization, climate of professionalism and time-tested caring culture constitute the framework for effectively channelizing corporate vitality.

The inspiration, courage and commitment derived from the vision and values has led to growing vitality. Such vitality in turn has led to enhanced market standing and profitability. A clutch of global honours and recognition for the company in the last several years testifies to the quality of transformation achieved.

In order to fulfill the vision of making a substantial and enduring contribution to the Indian society, the company consciously chose the more challenging strategy of pursuing multiple drivers of growth, against conventional wisdom. The company is engaged in managing diversity by creating synergies and special strengths based on cultivating a network of interdependencies within the organization. ITC's strategy of organization based on the governance principle of distributed leadership, seeks to derive the benefits of focus for each business while creatively deploying its diverse pool of core competencies to generate distinctive sources of sustainable competitive advantages.

Questions

1. How has developing value system in this organization helped it in achieving its objectives?

2. What values in business management can prosper its image in the market?

3. Should values remain confined within the company/industry or spread across the globe through trans-cultural value system?

CASE 3: CSR ENDEAVOURS OF TATA GROUP

TATA group of companies has been involved in social responsibilities since pre-independence era. It has identified itself with the growth and development of the nation and is committed to continue to improve the quality of life of people in the communities it serves.

J.R.D. Tata was the first leading businessman to explicitly recognize that business does not operate in isolation from society. He remarked, "The most significant contribution an organised industry can make is, by identifying itself with the life and the problems of the people of the community to which it belongs, and by applying its resources, skills and talents to serve and help them." Today, business is going even beyond the old-style approach of supporting deserving causes and individuals, and accepting its role as one of the driving forces of development.

Its socially responsible activities are listed below :

1. In the year 1892, Jamshedji Tata established the J.N. Tata Endowment Scheme

to provide higher education for deserving Indians. Since then, 3500 Tata scholarships have been awarded.

2. In 1975 the TATA group amended its Articles of Association and kept in consideration, its social and moral responsibilities to consumers, employees, shareholders, society and the local community.

3. In 1979 the Tata Steel Rural Development Society (TSRDS) was established for rural development. It covers around 700 villages in and around Tata Steel's business operations in the States of Jharkhand and Orissa. This society helps in improving the quality of life of communities by implementing a number of programmes in the areas of education, health care and sanitation services, family planning, drinking water, irrigation, agriculture, animal husbandry, vocational training, handicrafts, etc.

4. In July 2002, Tata formedgroup resource under the Tata Council for Community Initiatives (TCCI) on corporate sustainability and a strategy to assist other Tata Companies on corporate social responsibility.

5. The Tatas spent Rs.1.5 billion on social services, in the year 2001-02, on rural development, community health, basic education and vocational training.

6. Tatas also spend on basic infrastructure and disburse money through various charitable trusts and reconstruction societies.

Social values of TATA STEEL include the following:

I. Environment Management

1. Tata Steel is committed to continual improvement in environment performance by setting sound environmental objectives and targets and by integrating a process of review at every stage of production and services.

2. Tata Steel has received the prestigious Social Accountability (SA) 8000 Certification by Social Accountability International (SAI), CSA. This certification is a global verifiable standard for managing the workplace in the most effective manner by improving the workplace conditions. The Sukinda Chromite line has been conferred the SA: 8000:2001 Certification as a result of Tata Steel's practice of continuous improvement in the working conditions.

3. The company has submitted one of the strongest Corporate Sustainability Reports from emerging economies and is India's Top Reporter according to the United Nations Environment Programme (UNEP) and Standard & Poor's. Tata Steel was the first company in India to have started filing and publishing its performance as per Global Reporting Initiative.

4. Tata Steel achieved a reduction in Greenhouse Gas (GHG) emissions, specific energy consumption, suspended particulate matter (SPM) and in emissions of sulphur dioxide/nitrogen oxide. Consumption of raw material and water was brought down and waste reuse and recycling was increased during the year.

5. JUSCO (Jamshedpur Utilities and Services Company) has also initiated several schemes to harvest rainwater in building complexes in Jamshedpur with technical support from Center for Science & Environment, a New Delhi based expert agency. This effort will result in enormous savings in water abstracted from the rivers in Jamshedpur. In Noamundi, a large number of conservation measures have been undertaken to protect the environment. These include constant watering of mining areas to control dust, diversified afforestation and effective plant nurturing and tree protection measures.

II. Welfare

1. The Tata Relief Committee (TRC) has undertaken relief and rehabilitation work in the tsunami affected areas of South India. In the district of Tirunelveli, TRC was one of the first voluntary organizations to begin the relief operations for the displaced people and survivors of tsunami. Tata Relief Committee rushed with materials including household items, dry rations, roofing material and other basic emergency requirements. Base camps were set up at Tirunelveli, Tuticorin and Kanyakumari to provide relief materials.

2. The Lifeline Express Project is being sponsored by Tata Steel for the eighth time the maximum times by a single corporate house. This initiative has already reached out to nearly 30,000 patients.

3. An MOU (Memorandum of Understanding) is being signed with DuPont, known worldwide as world class Safety Performing Company for Safety Management. Plans for next five years will be formulated to enhance TATA Steel's safety performance. "Safety by Choice Tata Steel Voice" a safety Excellence initiative has been implemented.

4. The Tata Steel Rural Development Society (TSRDS) completed 25 years of service to rural communities around its operational units. Established in 1979, TSRDS works in the areas of education, sports and self-reliance to help improve the quality of life.

III. Sports

The JRD Tata Sports Complex houses a state-of-the-art stadium built to international standards. This year, several national and international sports events were hosted at the

JRD Sports Complex. Within the complex there are facilities for chess, courts for handball, tennis, volleyball and basketball; grounds for hockey and archery; etc.

India's First Planned Industrial City

Jamshedpur was named as a tribute to Tata Steel's founder, Jamshedji Tata. It is India's first planned industrial city and a model for the harmonious coexistence of industry and environment.

Jamshedpur Utilities and Services Company (JUSCO) is a wholly owned subsidiary of Tata Steel providing utilities and services to the city of Jamshedpur. It is responsible for housekeeping and hospitality, town planning and engineering, civil construction and maintenance, public health, education, horticulture, fleet management, water and wastewater management, power distribution and related activities. JUSCO is the country's only civic service provider.

1. The Jubilee park is spread over 225 acres of lush greenery dotted with flower beds and illuminated fountains. It was dedicated to the nation during Tata Steel's golden jubilee year.

2. The Tribal Culture Centre imparts vocational training, organizes tribal festivals and cultural programmes to empower the natives of the region.

3. The JRD Tata Technical Education Centre is the outcome of a tie-up between Tata Steel and Kettur Technical Training Foundation to improve the quality of technical education and cater to the requirements of industries in the region.

4. Centre for Excellence has the distinction of being the first centre in the country to provide a central facility where organizations can work together towards professional excellence. Promoters of science, technology, humanities, art, literature and culture can teach and endeavour to improve the overall productivity of national resources.

5. The Tata Steel Zoological Park houses a variety of fauna. The animals in the park are kept in conditions that are close to their natural habitat.

6. Treated waste water from the Town Sewage Treatment Plant is recycled and used in the Tata Steel Works.

7. The Shavak Nanavati Technical Institute is a 75-year old institute. It caters to Tata steel's technical requirements and imparts technical training to other professional institutions.

Questions

1. How important is it for a company to assume social responsibilities?

2. Tata Group of industries has made efforts to look after which sections of society?

3. State five areas of social responsiveness of TATA industries?

CASE – 4: WORLDCOM: ACTIONS LEAD TO CORPORATE REFORM

The story of WorldCom began in 1983 when businessmen Murray Waldron and William Rector sketched out a plan to create a long-distance telephone service provider on a napkin in a coffee shop in Hattiesburg, Mississippi. Their new company, Long Distance Discount Service (LDDS), began operating as a long distance reseller in 1984. Early investor Bernard Ebbers was named CEO the following year. Through acquisitions and mergers, LDDS grew quickly over the next fifteen years. It changed its name to WorldCom, achieved a worldwide presence, acquired telecommunications giant MCI, and eventually expanded beyond long-distance service to offer the whole range of telecommunications services. It seemed poised to become one of the largest telecommunications corporations in the world. Instead, it became the largest bankruptcy filing in U.S. history to date and another name on a long list of those disgraced by the accounting scandals of the early twenty-first century.

Financial Implications of Accounting Fraud

Unfortunately, for thousands of employees and shareholders, WorldCom used questionable accounting practices and improperly recorded $3.8 billion in capital expenditures, which boosted cash flows and profit over all four quarters in 2001 as well as the first quarter of 2002. This disguised the firm's actual net losses for the five quarters because capital expenditures can be deducted over a longer period of time, whereas expenses must be immediately subtracted from revenue. Investors, unaware of the alleged fraud, continued to buy the company's stock, which accelerated the stock's price. Internal investigations uncovered questionable accounting practices stretching as far back as 1999.

Even before the improper accounting practices were disclosed, however, by 2001 WorldCom was already in financial turmoil. Declining rates and revenues and an ambitious buying spree had pushed the company deeper into debt. In addition, chief executive Bernard Ebbers received a controversial $408 million loan to cover margin calls on loans that were secured by company stock in July 2001, WorldCom signed a credit agreement with multiple banks to borrow up to $2.65 billion and repay it within a year. According to the banks, WorldCom tapped the entire amount six weeks before the accounting irregularities were disclosed. The banks contend that if they had known WorldCom's true financial picture, they would not have extended the financing without demanding additional collateral.

On June 28, 2002, the Securities and Exchange Commission (SEC) directed WorldCom to detail the facts underlying the events the company had described in a June 25, 2002 press release. That press release had stated that WorldCom intended to restate its 2001 and first quarter 2002 financial statements and that Scott Sullivan who reported to Bernard Ebbers until he resigned in April 2002 had prepared the financial statements for 2001 and the first quarter of 2002. On February 6, 2002, a meeting had been held between the board's audit committee and Arthur Andersen, the firm's outside auditor, to discuss the audit for fiscal year 2001. Andersen assessed WorldCom's accounting practices to determine whether it had adequate controls to prevent material errors in its financial statements and attested that WorldCom's processes were, in fact, effective. When the committee asked Andersen whether its auditors had had any disagreements with WorldCom's management, Andersen replied that they had not; they were comfortable with the accounting positions WorldCom had taken.

WorldCom did not have the cash needed to pay $7.7 billion in debt, and therefore, filed for Chapter bankruptcy protection on July 21, 2002. In its bankruptcy filing, the firm listed $107 billion in assets and $41 billion in debt. WorldCom's bankruptcy filling allowed it to pay current employees, continue service to customers, retain possessing of assets, and gain a little breathing room to reorganize. However, the telecom giant credibility in the business of many large corporate and government clients, organizations did not fructify.

Who Is to Blame?

Naturally, no one has stepped forward to shoulder the blame for WorldCom's accounting scandal, neither its auditors, executives, board of directors, nor its analysts. As the primary outside auditor, Arthur Andersen also under fire for alleged mismanagement of many other large scandal plagued audits has been faulted for failing to uncover the accounting irregularities. In its defence, Andersen claimed that it could not have known about the improper accounting because former CFO Scott Sullivan never informed Andersen's auditors about the firm's questionable accounting practices. But, in WorldCom's statement to the SEC, the company claimed that Andersen did about these accounting practices, had no disagreement with WorldCom's management, and was not uncomfortable with any accounting positions taken by WorldCom. Several former WorldCom finance and accounting executives, including David Myers, Buford Yates, Betty Vinson, and Troy Normand, pleaded guilty to securities of charges. They claimed that they were directed by top managers to cover up WorldCom's worsening financial situation. Although they protested that these directions were improper, they said they agreed to follow orders after their superiors thinking that it was the only way to save the company. However, Scott Sullivan, studied that WorldCom showed signs of financial troubles: rates and revenues decline and debt rises.

WorldCom receives $2.65 billion in loans from twenty-six banks to be repaid by the end of 2001.

Arthur Andersen, LLP, and WorldCom's audit team meet to discuss the 2001 audit. Everything is deemed correct, and Andersen gives its approval. The U.S. Securities and Exchange Commission (SEC) requests more information concerning accounting procedures and loans to officers. Bernard Ebbers resigns as CEO of WorldCom and is replaced by vice chairman John Sidgmore.

CFO Scott Sullivan is fired after improper accounting of $3.8 billion in expenses is discovered, which covered up a net loss for 2001 and the first quarter of 2002.

WorldCom fires seventeen thousand employees to cut costs. John Sidgmore testifies before a congressional committee to explain how internal investigations uncovered the accounting problems. WorldCom files for reorganization under bankruptcy, an action that affects only the firm's U.S. operations, not its overseas subsidiaries. Continued internal investigations uncover an additional $3.8 billion in improperly reported earnings for 1999, 2000, 2001, and the first quarter of 2002, bringing the total amount of accounting errors to more than $7.6 billion. WorldCom names Greg Rayburn as chief restructuring officer and John Dubel as chief financial officer to lead the company through the reorganization process.

WorldCom formally announces it is seeking a permanent chief executive officer.

The U.S. Bankruptcy Court approves WorldCom's request to pay full severance and benefits to former employees, which had been limited under the company's. The U.S. Bankruptcy Court approves up to $1.1 billion in debtor-in-possession (DIP) financing for WorldCom while it undergoes reorganization. WorldCom files additional bankruptcy petitions for forty-three of its subsidiaries.

Michael D. CapeHas, former president of Hewlett-Packard Company, is named chairman and CEO.

WorldCom, now operating under the name MCI, agrees to pay $750 million to settle the SEC's civil fraud charges.

Many of these employees, pleaded not guilty. Chief executive Bernard Ebbers stated that he did nothing fraudulent and has nothing to hide. WorldCom's lawyers have said that Ebbers did not know of the money shifted into the capital expenditure accounts. However, the Wall Street Journal reported that an internal WorldCom report identified an e-mail and a voice mail that suggested otherwise. Former CEO Ebbers has not been charged with any crime as of this writing.

John Sidgmore, who briefly replaced Bernard Ebbers as CEO, blames WorldCom's former management for the company's woes. Richard Thornburgh, the independent investigator appointed by WorldCom's bankruptcy court, asserted that there was a "cause for substantial concern" regarding WorldCom's board of directors and independent auditors. The board has been accused of lax oversight, and the board's compensation committee has been attacked for approving Bernard Ebber's generous compensation package. Moreover, Thornburgh's report claimed that Ebbers and Sullivan ran WorldCom "with virtually no checks or restraints placed on their actions by the board of directors or other management." Another report, prepared for the firm's new board of directors, also criticized Ebbers for bucking efforts to develop a code of conduct, which Ebbers was said to have called a "colossal waste of time."

Additionally, Jack Grubman, a Wall Street analyst specializing in the telecommunications industry, who rated WorldCom's stock highly, has admitted he did so for too long. However, he insists he was unaware of the company's true financial shape. Grubman was later fired by Solomon Smith Barney because of accusations that he hyped telecommunications stocks, including Global Crossing and WorldCom, even after it became public that the stocks were poor investments.

Many people have blamed the rising number of telecommunication company failures and scandals on neophytes who had no experience in the telecommunication industry. Among these telecom outsiders are a junk bond financier (Gary Winnick of Global Crossing), a railroad baron (Phil Anschutz, who founded Qwest), and Bernard Ebbers, who operated a motel before he took the helm of WorldCom. They tried to transform their startups into gigantic full-service providers like AT&T, but in an increasingly competitive industry it was unlikely that so many large companies could survive.

Will WorldCom survive restructuring in bankruptcy court? Many signs point to yes, but there are skeptics who don't think the telecom giant can be pulled out of the hole it has found itself in. WorldCom's greatest problem is accountability. It has taken many steps toward a successful reorganization, including securing $1.1 billion in loans and appointing Michael Capellas as chairman and CEO. WorldCom has also taken many actions to restore confidence in the company, including replacing the board members who failed to prevent the accounting scandal, firing many managers, reorganizing its finance and accounting functions, and making other changes designed to help correct past problems and prevent them from recurring. The hiring of four line controllers to oversee revenue accounting, operational accounting, financial accounting and controls and procedures are also in process. Additionally, the audit department staff is being increased and will now report directly to the audit committee of the company's new board. "We are working to create a new WorldCom," John Sidgmore said, "We have developed and implemented new systems, policies, and procedures."

Questions

1. What are some things that WorldCom executives could have done to prevent Accounting scandal?

2. How could corporate ethics have played a part in this failure? How could they help to bring about a new and successful WorldCom?

3. Give a thorough update on WorldCom's struggle to reorganize. What have WorldCom executives paid for their part in the fiasco? Do you think these penalties are sufficient?

CASE – 5: ARTHUR ANDERSEN: QUESTIONABLE ACCOUNTING PRACTICES

Arthur Andersen LLP was founded in Chicago in 1913 by Arthur Andersen and partner Clarence Delany. Over a span of nearly ninety years, the Chicago accounting firm would become known as one of the "Big Five" largest accounting firms in the United States, together with Deloitte & Touche, PricewaterhouseCoopers, Ernst & Young, and KPMG. For most of those years, the firm's name was nearly synonymous with trust, integrity, and ethics. Such values are crucial for a firm charged with independently auditing and confirming the financial statements of public corporations, whose accuracy investors depend on for investment decisions.

In its earlier days, Andersen set standards for the accounting profession and advanced new initiatives on the strength of its then undeniable integrity. One example of Andersen's leadership in the profession occurred in the late 1970s when companies began acquiring IBM's new 360 mainframe computer system, the most expensive new computer technology available at the time. Many companies had been depreciating computer hardware on the basis of an assumed ten-year useful life. Andersen, under the leadership of Leonard Spacek, determined that a more realistic life span for the machines was five years. Andersen therefore advised its accounting clients to use the shorter time period for depreciation purposes, although this resulted in higher expenses charged against income and a smaller bottom line. Public corporations that failed to adopt the more conservative measure would receive an "adverse" opinion from Andersen's auditors, something they could ill afford.

Arthur Andersen once exemplified the rock-solid character and integrity that was synonymous with the accounting profession. But high-profile bankruptcies of clients such as Enron and WorldCom capped a string of accounting scandals that eventually cost investors nearly $300 billion and hundreds of thousands of people their jobs. As a result, the Chicago based accounting firm was forced to close its doors after ninety years of business.

The Advent of Consulting

Leonard Spacek joined the company in 1947 following the death of founder Arthur Andersen. He was perhaps best known for his uncompromising insistence on auditor independence, which was in stark contrast to the philosophy of combining auditing and consulting services that many firms, including Andersen itself, later adopted. Andersen began providing consulting services to large clients such as General Electric and Schlitz Brewing in the 1950s. Over the next thirty years, Andersen's consulting business became more profitable per partner than its core accounting and tax services businesses.

According to the American Institute of Certified Public Accountants (AICPA), the objective of an independent audit of a client's financial statements is "the expression of an opinion on the fairness with which [the financial statements] present, in all material respects, financial position, results of operations, and its cash flows in conformity with generally accepted accounting principles." The primary responsibility of an auditor is to express an opinion on a client firm's financial statements after conducting an audit to obtain reasonable assurance that the client's financial statements are free of material misstatement. It is important to note that financial statements are the responsibility of a company's management and not the outside auditor.

At Andersen, growth became the priority, and its emphasis on recruiting and retaining big clients perhaps came at the expense of quality and independent audits. The company linked its consulting business in a joint cooperative relationship with its audit arm, which compromised its auditors' independence, a quality crucial to the execution of a credible audit. The firm's focus on growth also generated a fundamental change in its corporate culture, one in which obtaining high-profit consulting business seems to have been regarded more highly than providing objective auditing services. Those individuals who could deliver the big accounts were often promoted ahead of the practitioners of quality audits.

Andersen's consulting business became recognized as one of the fastest grown and most profitable consulting networks in the world. Revenues from consulting began catching up with the auditing unit in the early 1980s, and surpassed them for the first time in 1984. Although Andersen's consulting business was growing at a rapid pace, its audit practice remained the company's bread and butter. Ten years later Arthur Andersen merged its operational and business systems consulting units and set up a separate business consulting practice in order to offer clients a broader range of integrated services. Throughout the 1990s, Andersen reaped huge profits by selling of consulting services to many clients whose financial statements it also audited. This lucrative full service strategy would later pose an ethical dilemma for some Anderson partners, who had to decide how to treat questionable accounting practices discovered at some of Andersen's largest clients.

Thanks to the growth of Andersen's consulting services, many viewed it as a successful model that other large accounting firms should emulate. However, this same model eventually raised alarm bells at the Securities and Exchange Commission (SEC), concerned over its potential for compromising the independence of audits, in 1998, then SEC chairman Arthur Levitt publicly voiced these concerns and recommended new rules that would restrict the non audit services that accounting firms could provide to their audit clientsa suggestion that Andersen vehemently opposed.

Nonetheless, in 1999 Andersen chose to split its accounting and consulting function into two separate and often competing units. Reportedly, under this arrangement, competition between the two units for accounts tended to discourage a team spirit and instead fostered secrecy and selfishness. Communication suffered, hampering the firm's ability to respond quickly and effectively to crises. As revenues grew, the consulting unit demanded greater compensation and recognition. Infighting between the consulting and auditing units grew until the company was essentially split into two opposing factions.

In August 2000, following an arbitration hearing, a judge ruled that Andersen's consulting arm could effectively divorce the accounting firm and operate independently. By that time, Andersen's consulting business consisted of about eleven thousand consultants and brought in global revenues of nearly $2 billion. Arthur Andersen as a whole employed more than eighty-five thousand people worldwide. The new consulting company promptly changed its name to Accenture the following January. The court later. ordered Arthur Andersen to change its name to Andersen Worldwide in order to better represent its new global brand of accounting services.

Meanwhile, in January 2001 Andersen named Joseph Berardino as the new CEO of the U.S. audit practice. His first task was to navigate the smaller company through a number of lawsuits that had developed in prior years. The company paid $110 million in May 2001 to settle claims brought by Sunbeam shareholders for accounting irregularities and $100 million to settle with Waste Management shareholders over similar charges a month later. In the meantime, news that Enron had overstated earnings became public, sending shock waves, through the financial markets. Over the following year, many companies, a number of them Andersen clients, were forced to restate earnings. The following sections describe a few of the cases that helped lead to Andersen's collapse.

Baptist Foundation of Arizona

In what would become the largest bankruptcy of a non-profit charity in U.S history, the Baptist Foundation of Arizona (BFA), which Andersen served as auditor bilked investors out of about $570 million. BFA, an agency of the Arizona Southern Baptist Convention, was founded in 1948 to raise and manage endowments for church work in Arizona. It operated like a bank, paying interest on deposits that were used mostly to invest in Arizona real estate. The foundation also offered estate and financial planning services to the state's more than four hundred Southern Baptist churches, and was one of the few foundations to offer investments to individuals.

Sunbeam

Andersen's troubles over Sunbeam Corp. began when its audits failed to address serious accounting errors that eventually led to a class action lawsuit by Sunbeam investors and the ouster of CEO Albert Dunlap in 1998.

Waste Management

Andersen also found itself in court over questionable accounting practices with regard to $1.4 billion of overstated earnings at Waste Management. A complaint filed by the SEC charged Waste Management with perpetrating a "massive" financial fraud over period of more than five years.

Trouble with Telecoms

Unfortunately for Andersen, the accusations of accounting fraud did not end with Enron. News soon surfaced that WorldCom, Andersen's largest client, had improperly accounted for nearly $3.9 billion of expenses and had overstated earnings in 2001 and the first part of 2002. After WorldCom restated its earnings, its stock price plummeted and investors launched a barrage of lawsuits that sent the telecom into bankruptcy court. WorldCom's bankruptcy filing eclipsed Enron's as the largest in U.S. history. Andersen blamed WorldCom for the scandal, insisting that the expense irregularities had not been disclosed to its auditors and that it had complied with SEC standards in its auditing of WorldCom. WorldCom, however, pointed the finger of blame not only at its former managers but also at Andersen for failing to find the accounting irregularities. The SEC filed fraud charges against WorldCom, which fired its CFO.

Corporate Culture and Ethical Ramifications

As the details of these investigations into accounting irregularities and fraud came to light, it became apparent that Andersen was more concerned about its own revenue growth than where the revenue came from or whether its independence as an auditor had been compromised. One of the reasons for this confusion in its corporate culture may have been that numerous inexperienced business consultants and untrained auditors were sent to client sites who were largely ignorant of company policies. Another factor may have been its partners' limited involvement in the process of issuing opinions. As the company grew, the number of partners stagnated, there is also evidence that Andersen had limited oversight over its audit, teams and that such visibility was impaired by a relative lack of checks and balances that could have identified when audit teams had strayed from accepted policies. Audit teams had great discretion in terms of issuing financials and restatements.

In February 2002, Andersen hired former Federal Reserve Board chairman Paul Volker to institute reform and help restore its reputation. Soon after Volker came on board, however, Andersen was indicted for obstruction of justice in connection with the shredding of Enron documents. During the investigations, Andersen had been trying to negotiate merger deals for its international partnerships and salvage what was left of its U.S operations. But amid a mass exodus of clients and partners and the resignation of Berardino, the company was forced to begin selling off various business units, and ultimately laid off more than seven thousand employees in the United States.

Implications for Regulation and Accounting Ethics

The accounting scandals of the early twenty-first century sent many Andersen clients into bankruptcy court and subjected even more to greater scrutiny. They also helped spur a new focus on business ethics, driven largely by public demands for greater corporate transparency and accountability. In response, Congress passed the Sarbanes-Oxley Act of 2002, which established new guidelines and direction for corporate and accounting responsibility. The act was enacted to combat securities and accounting fraud and includes, among other things, provisions for a new accounting oversight board, stiffer penalties for violators, and higher standards of corporate governance.

For the accounting profession, Sarbanes-Oxley emphasizes auditor independence and quality, restricts accounting firms' ability to provide both audit and non-audit services for the same clients, and requires periodic reviews of audit firms. All are provisions that the Arthur Andersen of the past would likely have supported wholeheartedly. Some are concerned, however, that such sweeping legislative and regulatory reform may be occurring too quickly in response to intense public and political pressure. The worry is that these reforms may not have been given enough forethought and cost benefit consideration for those public corporations.

Questions

1. Describe the legal and ethical issues surrounding Andersen's auditing of companies accused of accounting improprieties.

2. What evidence is there that Andersen's corporate culture contributed to its downfall?

3. How can the provisions of the Sarbanes-Oxley Act help minimize the likelihood of auditors failing to identify accounting irregularities?

CASE—6: THE COCA-COLA – HONEST TEA DEAL PROMOTING SUSTAINABILITY OR CORPORATE GREEN WASHING?

US-based Honest Tea Inc. (Honest Tea) was founded by Seth Goldman and Barry Nalebuff in 1998, as all natural, bottled iced tea company. From natural tea, Honest Tea went on to sell organic tea in 1999. Since its inception, social responsibility and environmental sustainability were of prime importance to Honest Tea and were a part of the company's identity and purpose. Honest Tea partnered with local communities to procure organic ingredients. As Honest Tea procured fair trade products, the workers on the firms benefited as they were provided with better wages and working conditions.

In a span of ten years, Honest Tea's revenue growth witnessed a CAGR of 66%. The company's revenues reached US$ 38 million by 2008 from US$ 250,000 in 1998. In February 2008, the Coca-Cola Company acquired an equity stake of 40% in Honest Tea. The agreement between the companies stated that Coca-Cola can acquire the remaining stake after three years. Seth Goldman maintained that Coca-Cola's acquisition of the stake would benefit Honest Tea, as it would strengthen the company's distribution capabilities.

However, customers who lent their support to Honest Tea over the years were not happy with this deal. They said that Honest Tea was selling out to Coca-Cola, which had a history of environmental and labour abuses in some of its international markets. Some of the industry experts were of the view that Coca-Cola may discontinue Honest Tea's sustainable business practices.

Questions

1. What is issue of sustainability as a corporate strategy with reference to the incorporation and operations of Honest Tea.

2. Are the challenges faced by startups that adopt sustainable practices and the ways to overcome these challenges.

CASE—7: THE BRIBERY SCANDAL AT SIEMENS AG

This case discusses the bribery scandals that were unearthed at Siemens AG (Siemens) in 2006 and 2007. These scandals involved some of the company's employees bribing foreign officials to gain contracts and creating slush funds for this purpose. In another case, the company was accused of bribing labour representatives on the supervisory board in order to gain their support for its policies. After the German authorities conducted raids on Siemens offices in Germany, investigations were initiated on Siemens in several other countries like the US, Greece, Italy and Switzerland for possible misconduct. As a fallout of this scandal, the CEO of the company and the chairman of the supervisory board had to resign even though they were not directly implicated, as the scandals had occurred during their tenure. With bribery scandals surfacing in Siemens and many other German companies like Volkswagen, questions were also raised about the effectiveness of the Co-determination law in Germany, which advocated a system where in a supervisory board governed the management board and at least half the supervisory board seats had to be filled by labour representatives. Critics contented that in such a system, the management always needed the labour representatives support for company policies, which could lead to a suspicious alliance between them. The case also highlights the opinions of several analysts on the issues related to bribing by the German companies and Siemens in particular and the challenges the new CEO is likely to face at Siemens.

Questions

1. Do issues of bribery scandals impact on Siemens AG on the company and the economic climate in Germany?

2. Analyze the steps taken by Siemens AG to prevent such incidents in future.

3. Discuss the role of the co-determination law in the bribery scandals that surfaced in German companies.

CASE—8: COCA-COLA INDIA'S CORPORATE SOCIAL RESPONSIBILITY STRATEGY

This case is about Coca-Cola's corporate social responsibility (CSR) initiatives in India. It details the activities taken up by Coca-Cola India's management and employees to contribute to the society and community in which the company operates. Coca-Cola India being one of the largest beverage companies in India, realized that CSR had to be an integral part of its corporate agenda. According to the company, it was aware of the environmental, social, and economic impact caused by a business of its scale and therefore it had decided to implement a wide range of initiatives to improve the quality of life of its customers, the workforce, and society at large. However, the company came in for severe criticism from activists and environmental experts who charged it with depleting groundwater resources in the areas in which its bottling plants were located, thereby affecting the livelihood of poor farmers, dumping toxic and hazardous waste materials near its bottling facilities, and discharging waste water into the agricultural lands of farmers. Moreover, its allegedly unethical business practices in developing countries led to its becoming one of the most boycotted companies in the world. Notwithstanding the criticisms, the company continued to champion various initiatives such as rainwater harvesting, restoring groundwater resources, going in for sustainable packaging and recycling, and serving the communities where it operated. Coca-Cola planned to become water neutral in India by 2009 as part of its global strategy of achieving water neutrality. However, criticism against the company refused to die down. Critics felt that Coca-Cola was spending millions of dollars to project a 'green' and 'environment-friendly' image of itself, while failing to make any change in its operations. They said this was an attempt at green washing as Coca-Cola's business practices in India had tarnished its brand image not only in India but also globally. The case discusses the likely challenges for Coca-Cola India as it prepares to implement its new CSR strategy in the country.

Questions

1. Analyze the CSR strategy adopted by Coca-Cola India.

2. State the issues and challenges faced by Coca-Cola with regard to its sustainability initiatives in India.

3. What are the underlying reasons for the growing criticism against Coca-Cola in India.

CASE—9: BUSINESS ETHICS AND GOVERNANCE ISSUES AT HP THE PRETEXTING CONTROVERSY

The case examines the business ethics and governance issues relating to the pretexting controversy that engulfed US based IIP during the second half of 2006. Though the civil claims arising out of the controversy were settled, it raised several other issues pertaining to invasion of privacy, identity theft, and using pretexting to obtain confidential information. When the Board of Directors at HP found that highly confidential information that was discussed among the Board members was being reported in detail in the press, an investigation was initiated. The investigation was carried out by a team constituted by Patricia Dunn, the then Chairperson of the Board. During the probe, it was found that Keyworth, one of the Directors was responsible for the information leaks. The matter was reported to the Board, and one of the Directors, Tom Perkins resigned from the Board, to express his displeasure about the way the investigation was carried out. He asked the HP Board to disclose the details of the investigation process. HP admitted that pretexting was used to obtain the information about the source of leaks. This led to a series of investigations by several governmental agencies and the Attorney General of California on the illegal methods used by HP to carry out the probe. As a result of these investigations, Dunn and four other persons were indicted and the company paid US$ 14.5 million to settle civil claims.

Questions

1. Understand the business ethics issues arising out of the pretexting controversy at HP.

2. Examine the corporate governance issues relating to the pretexting controversy at HP and study the investigation process employed by HP to find the source of confidential information leaks.

3. Examine the illegal/unlawful methods used during the investigations and analyze the implications of the pretexting controversy at HP.

CASE—10: THE BHEL DISINVESTMENT

The decision of the United Progressive Alliance Government to self off 10 per cent of public holding of equity in Bharat Heavy Electricals Limited {BHEL) is in direct contravention of its own promises. The National Common Minimum Programme made it clear that profitable public sector enterprises would not be privatized. In the case of "navratna" PSEs such as BHEL, it had further explicitly promised that while they would be allowed to raise resources from the capital market, the Government would devolve full managerial and commercial autonomy to such enterprises to strengthen them and enable them to become globally competitive players.

The decision to divest shares in BHEL just like the earlier sale by the UPA Government of some equity in NTPC completely disregards both of these important promises. This particular act of divestment would take place without the agreement of the parties providing outside support to the UPA, and indeed by overriding the decision-making power of the Board of BHEL which has clearly not been consulted either. Further, the expected proceeds from the sale of this equity around Rs.2000 crore is to go into the Government's offers, to the newly created National Investment Fund. This is supposed to finance health and education schemes and also provide some funds for the revival of public sector units. In other words, this money is to go towards spending by the Government, in a move that is both unnecessary and fiscally irresponsible.

Unnecessary, because this amount of public resources can easily be found if the Government really intends to. After all, more than Rs.30,000 crore was found as additional defence expenditure. And irresponsible, because the sale of equity from a profitable enterprise necessarily involves the loss of future revenue because of foregone profits. The sale of any asset implies a capital loss, which properly should only be balanced through some other change in assets or liabilities, such as through retiring debt.

Using such money as a form of current revenue, which is in effect what is being done here, is completely illegitimate.

In the case of BHEL, the amount of foregone profits is not small at all, since last year alone, the total operating profits of the company (before depreciation interest and taxes) amounted to more than Rs. 19,000 crore and even post-tax profits were more than Rs. 10,000 crore. Indeed, BHEL, like several other "navratnas" has been showing profits continuously for some years, but has not been allowed to spend the money for its own expansion and development. Instead, it has simply added to its reserves, such that its reserves currently amount to well over Rs.50,000 crore. This is extremely wasteful, and it also denies BHEL and other such companies the opportunity to invest and restructure in ways that would increase their external competitiveness. Many profitable PSEs are currently in this position, whereby they are simply holding huge reserves because of constraints put upon them by the Government. The fear is that if they are privatized, the private purchaser will then have access to the use of these huge reserves that it can use to its own ends rather than in socially desirable ways. This was true, for example, in the case of both VSNL and Balco, where privatisation allowed the private buyers access to the enormous built up reserves to further other investment plans of the purchasing company.

It should be obvious that such a strategy is not only unwise but also counter to any genuine attempt to strengthen the public sector. It is even contrary to proposals made by a committee set up by this self same UPA government to consider issues of autonomy, delegation of financial powers and empowerment of central PSEs, chaired by Dr. Arjun Sengupta. The Report of this Committee ("Group of Experts on Empowerment of Central Public Sector Enterprises") was submitted in April 2005. Many of the points made in this Report have a direct bearing on disinvestment of the type now being pursued. The Report points out that "the Ministry in charge of the company should recognise the fact that they are not the owners of the company but are only exercising the functions of ownership as a custodian on behalf of the Government and the public at large."

The Report recommends much greater autonomy and clearly defined powers for the Board of Directors, which should be fully responsible for the supervision and control over the management of the company. It should have full powers of pursuing new lines of business, deciding on suitable acquisitions and mergers, setting up subsidiaries, as well as other decisions regarding capital expenditures, joint ventures, and so on. What this means is that the Board of a PSE, rather than the Ministry or Cabinet, should decide in the best interests of the company and keeping in mind its social objectives, how to use its resources, whether to issues shares for expansion, and so on. The Report therefore proposes a whole institutional package towards strengthening the PSEs, making their management more professional and providing them some autonomy from their Ministries in order to allow them to invest and restructure in ways that would strengthen their competitive position over time.

It is obvious that the disinvestment proposed for BHEL runs completely contrary to the recommendations of this Report. Instead, once again the profitable PSEs are being treated as milch cows for a Government that has decided that it is strapped for cash, regardless of what is in the best interests of the company itself and the economic activities it is engaged in. Another very important recommendation of the Report is that any decision to reduce Government shareholding to less than 51 per cent should be taken only with the consent of Parliament in each case. But the recent action of the government suggests it is willing to ignore not only its own promises and the views of its allies and supporting parties, but also the recommendations made by Committees that it has set up, if it does not suit its own temporary and short-sighted goals.

Questions

1. Is it ethical for the UPA Government to go back on its promises and not taking the consent of Board of Directors and the views of its allies and supporting parties and the recommendations by its own committees set up by themselves, in selling off 10 per cent of public holding of equity in BHEL.

   Give your views putting yourself in the positions of:

   i. As the PM of Indian Government.

   ii. As Chairman of BHEL.

   iii. As a dignitary of this country.

   iv. As a common man of this country.

2. What made the Government to take this quick decision ignoring several decisions of its own. Do you find any motive behind it?

CASE—11: ENRON A SAGA OF BOOM AND BUST UNITED STATES

Enron was formed from the merger in 1985 of Houston Natural Gas with Intermonth gas co., Omaha.In just over 15 years, it transformed itself from a regulated natural gas company into one of the world's largest energy traders. With more than 21,000 employees around the world, its revenues were over $100 bn in 2000. Enron grew rapidly, containing three businesses energy, wholesale and global services. The size of dealings made Enron briefly one of the biggest energy companies in the world, with sale of $101 bn in 2000, rivaling venerable names such as Shell and Exxon.

Enron's prominence came not only from the key role it played in world energy markets but also because under President Bush, the US administration depended on its chairman Kenneth Lay for advice on energy. By some estimates, it had many lakhs of investors through the holdings of pension funds across the US.

Enron established itself in the UK at the first signs of energy liberalization, becoming the first company to begin construction of a power plant after the electric industry was privatized. For a decade or so, Enron's revolutionary approach was universally applauded.

Innovation

The genius behind Enron was the realization that energy, water, and even products such as telecom bandwidth were essentially commodities that company write-up is based on several articles, papers, in leading economic papers such as Economic Times, Financial, Business Standard and an article March 2002 edition of the Analysts. Enron was a huge "market-maker" in it acted as the main broker in energy products. Among its innovations, it has prized the German power and gas markets, created a virtual gas storage facility in the UK, pioneered the world's largest online commodity trading site.

Earned Eminence and Awards

Fortune magazine tipped the firm as one of the 10 growth to last the decade. The company has won a string of awards, including Fortune it's most innovative company award for an unprecedented six years between 1996 to 2001. In 2000, it won the Financial Times "energy company of the year" award and successful investment decision.

Biggest Corporate Failure

Enron has filed bankruptcy which allows a company to continue trading protection from its creditors. The firm had already confessed to having profits. It leaves debts behind of about $ 15 bn. The collapse of Enron is the failure in US history.

How the Collapse Happened

The firm reported its third quarter results in October, 2001; it revealed a large, lined hole that sent its share price tumbling. The US financial regulator the Exchange Commission (SEC) launched an investigation into the firm and its Enron then admitted it had inflated its profits, sending shares even lower.

A potential buyer for Enron, shied away from the company, leaving it no choice but to file for bankruptcy on 2 December, 2001.

Outcome of Enron failure

The fallout is immense. Enron has left behind $15 bn of debts. And many banks around the world are exposed to the firm, from lending it money and trading with it. Amongst others, JP Morgan has admitted to $900m of exposure, the Citigroup to up to $800m. Some banks are already proceeding with legal action, and the New York based Amalgamated Bank is suing Enron top executives for $15 bn. It seems likely that a few smaller firms that dealt extensively with Enron could go bust soon, too. On an individual level, many employees have lost their jobs and seen the value of their pensions which had been invested heavily in Enron's own stock-wiped away. And Enron's shareholders have seen shares which were worth $85 just a year ago become virtually worthless.

In the longer term, the failure of Enron makes all-out market deregulation look a lot less attractive. Policy-makers used to be keen on the idea of applying Wall Street techniques to energy markets, hoping that it might result in greater efficiency and cheaper prices. Until recently, that hope seemed to be coming true. Now investors will also be more cautious about putting their money into companies that they do not understand.

As criminal investigation has been launched, it is a possibility their senior executives at the firm were involved in fraud. In order to fiddle its balance sheets, the firm used complex financial partnership in order to conceal debt. And many of the company's executives allegedly raked in massive profits, selling their shares before the stock collapsed. But Enron's 20,000 employees lost billions of dollars in their pension plans, after they were barred by the company from selling shares when their value plummeted.

Regulators have started to take action, too. The American Institute of Certified Public Accountants (AICPA), which issues auditing standards through its Auditing Standards Board, said it would propose a new auditing standard early next year for detecting fraud.

The AICPA plans to issue guidance for company management and audit committees, as well as revised auditor standards on the review of quarterly financial statements. It also said it would make recommendations to the SEC on disclosing special purpose entities, complex financial vehicles often kept off a firm's balance sheet precisely the sort of structure that may have played a role in Enron's downfall. Now, the accounting industry is taking a long hard look at itself.

Some More Facts

1. The real accounting scandal is not that handful of over-ambitious companies like Enron, WorldCom, Parmalat, etc. broke accounting rules to inflate their earnings, but

that a majority of companies are inflating their accounting profit by window dressing the figure one way or the other, which is now popularly known as 'creative accounting'.

2. Many companies in almost all the countries do indulge in window dressing their figures by inflating incomes or deflating expenses to beef up profit. But those businesses which do so during depression phase of business cycle, with no intention to cheat but to remain afloat in the eyes of people is different from those which indulge in creative accounting by deflating expenses (under provisioning) or inflating incomes when there is none. When accountants start doing the latter, it becomes accounting myth or financial scandal.

3. Consider one of the actions relating to Enron. In June 1999, some top executives at investment bank Credit Suisse First Boston (CSFB) discussed a controversial investment proposal with Enron. Andrew Fastow, the chief financial officer of the Houston-based energy company, had approached CSFB about joining hands with him in a new off balance sheet partnership known as LMJI that was intended to hedge the value of Enron's investment in an internet company.

Any deal that offered a chance to come closer to Mr. Fastow and Enron was intriguing. The company was one of Wall Street's cash cows when it came to investment banking fees. But all the Bankers like at least one investment CSFB banker were alarmed. Robert Jeffe, a managing director, later told the investigators that he was extremely upset that Mr. Fastow would represent both Enron and LJMI in future dealings as there will definitely be conflict of interest. Mr. Jeffe also saw that Mr. Fastow would earn more than $ 20 million from this arrangement. He further said that "This was something that I would never do even if I have approval from the President of the United States or the Supreme Court." Despite this, a group of CSFB executives including Chuck Ward, the former head of investment banking, Mark Paterson, former head of Leveraged Finance, and Richard Thornburgh, the former Chief Financial Officer, signed off on the deal. They were, it seems, persuaded by assurances from Enron's Board and opinions from its lowers and accountants. On 26 November, 2003, the Bankruptcy examiner, Neal Batson, concluded that SFB's participation in LJMI enabled Enron to book $ 95 million of questionable profits, forming 10% of total profit, while allowing improper profit of $ 40 million to themselves. This myth was exposed by a very bold lady employee (Watkin) who worked directly under the Chief Financial officer Mr. Fastow, and the call partnership firm 'LMJI' was really owned by him only.

Similarly, Enron managed billions of dollars of debt that JP Morgan and Citigroup helped Enron disguise as commodity trades.

Professor Paul Krugman said Enron would be a greater disaster for America than the lesson for all of us in general and corporates and regulators in particular from the Enron saga is: some controls are necessary to check creative accounting, as there are many creative minds in corporates who are prepared to take the investors for a ride. Another important lesson for the government and regulators of all the countries is - (i) Pack the corporate Boards with independent directors and (ii) Every company with a paid-up capital and free reserves exceeding Rs. 10 crores and turnover per annum exceeding Rs.50 crores must have (a) Nomination Committee, (b) Remuneration Committee and (c) Audit Committees (all 3 committees), without exception.

Questions

1. What do you think are the reasons for the sudden collapse of Enron?

2. Was Enron having proper corporate governance in place? If not, what precisely was lacking?

3. What lessons can be learned from the Enron saga?

4. Comment on the risks involved in "related party transitions" and "collusion" of CEO, CFO, Banks and Auditors from the experiences learned from "Enron Episode".

5. Had there been more than fifty percent independent directors on the Board of Enron, and a reasonably strong "Audit Committee" in place, could the fate of the company have been different?

INDEX